CRYPTOGRAPHY
Theory and Practice

SECOND EDITION

The CRC Press Series on

DISCRETE
MATHEMATICS
AND
ITS APPLICATIONS

Series Editor
Kenneth H. Rosen, Ph.D.
AT&T Bell Laboratories

CRYPTOGRAPHY
Theory and Practice

SECOND EDITION

DOUGLAS R. STINSON

Department of Combinatorics and Optimization
University of Waterloo
Waterloo, Ontario
Canada

CHAPMAN & HALL/CRC

A CRC Press Company
Boca Raton   London   New York   Washington, D.C.

Library of Congress Cataloging-in-Publication Data

Stinson, Douglas R. (Douglas Robert), 1956–
   Cryptography : theory and practice / Douglas R. Stinson. — 2nd ed.
     p. cm.
   Includes bibliographical references and index.
   ISBN 1-58488-206-9
   1. Coding theory. 2. Cryptography. I. Title.

QA268 .S75 2002
005.8′2—dc21

2002017476
CIP

Visit the CRC Press Web site at www.crcpress.com

© 2002 by Chapman & Hall/CRC

No claim to original U.S. Government works
International Standard Book Number 1-58488-206-9
Library of Congress Card Number 2002017476
Printed in the United States of America   3 4 5 6 7 8 9 0
Printed on acid-free paper

Preface

The first edition of this book was published in March 1995. At that time, my objective was to produce a general textbook that treated all the essential core areas of cryptography, as well as a selection of more advanced topics. In writing the book, I tried to design it to be flexible enough to permit a wide variety of approaches to the subject, so that it could be used for both undergraduate and graduate university courses in cryptography in mathematics, computer science or engineering.

The following were some of the important features of the first edition of the book, all of which are retained in the second edition.

- Mathematical background was provided where it was needed, in a "just-in-time" fashion.

- Informal descriptions of the cryptosystems were given along with more precise pseudo-code descriptions.

- Examples were presented to illustrate the workings of the algorithms.

- The mathematical underpinnings of the algorithms and cryptosystems were explained carefully and rigorously.

- Numerous exercises were included, some of them quite challenging.

The first edition of the book contained thirteen chapters. When I began to work on this revision a couple of years ago, it soon became clear that there was a wealth of new material I wanted to incorporate into the second edition. In order to keep the book from growing too large, and, also, to allow me to finish the book in a reasonable amount of time, I decided to focus the second edition more tightly on the core areas of cryptography that are most likely to be covered in a course. The result is that the second edition of the book consists of updated, revised, expanded and reorganized versions of the first seven chapters of the first edition. I plan to begin writing a companion volume quite soon, which will contain updated treatments of other chapters from the first edition, as well as chapters covering new topics.

Here is a brief synopsis of the seven chapters in this second edition of "Cryptography Theory and Practice":

• Chapter 1 remains a fairly elementary introduction to simple "classical" cryptosystems. Some topics have been updated or improved, e.g., a simplified cryptanalysis of the *Vigenère Cipher*, based on a suggestion of Dan Velleman, is given.

• Chapter 2 covers the main elements of Shannon's approach to cryptography, including the concept of perfect secrecy and the use of information theory in cryptography. It has not been changed significantly; however, it includes a more careful treatment of elementary probability theory than the first edition did.

• Chapter 3 has been almost completely rewritten. The corresponding chapter in the first edition dealt almost exclusively with the *Data Encryption Standard*, which is now obsolete. I decided to use substitution-permutation networks as a mathematical model to introduce many of the concepts of modern block cipher design and analysis, including differential and linear cryptanalysis. There is more emphasis on general principles than before, and the specific cryptosystems that are discussed (*DES* and the new *Advanced Encryption Standard*) serve to illustrate these general principles.

• Chapter 4 is a significantly improved version of the old Chapter 7. This chapter now contains a unified treatment of keyed and unkeyed hash functions and their application to the construction of message authentication codes. There is an emphasis on mathematical analysis and security proofs. This chapter includes a description of the *Secure Hash Algorithm*.

• Chapter 5 concerns the *RSA Cryptosystem*, together with a considerable amount of background on number-theoretic topics such as primality testing and factoring. It has been expanded to include several new sections, including Pollard's rho algorithm, Wiener's low decryption exponent attack, and semantically secure RSA-based cryptosystems.

• Chapter 6 discusses public-key cryptosystems, such as the *ElGamal Cryptosystem*, that are based on the Discrete Logarithm problem. This chapter also includes a considerable amount of new material, such as the Pollard rho algorithm, lower bounds on the complexity of generic algorithms, an expanded discussion of elliptic curves, semantic security of discrete logarithm cryptosystems and the Diffie-Hellman problems. There is no longer any discussion of knapsack cryptosystems or the *McEliece Cryptosystem* in this chapter.

• Chapter 7 deals with signature schemes. As before, it presents schemes such as the *Digital Signature Algorithm*, and it includes treatment of special types of signature schemes such as undeniable and fail-stop signature schemes. New material includes a careful discussion of security definitions, variants of the *ElGamal Signature Scheme* (such as the *Schnorr Signature Scheme* and the *Elliptic Curve Digital Signature Algorithm*) and provably secure signature schemes such as *Full Domain Hash*.

One of the most difficult things about writing any book in cryptography is deciding how much mathematical background to include. Cryptography is a broad subject, and it requires knowledge of several areas of mathematics, including number theory, groups, rings and fields, linear algebra, probability and information theory. As well, some familiarity with computational complexity, algorithms and NP-completeness theory is useful. In my opinion, it is the breadth of mathematical background required that often creates difficulty for students studying cryptography for the first time.

I tried not to assume too much mathematical background, and thus I developed mathematical tools as they are needed, for the most part. But it would certainly be helpful for the reader to have some familiarity with basic linear algebra and modular arithmetic. On the other hand, a more specialized topic, such as the concept of entropy from information theory, is introduced from scratch.

I apologize to anyone who does not agree with the phrase "Theory and Practice" in the title. I admit that the book is more theory than practice. What I mean by this phrase is that I have tried to select the material to be included in the book both on the basis of theoretical interest and practical importance. Therefore, I include systems that are not of practical use if they are mathematically elegant or illustrate an important concept or technique. But, on the other hand, I do present the most important systems that are used in practice, including several U.S. cryptographic standards.

Many people provided encouragement while I wrote this book, pointed out typos and errors in draft versions of this second edition, and gave me useful suggestions on new material to include and how various topics should be treated. In particular, I would like to thank Howard Heys, Alfred Menezes and Edlyn Teske.

Douglas R. Stinson
Waterloo, Ontario

To my children, Michela and Aiden

Contents

1

Classical Cryptography

In this chapter, we provide a gentle introduction to cryptography and cryptanalysis. We present several simple systems, and describe how they can be "broken." Along the way, we discuss various mathematical techniques that will be used throughout the book.

1.1 Introduction: Some Simple Cryptosystems

The fundamental objective of cryptography is to enable two people, usually referred to as Alice and Bob, to communicate over an insecure channel in such a way that an opponent, Oscar, cannot understand what is being said. This channel could be a telephone line or computer network, for example. The information that Alice wants to send to Bob, which we call "plaintext," can be English text, numerical data, or anything at all — its structure is completely arbitrary. Alice encrypts the plaintext, using a predetermined key, and sends the resulting ciphertext over the channel. Oscar, upon seeing the ciphertext in the channel by eavesdropping, cannot determine what the plaintext was; but Bob, who knows the encryption key, can decrypt the ciphertext and reconstruct the plaintext.

These ideas are described formally using the following mathematical notation.

Definition 1.1: A *cryptosystem* is a five-tuple $(\mathcal{P}, \mathcal{C}, \mathcal{K}, \mathcal{E}, \mathcal{D})$, where the following conditions are satisfied:

1. $\mathcal{P}$ is a finite set of possible *plaintexts*
2. $\mathcal{C}$ is a finite set of possible *ciphertexts*
3. $\mathcal{K}$, the *keyspace*, is a finite set of possible *keys*
4. For each $K \in \mathcal{K}$, there is an *encryption rule* $e_K \in \mathcal{E}$ and a corresponding *decryption rule* $d_K \in \mathcal{D}$. Each $e_K : \mathcal{P} \to \mathcal{C}$ and $d_K : \mathcal{C} \to \mathcal{P}$ are functions such that $d_K(e_K(x)) = x$ for every plaintext element $x \in \mathcal{P}$.

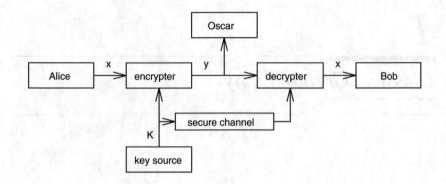

FIGURE 1.1
The communication channel

The main property is property 4. It says that if a plaintext x is encrypted using e_K, and the resulting ciphertext is subsequently decrypted using d_K, then the original plaintext x results.

Alice and Bob will employ the following protocol to use a specific cryptosystem. First, they choose a random key $K \in \mathcal{K}$. This is done when they are in the same place and are not being observed by Oscar, or, alternatively, when they do have access to a secure channel, in which case they can be in different places. At a later time, suppose Alice wants to communicate a message to Bob over an insecure channel. We suppose that this message is a string

$$\mathbf{x} = x_1 x_2 \cdots x_n$$

for some integer $n \geq 1$, where each plaintext symbol $x_i \in \mathcal{P}$, $1 \leq i \leq n$. Each x_i is encrypted using the encryption rule e_K specified by the predetermined key K. Hence, Alice computes $y_i = e_K(x_i)$, $1 \leq i \leq n$, and the resulting ciphertext string

$$\mathbf{y} = y_1 y_2 \cdots y_n$$

is sent over the channel. When Bob receives $y_1 y_2 \cdots y_n$, he decrypts it using the decryption function d_K, obtaining the original plaintext string, $x_1 x_2 \cdots x_n$. See Figure 1.1 for an illustration of the communication channel.

Clearly, it must be the case that each encryption function e_K is an *injective function* (i.e., one-to-one); otherwise, decryption could not be accomplished in an unambiguous manner. For example, if

$$y = e_K(x_1) = e_K(x_2)$$

where $x_1 \neq x_2$, then Bob has no way of knowing whether y should decrypt to x_1 or x_2. Note that if $\mathcal{P} = \mathcal{C}$, it follows that each encryption function is a permutation. That is, if the set of plaintexts and ciphertexts are identical, then each encryption function just rearranges (or permutes) the elements of this set.

1.1.1   The Shift Cipher

In this section, we will describe the *Shift Cipher*, which is based on modular arithmetic. But first we review some basic definitions of modular arithmetic.

Definition 1.2:   Suppose a and b are integers, and m is a positive integer. Then we write $a \equiv b \pmod{m}$ if m divides $b - a$. The phrase $a \equiv b \pmod{m}$ is called a *congruence*, and it is read as "a is *congruent* to b modulo m." The integer m is called the *modulus*.

Suppose we divide a and b by m, obtaining integer quotients and remainders, where the remainders are between 0 and $m - 1$. That is, $a = q_1 m + r_1$ and $b = q_2 m + r_2$, where $0 \le r_1 \le m - 1$ and $0 \le r_2 \le m - 1$. Then it is not difficult to see that $a \equiv b \pmod{m}$ if and only if $r_1 = r_2$. We will use the notation $a \bmod m$ (without parentheses) to denote the remainder when a is divided by m, i.e., the value r_1 above. Thus $a \equiv b \pmod{m}$ if and only if $a \bmod m = b \bmod m$. If we replace a by $a \bmod m$, we say that a is *reduced* modulo m.

We give a couple of examples. To compute $101 \bmod 7$, we write $101 = 7 \times 14 + 3$. Since $0 \le 3 \le 6$, it follows that $101 \bmod 7 = 3$. As another example, suppose we want to compute $(-101) \bmod 7$. In this case, we write $-101 = 7 \times (-15) + 4$. Since $0 \le 4 \le 6$, it follows that $(-101) \bmod 7 = 4$.

REMARK   Many computer programming languages define $a \bmod m$ to be the remainder in the range $-m + 1, \ldots, m - 1$ having the same sign as a. For example, $(-101) \bmod 7$ would be -3, rather than 4 as we defined it above. But for our purposes, it is much more convenient to define $a \bmod m$ always to be non-negative. ∎

We can now define arithmetic modulo m: $\mathbb{Z}_m$ is defined to be the set $\{0, \ldots, m - 1\}$, equipped with two operations, $+$ and $\times$. Addition and multiplication in $\mathbb{Z}_m$ work exactly like real addition and multiplication, except that the results are reduced modulo m.

For example, suppose we want to compute 11×13 in $\mathbb{Z}_{16}$. As integers, we have $11 \times 13 = 143$. Then we reduce 143 modulo 16 as described above: $143 = 8 \times 16 + 15$, so $143 \bmod 16 = 15$, and hence $11 \times 13 = 15$ in $\mathbb{Z}_{16}$.

These definitions of addition and multiplication in $\mathbb{Z}_m$ satisfy most of the familiar rules of arithmetic. We will list these properties now, without proof:

1. addition is *closed*, i.e., for any $a, b \in \mathbb{Z}_m$, $a + b \in \mathbb{Z}_m$

2. addition is *commutative*, i.e., for any $a, b \in \mathbb{Z}_m$, $a + b = b + a$

3. addition is *associative*, i.e., for any $a, b, c \in \mathbb{Z}_m$, $(a + b) + c = a + (b + c)$

4. 0 is an *additive identity*, i.e., for any $a \in \mathbb{Z}_m$, $a + 0 = 0 + a = a$

5. the *additive inverse* of any $a \in \mathbb{Z}_m$ is $m-a$, i.e., $a+(m-a) = (m-a)+a = 0$ for any $a \in \mathbb{Z}_m$

6. multiplication is *closed*, i.e., for any $a, b \in \mathbb{Z}_m$, $ab \in \mathbb{Z}_m$

7. multiplication is *commutative*, i.e., for any $a, b \in \mathbb{Z}_m$, $ab = ba$

8. multiplication is *associative*, i.e., for any $a, b, c \in \mathbb{Z}_m$, $(ab)c = a(bc)$

9. 1 is a *multiplicative identity*, i.e., for any $a \in \mathbb{Z}_m$, $a \times 1 = 1 \times a = a$

10. the *distributive property* is satisfied, i.e., for any $a, b, c \in \mathbb{Z}_m$, $(a + b)c = (ac) + (bc)$ and $a(b + c) = (ab) + (ac)$.

Properties 1, 3–5 say that $\mathbb{Z}_m$ forms an algebraic structure called a *group* with respect to the addition operation. Since property 2 also holds, the group is said to be an *abelian group*.

Properties 1–10 establish that $\mathbb{Z}_m$ is, in fact, a *ring*. We will see many other examples of groups and rings in this book. Some familiar examples of rings include the integers, $\mathbb{Z}$; the real numbers, $\mathbb{R}$; and the complex numbers, $\mathbb{C}$. However, these are all infinite rings, and our attention will be confined almost exclusively to finite rings.

Since additive inverses exist in $\mathbb{Z}_m$, we can also subtract elements in $\mathbb{Z}_m$. We define $a - b$ in $\mathbb{Z}_m$ to be $(a - b) \bmod m$. That is, we compute the integer $a - b$ and then reduce it modulo m. For example, to compute $11 - 18$ in $\mathbb{Z}_{31}$, we first subtract 18 from 11, obtaining -7, and then compute $(-7) \bmod 31 = 24$.

We present the *Shift Cipher* as Cryptosystem 1.1. It is defined over $\mathbb{Z}_{26}$ since there are 26 letters in the English alphabet, though it could be defined over $\mathbb{Z}_m$ for any modulus m. It is easy to see that the *Shift Cipher* forms a cryptosystem as defined above, i.e., $d_K(e_K(x)) = x$ for every $x \in \mathbb{Z}_{26}$.

Cryptosystem 1.1: *Shift Cipher*

Let $\mathcal{P} = \mathcal{C} = \mathcal{K} = \mathbb{Z}_{26}$. For $0 \le K \le 25$, define

$$e_K(x) = (x + K) \bmod 26$$

and

$$d_K(y) = (y - K) \bmod 26$$

$(x, y \in \mathbb{Z}_{26})$.

REMARK   For the particular key $K = 3$, the cryptosystem is often called the *Caesar Cipher*, which was purportedly used by Julius Caesar.   ∎

We would use the *Shift Cipher* (with a modulus of 26) to encrypt ordinary English text by setting up a correspondence between alphabetic characters and residues modulo 26 as follows: $A \leftrightarrow 0$, $B \leftrightarrow 1, \ldots, Z \leftrightarrow 25$. Since we will be using this correspondence in several examples, let's record it for future use:

A	B	C	D	E	F	G	H	I	J	K	L	M
0	1	2	3	4	5	6	7	8	9	10	11	12

N	O	P	Q	R	S	T	U	V	W	X	Y	Z
13	14	15	16	17	18	19	20	21	22	23	24	25

A small example will illustrate.

Example 1.1 Suppose the key for a *Shift Cipher* is $K = 11$, and the plaintext is

> wewillmeetatmidnight.

We first convert the plaintext to a sequence of integers using the specified correspondence, obtaining the following:

$$
\begin{array}{cccccccccc}
22 & 4 & 22 & 8 & 11 & 11 & 12 & 4 & 4 & 19 \\
0 & 19 & 12 & 8 & 3 & 13 & 8 & 6 & 7 & 19
\end{array}
$$

Next, we add 11 to each value, reducing each sum modulo 26:

$$
\begin{array}{cccccccccc}
7 & 15 & 7 & 19 & 22 & 22 & 23 & 15 & 15 & 4 \\
11 & 4 & 23 & 19 & 14 & 24 & 19 & 17 & 18 & 4
\end{array}
$$

Finally, we convert the sequence of integers to alphabetic characters, obtaining the ciphertext:

> HPHTWWXPPELEXTOYTRSE.

To decrypt the ciphertext, Bob will first convert the ciphertext to a sequence of integers, then subtract 11 from each value (reducing modulo 26), and finally convert the sequence of integers to alphabetic characters. $\quad\square$

REMARK In the above example we are using upper case letters for ciphertext and lower case letters for plaintext, in order to improve readability. We will do this elsewhere as well. ∎

If a cryptosystem is to be of practical use, it should satisfy certain properties. We informally enumerate two of these properties now.

1. Each encryption function e_K and each decryption function d_K should be efficiently computable.

2. An opponent, upon seeing a ciphertext string **y**, should be unable to determine the key K that was used, or the plaintext string **x**.

The second property is defining, in a very vague way, the idea of "security." The process of attempting to compute the key K, given a string of ciphertext **y**, is called *cryptanalysis*. (We will make these concepts more precise as we proceed.) Note that, if Oscar can determine K, then he can decrypt **y** just as Bob would, using d_K. Hence, determining K is at least as difficult as determining the plaintext string **x**, given the ciphertext string **y**.

We observe that the *Shift Cipher* (modulo 26) is not secure, since it can be cryptanalyzed by the obvious method of *exhaustive key search*. Since there are only 26 possible keys, it is easy to try every possible decryption rule d_K until a "meaningful" plaintext string is obtained. This is illustrated in the following example.

Example 1.2   Given the ciphertext string

JBCRCLQRWCRVNBJENBWRWN,

we successively try the decryption keys d_0, d_1, etc. The following is obtained:

jbcrclqrwcrvnbjenbwrwn
iabqbkpqvbqumaidmavqvm
hzapajopuaptlzhclzupul
gyzozinotzoskygbkytotk
fxynyhmnsynrjxfajxsnsj
ewxmxglmrxmqiweziwrmri
dvwlwfklqwlphvdyhvqlqh
cuvkvejkpvkogucxgupkpg
btujudijoujnftbwftojof
astitchintimesavesnine

At this point, we have determined the plaintext and we can stop. The key is $K = 9$.   ☐

On average, a plaintext will be computed using this method after trying $26/2 = 13$ decryption rules.

As the above example indicates, a necessary condition for a cryptosystem to be secure is that an exhaustive key search should be infeasible; i.e., the keyspace should be very large. As might be expected, however, a large keyspace is not sufficient to guarantee security.

1.1.2 The Substitution Cipher

Another well-known cryptosystem is the *Substitution Cipher*, which we define now. This cryptosystem has been used for hundreds of years. Puzzle "cryptograms" in newspapers are examples of *Substitution Ciphers*.

Cryptosystem 1.2: *Substitution Cipher*

Let $\mathcal{P} = \mathcal{C} = \mathbb{Z}_{26}$. $\mathcal{K}$ consists of all possible permutations of the 26 symbols $0, 1, \ldots, 25$. For each permutation $\pi \in \mathcal{K}$, define

$$e_\pi(x) = \pi(x),$$

and define

$$d_\pi(y) = \pi^{-1}(y),$$

where π^{-1} is the inverse permutation to π.

Actually, in the case of the *Substitution Cipher*, we might as well take $\mathcal{P}$ and $\mathcal{C}$ both to be the 26-letter English alphabet. We used $\mathbb{Z}_{26}$ in the *Shift Cipher* because encryption and decryption were algebraic operations. But in the *Substitution Cipher*, it is more convenient to think of encryption and decryption as permutations of alphabetic characters.

Here is an example of a "random" permutation, π, which could comprise an encryption function. (As before, plaintext characters are written in lower case and ciphertext characters are written in upper case.)

a	b	c	d	e	f	g	h	i	j	k	l	m
X	N	Y	A	H	P	O	G	Z	Q	W	B	T

n	o	p	q	r	s	t	u	v	w	x	y	z
S	F	L	R	C	V	M	U	E	K	J	D	I

Thus, $e_\pi(a) = X$, $e_\pi(b) = N$, etc. The decryption function is the inverse permutation. This is formed by writing the second lines first, and then sorting in alphabetical order. The following is obtained:

A	B	C	D	E	F	G	H	I	J	K	L	M
d	l	r	y	v	o	h	e	z	x	w	p	t

N	O	P	Q	R	S	T	U	V	W	X	Y	Z
b	g	f	j	q	n	m	u	s	k	a	c	i

Hence, $d_\pi(A) = d$, $d_\pi(B) = l$, etc.

As an exercise, the reader might decrypt the following ciphertext using this decryption function:

MGZVYZLGHCMHJMYXSSFMNHAHYCDLMHA.

A key for the *Substitution Cipher* just consists of a permutation of the 26 alphabetic characters. The number of possible permutations is 26!, which is more than 4.0×10^{26}, a very large number. Thus, an exhaustive key search is infeasible, even for a computer. However, we shall see later that a *Substitution Cipher* can easily be cryptanalyzed by other methods.

1.1.3   The Affine Cipher

The *Shift Cipher* is a special case of the *Substitution Cipher* which includes only 26 of the 26! possible permutations of 26 elements. Another special case of the *Substitution Cipher* is the *Affine Cipher*, which we describe now. In the *Affine Cipher*, we restrict the encryption functions to functions of the form

$$e(x) = (ax + b) \bmod 26,$$

$a, b \in \mathbb{Z}_{26}$. These functions are called *affine functions*, hence the name *Affine Cipher*. (Observe that when $a = 1$, we have a *Shift Cipher*.)

In order that decryption is possible, it is necessary to ask when an affine function is injective. In other words, for any $y \in \mathbb{Z}_{26}$, we want the congruence

$$ax + b \equiv y \pmod{26}$$

to have a unique solution for x. This congruence is equivalent to

$$ax \equiv y - b \pmod{26}.$$

Now, as y varies over $\mathbb{Z}_{26}$, so, too, does $y - b$ vary over $\mathbb{Z}_{26}$. Hence, it suffices to study the congruence $ax \equiv y \pmod{26}$ $(y \in \mathbb{Z}_{26})$.

We claim that this congruence has a unique solution for every y if and only if $\gcd(a, 26) = 1$ (where the gcd function denotes the greatest common divisor of its arguments). First, suppose that $\gcd(a, 26) = d > 1$. Then the congruence $ax \equiv 0 \pmod{26}$ has (at least) two distinct solutions in $\mathbb{Z}_{26}$, namely $x = 0$ and $x = 26/d$. In this case $e(x) = (ax + b) \bmod 26$ is not an injective function and hence not a valid encryption function.

For example, since $\gcd(4, 26) = 2$, it follows that $4x + 7$ is not a valid encryption function: x and $x + 13$ will encrypt to the same value, for any $x \in \mathbb{Z}_{26}$.

Let's next suppose that $\gcd(a, 26) = 1$. Suppose for some x_1 and x_2 that

$$ax_1 \equiv ax_2 \pmod{26}.$$

Then

$$a(x_1 - x_2) \equiv 0 \pmod{26},$$

and thus

$$26 \mid a(x_1 - x_2).$$

We now make use of a fundamental property of integer division: if $\gcd(a, b) = 1$ and $a \mid bc$, then $a \mid c$. Since $26 \mid a(x_1 - x_2)$ and $\gcd(a, 26) = 1$, we must therefore have that

$$26 \mid (x_1 - x_2),$$

i.e., $x_1 \equiv x_2 \pmod{26}$.

At this point we have shown that, if $\gcd(a, 26) = 1$, then a congruence of the form $ax \equiv y \pmod{26}$ has, at most, one solution in $\mathbb{Z}_{26}$. Hence, if we let x vary over $\mathbb{Z}_{26}$, then $ax \bmod 26$ takes on 26 distinct values modulo 26. That is, it takes on every value exactly once. It follows that, for any $y \in \mathbb{Z}_{26}$, the congruence $ax \equiv y \pmod{26}$ has a unique solution for x.

There is nothing special about the number 26 in this argument. The following result can be proved in an analogous fashion.

THEOREM 1.1 *The congruence $ax \equiv b \pmod{m}$ has a unique solution $x \in \mathbb{Z}_m$ for every $b \in \mathbb{Z}_m$ if and only if $\gcd(a, m) = 1$.*

Since $26 = 2 \times 13$, the values of $a \in \mathbb{Z}_{26}$ such that $\gcd(a, 26) = 1$ are $a = 1$, 3, 5, 7, 9, 11, 15, 17, 19, 21, 23, and 25. The parameter b can be any element in $\mathbb{Z}_{26}$. Hence the *Affine Cipher* has $12 \times 26 = 312$ possible keys. (Of course, this is much too small to be secure.)

Let's now consider the general setting where the modulus is m. We need another definition from number theory.

Definition 1.3: Suppose $a \geq 1$ and $m \geq 2$ are integers. If $\gcd(a, m) = 1$, then we say that a and m are *relatively prime*. The number of integers in $\mathbb{Z}_m$ that are relatively prime to m is often denoted by $\phi(m)$ (this function is called the *Euler phi-function*).

A well-known result from number theory gives the value of $\phi(m)$ in terms of the prime power factorization of m. (An integer $p > 1$ is *prime* if it has no positive divisors other than 1 and p. Every integer $m > 1$ can be factored as a product of powers of primes in a unique way. For example, $60 = 2^2 \times 3 \times 5$ and $98 = 2 \times 7^2$.)

We record the formula for $\phi(m)$ in the following theorem.

THEOREM 1.2 *Suppose*

$$m = \prod_{i=1}^{n} p_i^{e_i},$$

where the p_i's are distinct primes and $e_i > 0$, $1 \leq i \leq n$. Then

$$\phi(m) = \prod_{i=1}^{n} (p_i^{e_i} - p_i^{e_i - 1}).$$

It follows that the number of keys in the *Affine Cipher* over $\mathbb{Z}_m$ is $m\phi(m)$, where $\phi(m)$ is given by the formula above. (The number of choices for b is m, and the number of choices for a is $\phi(m)$, where the encryption function is $e(x) = ax + b$.) For example, when $m = 60$, $\phi(60) = 2 \times 2 \times 4 = 16$ and the number of keys in the *Affine Cipher* is 960.

Let's now consider the decryption operation in the *Affine Cipher* with modulus $m = 26$. Suppose that $\gcd(a, 26) = 1$. To decrypt, we need to solve the congruence $y \equiv ax + b \pmod{26}$ for x. The discussion above establishes that the congruence will have a unique solution in $\mathbb{Z}_{26}$, but it does not give us an efficient method of finding the solution. What we require is an efficient algorithm to do this. Fortunately, some further results on modular arithmetic will provide us with the efficient decryption algorithm we seek.

We require the idea of a multiplicative inverse.

Definition 1.4:   Suppose $a \in \mathbb{Z}_m$. The *multiplicative inverse* of a modulo m, denoted $a^{-1} \bmod m$, is an element $a' \in \mathbb{Z}_m$ such that $aa' \equiv a'a \equiv 1 \pmod{m}$. If m is fixed, we sometimes write a^{-1} for $a^{-1} \bmod m$.

By similar arguments to those used above, it can be shown that a has a multiplicative inverse modulo m if and only if $\gcd(a, m) = 1$; and if a multiplicative inverse exists, it is unique modulo m. Also, observe that if $b = a^{-1}$, then $a = b^{-1}$. If p is prime, then every non-zero element of $\mathbb{Z}_p$ has a multiplicative inverse. A ring in which this is true is called a *field*.

In a later section, we will describe an efficient algorithm for computing multiplicative inverses in $\mathbb{Z}_m$ for any m. However, in $\mathbb{Z}_{26}$, trial and error suffices to find the multiplicative inverses of the elements relatively prime to 26:

$$1^{-1} = 1,$$
$$3^{-1} = 9,$$
$$5^{-1} = 21,$$
$$7^{-1} = 15,$$
$$11^{-1} = 1,$$
$$17^{-1} = 23, \text{ and}$$
$$25^{-1} = 25.$$

(All of these can be verified easily. For example, $7 \times 15 = 105 \equiv 1 \pmod{26}$, so $7^{-1} = 15$ and $15^{-1} = 7$.)

Consider our congruence $y \equiv ax + b \pmod{26}$. This is equivalent to

$$ax \equiv y - b \pmod{26}.$$

Since $\gcd(a, 26) = 1$, a has a multiplicative inverse modulo 26. Multiplying both sides of the congruence by a^{-1}, we obtain

$$a^{-1}(ax) \equiv a^{-1}(y - b) \pmod{26}.$$

By associativity of multiplication modulo 26, we have that

$$a^{-1}(ax) \equiv (a^{-1}a)x \equiv 1x \equiv x \pmod{26}.$$

Consequently, $x = a^{-1}(y - b) \bmod 26$. This is an explicit formula for x, that is, the decryption function is

$$d(y) = a^{-1}(y - b) \bmod 26.$$

So, finally, the complete description of the *Affine Cipher* is given as Cryptosystem 1.3.

Cryptosystem 1.3: *Affine Cipher*

Let $\mathcal{P} = \mathcal{C} = \mathbb{Z}_{26}$ and let

$$\mathcal{K} = \{(a, b) \in \mathbb{Z}_{26} \times \mathbb{Z}_{26} : \gcd(a, 26) = 1\}.$$

For $K = (a, b) \in \mathcal{K}$, define

$$e_K(x) = (ax + b) \bmod 26$$

and

$$d_K(y) = a^{-1}(y - b) \bmod 26$$

$(x, y \in \mathbb{Z}_{26})$.

Let's do a small example.

Example 1.3 Suppose that $K = (7, 3)$. As noted above, $7^{-1} \bmod 26 = 15$. The encryption function is

$$e_K(x) = 7x + 3,$$

and the corresponding decryption function is

$$d_K(y) = 15(y - 3) = 15y - 19,$$

where all operations are performed in $\mathbb{Z}_{26}$. It is a good check to verify that $d_K(e_K(x)) = x$ for all $x \in \mathbb{Z}_{26}$. Computing in $\mathbb{Z}_{26}$, we get

$$d_K(e_K(x)) = d_K(7x + 3)$$
$$= 15(7x + 3) - 19$$
$$= x + 45 - 19$$
$$= x.$$

To illustrate, let's encrypt the plaintext *hot*. We first convert the letters h, o, t to residues modulo 26. These are respectively 7, 14, and 19. Now, we encrypt:

$$
\begin{array}{rcl}
(7 \times 7 + 3) \bmod 26 & = & 52 \bmod 26 & = & 0 \\
(7 \times 14 + 3) \bmod 26 & = & 101 \bmod 26 & = & 23 \\
(7 \times 19 + 3) \bmod 26 & = & 136 \bmod 26 & = & 6.
\end{array}
$$

So the three ciphertext characters are 0, 23, and 6, which corresponds to the alphabetic string AXG. We leave the decryption as an exercise for the reader. ▯

1.1.4 The Vigenère Cipher

In both the *Shift Cipher* and the *Substitution Cipher*, once a key is chosen, each alphabetic character is mapped to a unique alphabetic character. For this reason, these cryptosystems are called *monoalphabetic cryptosystems*. We now present a cryptosystem which is not monoalphabetic, the well-known *Vigenère Cipher*. This cipher is named after Blaise de Vigenère, who lived in the sixteenth century.

Cryptosystem 1.4: *Vigenère Cipher*

Let m be a positive integer. Define $\mathcal{P} = \mathcal{C} = \mathcal{K} = (\mathbb{Z}_{26})^m$. For a key $K = (k_1, k_2, \ldots, k_m)$, we define

$$e_K(x_1, x_2, \ldots, x_m) = (x_1 + k_1, x_2 + k_2, \ldots, x_m + k_m)$$

and

$$d_K(y_1, y_2, \ldots, y_m) = (y_1 - k_1, y_2 - k_2, \ldots, y_m - k_m),$$

where all operations are performed in $\mathbb{Z}_{26}$.

Using the correspondence $A \leftrightarrow 0$, $B \leftrightarrow 1, \ldots, Z \leftrightarrow 25$ described earlier, we can associate each key K with an alphabetic string of length m, called a *keyword*. The *Vigenère Cipher* encrypts m alphabetic characters at a time: each plaintext element is equivalent to m alphabetic characters.

Let's do a small example.

Example 1.4 Suppose $m = 6$ and the keyword is $CIPHER$. This corresponds to the numerical equivalent $K = (2, 8, 15, 7, 4, 17)$. Suppose the plaintext is the string

                      thiscryptosystemisnotsecure.

We convert the plaintext elements to residues modulo 26, write them in groups of six, and then "add" the keyword modulo 26, as follows:

19	7	8	18	2	17	24	15	19	14	18	24
2	8	15	7	4	17	2	8	15	7	4	17
21	15	23	25	6	8	0	23	8	21	22	15

$$
\begin{array}{rrrrrrrrrrr}
18 & 19 & 4 & 12 & 8 & 18 & 13 & 14 & 19 & 18 & 4 & 2 \\
2 & 8 & 15 & 7 & 4 & 17 & 2 & 8 & 15 & 7 & 4 & 17 \\
\hline
20 & 1 & 19 & 19 & 12 & 9 & 15 & 22 & 8 & 25 & 8 & 19
\end{array}
$$

$$
\begin{array}{rrr}
20 & 17 & 4 \\
2 & 8 & 15 \\
\hline
22 & 25 & 19
\end{array}
$$

The alphabetic equivalent of the ciphertext string would thus be:

$$\text{VPXZGIAXIVWPUBTTMJPWIZITWZT.}$$

To decrypt, we can use the same keyword, but we would subtract it modulo 26 from the ciphertext, instead of adding. □

Observe that the number of possible keywords of length m in a *Vigenère Cipher* is 26^m, so even for relatively small values of m, an exhaustive key search would require a long time. For example, if we take $m = 5$, then the keyspace has size exceeding 1.1×10^7. This is already large enough to preclude exhaustive key search by hand (but not by computer).

In a *Vigenère Cipher* having keyword length m, an alphabetic character can be mapped to one of m possible alphabetic characters (assuming that the keyword contains m distinct characters). Such a cryptosystem is called a *polyalphabetic cryptosystem*. In general, cryptanalysis is more difficult for polyalphabetic than for monoalphabetic cryptosystems.

1.1.5 The Hill Cipher

In this section, we describe another polyalphabetic cryptosystem called the *Hill Cipher*. This cipher was invented in 1929 by Lester S. Hill. Let m be a positive integer, and define $\mathcal{P} = \mathcal{C} = (\mathbb{Z}_{26})^m$. The idea is to take m linear combinations of the m alphabetic characters in one plaintext element, thus producing the m alphabetic characters in one ciphertext element.

For example, if $m = 2$, we could write a plaintext element as $x = (x_1, x_2)$ and a ciphertext element as $y = (y_1, y_2)$. Here, y_1 would be a linear combination of x_1 and x_2, as would y_2. We might take

$$y_1 = (11x_1 + 3x_2) \bmod 26$$

$$y_2 = (8x_1 + 7x_2) \bmod 26.$$

Of course, this can be written more succinctly in matrix notation as follows:

$$(y_1, y_2) = (x_1, x_2) \begin{pmatrix} 11 & 8 \\ 3 & 7 \end{pmatrix},$$

where all operations are performed in $\mathbb{Z}_{26}$. In general, we will take an $m \times m$ matrix K as our key. If the entry in row i and column j of K is $k_{i,j}$, then we write

$K = (k_{i,j})$. For $x = (x_1, \ldots, x_m) \in \mathcal{P}$ and $K \in \mathcal{K}$, we compute $y = e_K(x) = (y_1, \ldots, y_m)$ as follows:

$$(y_1, y_2, \ldots, y_m) = (x_1, x_2, \ldots, x_m) \begin{pmatrix} k_{1,1} & k_{1,2} & \cdots & k_{1,m} \\ k_{2,1} & k_{2,2} & \cdots & k_{2,m} \\ \vdots & \vdots & & \vdots \\ k_{m,1} & k_{m,2} & \cdots & k_{m,m} \end{pmatrix}.$$

In other words, using matrix notation, $y = xK$.

We say that the ciphertext is obtained from the plaintext by means of a *linear transformation*. We have to consider how decryption will work, that is, how x can be computed from y. Readers familiar with linear algebra will realize that we will use the inverse matrix K^{-1} to decrypt. The ciphertext is decrypted using the matrix equation $x = yK^{-1}$.

Here are the definitions of necessary concepts from linear algebra. If $A = (a_{i,j})$ is an $\ell \times m$ matrix and $B = (b_{j,k})$ is an $m \times n$ matrix, then we define the *matrix product* $AB = (c_{i,k})$ by the formula

$$c_{i,k} = \sum_{j=1}^{m} a_{i,j} b_{j,k}$$

for $1 \leq i \leq \ell$ and $1 \leq k \leq n$. That is, the entry in row i and column k of AB is formed by taking the ith row of A and the kth column of B, multiplying corresponding entries together, and summing. Note that AB is an $\ell \times n$ matrix.

Matrix multiplication is associative (that is, $(AB)C = A(BC)$) but not, in general, commutative (it is not always the case that $AB = BA$, even for square matrices A and B).

The $m \times m$ *identity matrix*, denoted by I_m, is the $m \times m$ matrix with 1's on the main diagonal and 0's elsewhere. Thus, the 2×2 identity matrix is

$$I_2 = \begin{pmatrix} 1 & 0 \\ 0 & 1 \end{pmatrix}.$$

I_m is termed an identity matrix since $AI_m = A$ for any $\ell \times m$ matrix A and $I_m B = B$ for any $m \times n$ matrix B. Now, the *inverse matrix* of an $m \times m$ matrix A (if it exists) is the matrix A^{-1} such that $AA^{-1} = A^{-1}A = I_m$. Not all matrices have inverses, but if an inverse exists, it is unique.

With these facts at hand, it is easy to derive the decryption formula given above: since $y = xK$, we can multiply both sides of the formula by K^{-1}, obtaining

$$yK^{-1} = (xK)K^{-1} = x(KK^{-1}) = xI_m = x.$$

(Note the use of the associativity property.)

We can verify that the encryption matrix above has an inverse in $\mathbb{Z}_{26}$:

$$\begin{pmatrix} 11 & 8 \\ 3 & 7 \end{pmatrix}^{-1} = \begin{pmatrix} 7 & 18 \\ 23 & 11 \end{pmatrix}.$$

since

$$\begin{pmatrix} 11 & 8 \\ 3 & 7 \end{pmatrix} \begin{pmatrix} 7 & 18 \\ 23 & 11 \end{pmatrix} = \begin{pmatrix} 11 \times 7 + 8 \times 23 & 11 \times 18 + 8 \times 11 \\ 3 \times 7 + 7 \times 23 & 3 \times 18 + 7 \times 11 \end{pmatrix}$$

$$= \begin{pmatrix} 261 & 286 \\ 182 & 131 \end{pmatrix}$$

$$= \begin{pmatrix} 1 & 0 \\ 0 & 1 \end{pmatrix}.$$

(Remember that all arithmetic operations are done modulo 26.)

Let's now do an example to illustrate encryption and decryption in the *Hill Cipher*.

Example 1.5 Suppose the key is

$$K = \begin{pmatrix} 11 & 8 \\ 3 & 7 \end{pmatrix}.$$

From the computations above, we have that

$$K^{-1} = \begin{pmatrix} 7 & 18 \\ 23 & 11 \end{pmatrix}.$$

Suppose we want to encrypt the plaintext *july*. We have two elements of plaintext to encrypt: $(9, 20)$ (corresponding to *ju*) and $(11, 24)$ (corresponding to *ly*). We compute as follows:

$$(9, 20) \begin{pmatrix} 11 & 8 \\ 3 & 7 \end{pmatrix} = (99 + 60, 72 + 140) = (3, 4)$$

and

$$(11, 24) \begin{pmatrix} 11 & 8 \\ 3 & 7 \end{pmatrix} = (121 + 72, 88 + 168) = (11, 22).$$

Hence, the encryption of *july* is *DELW*. To decrypt, Bob would compute:

$$(3, 4) \begin{pmatrix} 7 & 18 \\ 23 & 11 \end{pmatrix} = (9, 20)$$

and

$$(11, 22) \begin{pmatrix} 7 & 18 \\ 23 & 11 \end{pmatrix} = (11, 24).$$

Hence, the correct plaintext is obtained. ∎

At this point, we have shown that decryption is possible if K has an inverse. In fact, for decryption to be possible, it is necessary that K has an inverse. (This follows fairly easily from elementary linear algebra, but we will not give a proof here.) So we are interested precisely in those matrices K that are invertible.

The invertibility of a (square) matrix depends on the value of its determinant, which we define now.

Definition 1.5:   Suppose that $A = (a_{i,j})$ is an $m \times m$ matrix. For $1 \leq i \leq m$, $1 \leq j \leq m$, define A_{ij} to be the matrix obtained from A by deleting the ith row and the jth column. The *determinant* of A, denoted $\det A$, is the value $a_{1,1}$ if $m = 1$. If $m > 1$, then $\det A$ is computed recursively from the formula

$$\det A = \sum_{j=1}^{m} (-1)^{i+j} a_{i,j} \det A_{ij},$$

where i is any fixed integer between 1 and m.

It is not at all obvious that the value of $\det A$ is independent of the choice of i in the formula given above, but it can be proved that this is indeed the case. It will be useful to write out the formulas for determinants of 2×2 and 3×3 matrices. If $A = (a_{i,j})$ is a 2×2 matrix, then

$$\det A = a_{1,1}a_{2,2} - a_{1,2}a_{2,1}.$$

If $A = (a_{i,j})$ is a 3×3 matrix, then

$$\det A = a_{1,1}a_{2,2}a_{3,3} + a_{1,2}a_{2,3}a_{3,1} + a_{1,3}a_{2,1}a_{3,2}$$

$$-(a_{1,1}a_{2,3}a_{3,2} + a_{1,2}a_{2,1}a_{3,3} + a_{1,3}a_{2,2}a_{3,1}).$$

For large m, the recursive formula given in the definition above is not usually a very efficient method of computing the determinant of an $m \times m$ square matrix. A preferred method is to compute the determinant using so-called "elementary row operations"; see any text on linear algebra.

Two important properties of determinants that we will use are $\det I_m = 1$; and the multiplication rule $\det(AB) = \det A \times \det B$.

A real matrix K has an inverse if and only if its determinant is non-zero. However, it is important to remember that we are working over $\mathbb{Z}_{26}$. The relevant result for our purposes is that a matrix K has an inverse modulo 26 if and only if $\gcd(\det K, 26) = 1$. To see that this condition is necessary, suppose K has an inverse, denoted K^{-1}. By the multiplication rule for determinants, we have

$$1 = \det I = \det(KK^{-1}) = \det K \det K^{-1}.$$

Hence, $\det K$ is invertible in $\mathbb{Z}_{26}$, which is true if and only if $\gcd(\det K, 26) = 1$.

Sufficiency of this condition can be established in several ways. We will give an explicit formula for the inverse of the matrix K. Define a matrix K^* to have as its (i, j)-entry the value $(-1)^{i+j} \det K_{ji}$. (Recall that K_{ji} is obtained from K

by deleting the jth row and the ith column.) K^* is called the *adjoint matrix* of K. We state the following theorem, concerning inverses of matrices over $\mathbb{Z}_n$, without proof.

THEOREM 1.3 *Suppose $K = (k_{i,j})$ is an $m \times m$ matrix over $\mathbb{Z}_n$ such that $\det K$ is invertible in $\mathbb{Z}_n$. Then $K^{-1} = (\det K)^{-1}K^*$, where K^* is the adjoint matrix of K.*

REMARK The above formula for K^{-1} is not very efficient computationally, except for small values of m (e.g., $m = 2, 3$). For larger m, the preferred method of computing inverse matrices would involve elementary row operations. ∎

In the 2×2 case, we have the following formula, which is an immediate corollary of Theorem 1.3.

COROLLARY 1.4 *Suppose*

$$K = \begin{pmatrix} k_{1,1} & k_{1,2} \\ k_{2,1} & k_{2,2} \end{pmatrix}$$

is a matrix having entries in $\mathbb{Z}_n$, and $\det K = k_{1,1}k_{2,2} - k_{1,2}k_{2,1}$ is invertible in $\mathbb{Z}_n$. Then

$$K^{-1} = (\det K)^{-1} \begin{pmatrix} k_{2,2} & -k_{1,2} \\ -k_{2,1} & k_{1,1} \end{pmatrix}.$$

Let's look again at the example considered earlier. First, we have

$$\det \begin{pmatrix} 11 & 8 \\ 3 & 7 \end{pmatrix} = 11 \times (7 - 8 \times 3) \bmod 26$$

$$= (77 - 24) \bmod 26$$

$$= 53 \bmod 26$$

$$= 1.$$

Now, $1^{-1} \bmod 26 = 1$, so the inverse matrix is

$$\begin{pmatrix} 11 & 8 \\ 3 & 7 \end{pmatrix}^{-1} = \begin{pmatrix} 7 & 18 \\ 23 & 11 \end{pmatrix},$$

as we verified earlier.

Here is another example, using a 3×3 matrix.

Example 1.6 Suppose that

$$K = \begin{pmatrix} 10 & 5 & 12 \\ 3 & 14 & 21 \\ 8 & 9 & 11 \end{pmatrix},$$

where all entries are in $\mathbb{Z}_{26}$. The reader can verify that $\det K = 7$. In $\mathbb{Z}_{26}$, we have that $7^{-1} \bmod 26 = 15$. The adjoint matrix is

$$K^* = \begin{pmatrix} 17 & 1 & 15 \\ 5 & 14 & 8 \\ 19 & 2 & 21 \end{pmatrix}.$$

Finally, the inverse matrix is

$$K^{-1} = 15K^* = \begin{pmatrix} 21 & 15 & 17 \\ 23 & 2 & 16 \\ 25 & 4 & 3 \end{pmatrix}.$$

$\square$

We now give a precise description of the *Hill Cipher* over $\mathbb{Z}_{26}$.

Cryptosystem 1.5: *Hill Cipher*

Let $m \geq 2$ be an integer. Let $\mathcal{P} = \mathcal{C} = (\mathbb{Z}_{26})^m$ and let

$$\mathcal{K} = \{m \times m \text{ invertible matrices over } \mathbb{Z}_{26}\}.$$

For a key K, we define

$$e_K(x) = xK$$

and

$$d_K(y) = yK^{-1},$$

where all operations are performed in $\mathbb{Z}_{26}$.

1.1.6 The Permutation Cipher

All of the cryptosystems we have discussed so far involve substitution: plaintext characters are replaced by different ciphertext characters. The idea of a permutation cipher is to keep the plaintext characters unchanged, but to alter their positions by rearranging them using a permutation.

A *permutation* of a finite set X is a bijective function $\pi : X \to X$. In other words, π is one-to-one (injective) and onto (*surjective*). It follows that, for every $x \in X$, there is a unique element $x' \in X$ such that $\pi(x') = x$. This allows us to define the *inverse permutation*, $\pi^{-1} : X \to X$ by the rule

$$\pi^{-1}(x) = x' \quad \text{if and only if} \quad \pi(x') = x.$$

Then π^{-1} is also a permutation of X.

The *Permutation Cipher* (also known as the *Transposition Cipher*) is defined formally as Cryptosystem 1.6. This cryptosystem has been in use for hundreds of years. In fact, the distinction between the *Permutation Cipher* and the *Substitution Cipher* was pointed out as early as 1563 by Giovanni Porta.

Cryptosystem 1.6: *Permutation Cipher*

Let m be a positive integer. Let $\mathcal{P} = \mathcal{C} = (\mathbb{Z}_{26})^m$ and let $\mathcal{K}$ consist of all permutations of $\{1, \ldots, m\}$. For a key (i.e., a permutation) π, we define

$$e_\pi(x_1, \ldots, x_m) = (x_{\pi(1)}, \ldots, x_{\pi(m)})$$

and

$$d_\pi(y_1, \ldots, y_m) = (y_{\pi^{-1}(1)}, \ldots, y_{\pi^{-1}(m)}),$$

where π^{-1} is the inverse permutation to π.

As with the *Substitution Cipher*, it is more convenient to use alphabetic characters as opposed to residues modulo 26, since there are no algebraic operations being performed in encryption or decryption.

Here is an example to illustrate:

Example 1.7 Suppose $m = 6$ and the key is the following permutation π:

x	1	2	3	4	5	6
$\pi(x)$	3	5	1	6	4	2

Note that the first row of the above diagram lists the values of x, $1 \leq x \leq 6$, and the second row lists the corresponding values of $\pi(x)$. Then the inverse permutation π^{-1} can be constructed by interchanging the two rows, and rearranging the columns so that the first row is in increasing order. Carrying out these operations, we see that the permutation π^{-1} is the following:

x	1	2	3	4	5	6
$\pi^{-1}(x)$	3	6	1	5	2	4

Now, suppose we are given the plaintext

shesellsseashellsbytheseashore.

We first partition the plaintext into groups of six letters:

shesel | lsseas | hellsb | ythese | ashore

Now each group of six letters is rearranged according to the permutation π, yielding the following:

EESLSH | SALSES | LSHBLE | HSYEET | HRAEOS

So, the ciphertext is:

EESLSHSALSESLSHBLEHSYEETHRAEOS.

The ciphertext can be decrypted in a similar fashion, using the inverse permutation π^{-1}.   □

We now show that the *Permutation Cipher* is a special case of the *Hill Cipher*. Given a permutation π of the set $\{1, \ldots, m\}$, we can define an associated $m \times m$ permutation matrix $K_\pi = (k_{i,j})$ according to the formula

$$k_{i,j} = \begin{cases} 1 & \text{if } i = \pi(j) \\ 0 & \text{otherwise.} \end{cases}$$

(A *permutation matrix* is a matrix in which every row and column contains exactly one "1," and all other values are "0." A permutation matrix can be obtained from an identity matrix by permuting rows or columns.)

It is not difficult to see that Hill encryption using the matrix K_π is, in fact, equivalent to permutation encryption using the permutation π. Moreover, $K_\pi^{-1} = K_{\pi^{-1}}$, i.e., the inverse matrix to K_π is the permutation matrix defined by the permutation π^{-1}. Thus, Hill decryption is equivalent to permutation decryption.

For the permutation π used in the example above, the associated permutation matrices are

$$K_\pi = \begin{pmatrix} 0 & 0 & 1 & 0 & 0 & 0 \\ 0 & 0 & 0 & 0 & 0 & 1 \\ 1 & 0 & 0 & 0 & 0 & 0 \\ 0 & 0 & 0 & 0 & 1 & 0 \\ 0 & 1 & 0 & 0 & 0 & 0 \\ 0 & 0 & 0 & 1 & 0 & 0 \end{pmatrix}$$

and

$$K_\pi^{-1} = \begin{pmatrix} 0 & 0 & 1 & 0 & 0 & 0 \\ 0 & 0 & 0 & 0 & 1 & 0 \\ 1 & 0 & 0 & 0 & 0 & 0 \\ 0 & 0 & 0 & 0 & 0 & 1 \\ 0 & 0 & 0 & 1 & 0 & 0 \\ 0 & 1 & 0 & 0 & 0 & 0 \end{pmatrix}.$$

The reader can verify that the product of these two matrices is the identity matrix.

1.1.7   Stream Ciphers

In the cryptosystems we have studied so far, successive plaintext elements are encrypted using the same key, K. That is, the ciphertext string $\mathbf{y}$ is obtained as follows:

$$\mathbf{y} = y_1 y_2 \cdots = e_K(x_1) e_K(x_2) \cdots.$$

Cryptosystems of this type are often called *block ciphers*.

An alternative approach is to use what are called stream ciphers. The basic idea is to generate a keystream $\mathbf{z} = z_1 z_2 \cdots$, and use it to encrypt a plaintext string $\mathbf{x} = x_1 x_2 \cdots$ according to the rule

$$\mathbf{y} = y_1 y_2 \cdots = e_{z_1}(x_1) e_{z_2}(x_2) \cdots .$$

The simplest type of stream cipher is one in which the keystream is constructed from the key, independent of the plaintext string, using some specified algorithm. This type of stream cipher is called "synchronous" and can be defined formally as follows:

Definition 1.6: A *synchronous stream cipher* is a tuple $(\mathcal{P}, \mathcal{C}, \mathcal{K}, \mathcal{L}, \mathcal{E}, \mathcal{D})$, together with a function g, such that the following conditions are satisfied:

1. $\mathcal{P}$ is a finite set of possible *plaintexts*

2. $\mathcal{C}$ is a finite set of possible *ciphertexts*

3. $\mathcal{K}$, the *keyspace*, is a finite set of possible *keys*

4. $\mathcal{L}$ is a finite set called the *keystream alphabet*

5. g is the *keystream generator*. g takes a key K as input, and generates an infinite string $z_1 z_2 \cdots$ called the *keystream*, where $z_i \in \mathcal{L}$ for all $i \geq 1$.

6. For each $z \in \mathcal{L}$, there is an *encryption rule* $e_z \in \mathcal{E}$ and a corresponding *decryption rule* $d_z \in \mathcal{D}$. $e_z : \mathcal{P} \to \mathcal{C}$ and $d_z : \mathcal{C} \to \mathcal{P}$ are functions such that $d_z(e_z(x)) = x$ for every plaintext element $x \in \mathcal{P}$.

To illustrate this definition, we show how the *Vigenère Cipher* can be defined as a synchronous stream cipher. Suppose that m is the keyword length of a *Vigenère Cipher*. Define $\mathcal{K} = (\mathbb{Z}_{26})^m$ and $\mathcal{P} = \mathcal{C} = \mathcal{L} = \mathbb{Z}_{26}$; and define $e_z(x) = (x + z) \bmod 26$ and $d_z(y) = (y - z) \bmod 26$. Finally, define the keystream $z_1 z_2 \cdots$ as follows:

$$z_i = \begin{cases} k_i & \text{if } 1 \leq i \leq m \\ z_{i-m} & \text{if } i \geq m + 1, \end{cases}$$

where $K = (k_1, \ldots, k_m)$. This generates the keystream

$$k_1 k_2 \cdots k_m k_1 k_2 \cdots k_m k_1 k_2 \cdots$$

from the key $K = (k_1, k_2, \ldots, k_m)$.

REMARK We can think of a block cipher as a special case of a stream cipher where the keystream is constant: $z_i = K$ for all $i \geq 1$. ∎

A stream cipher is a *periodic stream cipher* with period d if $z_{i+d} = z_i$ for all integers $i \geq 1$. The *Vigenère Cipher* with keyword length m, as described above, can be thought of as a periodic stream cipher with period m.

Stream ciphers are often described in terms of binary alphabets, i.e., $\mathcal{P} = \mathcal{C} = \mathcal{L} = \mathbb{Z}_2$. In this situation, the encryption and decryption operations are just addition modulo 2:

$$e_z(x) = (x + z) \bmod 2$$

and

$$d_z(y) = (y + z) \bmod 2.$$

If we think of "0" as representing the boolean value "false" and "1" as representing "true," then addition modulo 2 corresponds to the *exclusive-or* operation. Hence, encryption (and decryption) can be implemented very efficiently in hardware.

Let's look at another method of generating a (synchronous) keystream. We will work over binary alphabets. Suppose we start with a binary m-tuple $(k_1, \ldots, k_m)$ and let $z_i = k_i$, $1 \leq i \leq m$ (as before). Now we generate the keystream using a *linear recurrence* of degree m:

$$z_{i+m} = \sum_{j=0}^{m-1} c_j z_{i+j} \bmod 2,$$

for all $i \geq 1$, where $c_0, \ldots, c_{m-1} \in \mathbb{Z}_2$ are specified constants.

REMARK   This recurrence is said to have *degree* m since each term depends on the previous m terms. It is *linear* because z_{i+m} is a linear function of previous terms. Note that we can take $c_0 = 1$ without loss of generality, for otherwise the recurrence will be of degree (at most) $m - 1$.   ∎

Here, the key K consists of the $2m$ values $k_1, \ldots, k_m, c_0, \ldots, c_{m-1}$. If

$$(k_1, \ldots, k_m) = (0, \ldots, 0),$$

then the keystream consists entirely of 0's. Of course, this should be avoided, as the ciphertext will then be identical to the plaintext. However, if the constants $c_0, \ldots, c_{m-1}$ are chosen in a suitable way, then any other initialization vector $(k_1, \ldots, k_m)$ will give rise to a periodic keystream having period $2^m - 1$. So a "short" key can give rise to a keystream having a very long period. This is certainly a desirable property: we will see in a later section how the *Vigenère Cipher* can be cryptanalyzed by exploiting the fact that the keystream has a short period.

Here is an example to illustrate.

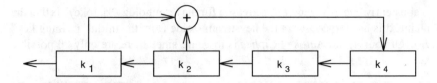

FIGURE 1.2
A linear feedback shift register

Example 1.8 Suppose $m = 4$ and the keystream is generated using the linear recurrence

$$z_{i+4} = (z_i + z_{i+1}) \bmod 2,$$

$i \geq 1$. If the keystream is initialized with any vector other than $(0, 0, 0, 0)$, then we obtain a keystream of period 15. For example, starting with $(1, 0, 0, 0)$, the keystream is

$$1\,0\,0\,0\,1\,0\,0\,1\,1\,0\,1\,0\,1\,1\,1\cdots.$$

Any other non-zero initialization vector will give rise to a cyclic permutation of the same keystream. $\quad\Box$

Another appealing aspect of this method of keystream generation is that the keystream can be produced efficiently in hardware using a *linear feedback shift register*, or LFSR. We would use a shift register with m stages. The vector $(k_1, \ldots, k_m)$ would be used to initialize the shift register. At each time unit, the following operations would be performed concurrently:

1. k_1 would be tapped as the next keystream bit
2. $k_2, \ldots, k_m$ would each be shifted one stage to the left
3. the "new" value of k_m would be computed to be

$$\sum_{j=0}^{m-1} c_j k_{j+1}$$

(this is the "linear feedback").

Observe that the linear feedback is carried out by tapping certain stages of the register (as specified by the constants c_j having the value "1") and computing a sum modulo 2 (which is an exclusive-or). This is illustrated in Figure 1.2, where we depict the LFSR that will generate the keystream of Example 1.8.

A *non-synchronous stream cipher* is a stream cipher in which each keystream element z_i depends on previous plaintext or ciphertext elements $(x_1, \ldots, x_{i-1}$ and/or $y_1, \ldots, y_{i-1})$ as well as the key K. A simple type of non-synchronous stream cipher, known as the *Autokey Cipher*, is presented as Cryptosystem 1.7. It is apparently due to Vigenère. The reason for the terminology "autokey" is that the plaintext is used to construct the keystream (aside from the initial "priming key" K). Of course, the *Autokey Cipher* is insecure since there are only 26 possible keys.

Cryptosystem 1.7: *Autokey Cipher*

Let $\mathcal{P} = \mathcal{C} = \mathcal{K} = \mathcal{L} = \mathbb{Z}_{26}$. Let $z_1 = K$, and define $z_i = x_{i-1}$ for all $i \geq 2$. For $0 \leq z \leq 25$, define

$$e_z(x) = (x + z) \bmod 26$$

and

$$d_z(y) = (y - z) \bmod 26$$

$(x, y \in \mathbb{Z}_{26})$.

Here is an example to illustrate:

Example 1.9 Suppose the key is $K = 8$, and the plaintext is

rendezvous.

We first convert the plaintext to a sequence of integers:

$$17 \quad 4 \quad 13 \quad 3 \quad 4 \quad 25 \quad 21 \quad 14 \quad 20 \quad 18$$

The keystream is as follows:

$$8 \quad 17 \quad 4 \quad 13 \quad 3 \quad 4 \quad 25 \quad 21 \quad 14 \quad 20$$

Now we add corresponding elements, reducing modulo 26:

$$25 \quad 21 \quad 17 \quad 16 \quad 7 \quad 3 \quad 20 \quad 9 \quad 8 \quad 12$$

In alphabetic form, the ciphertext is:

VRQHDUJIM.

Now let's look at how Alice decrypts the ciphertext. She will first convert the alphabetic string to the numeric string

$$25 \quad 21 \quad 17 \quad 16 \quad 7 \quad 3 \quad 20 \quad 9 \quad 8 \quad 12$$

Then she can compute

$$x_1 = d_8(25) = (25 - 8) \bmod 26 = 17.$$

Next,
$$x_2 = d_{17}(21) = (21 - 17) \bmod 26 = 4,$$

and so on. Each time she obtains another plaintext character, she also uses it as
the next keystream element. ▯

In the next section, we discuss methods that can be used to cryptanalyze the
various cryptosystems we have presented.

1.2 Cryptanalysis

In this section, we discuss some techniques of cryptanalysis. The general assumption that is usually made is that the opponent, Oscar, knows the cryptosystem being used. This is usually referred to as *Kerckhoffs' principle*. Of course, if Oscar
does not know the cryptosystem being used, that will make his task more difficult. But we do not want to base the security of a cryptosystem on the (possibly
shaky) premise that Oscar does not know what system is being employed. Hence,
our goal in designing a cryptosystem will be to obtain security under Kerckhoffs'
principle.

First, we want to differentiate between different attack models on cryptosystems. The *attack model* specifies the information available to the adversary when
he mounts his attack. The most common types of attack models are enumerated
as follows.

ciphertext only attack
> The opponent possesses a string of ciphertext, **y**.

known plaintext attack
> The opponent possesses a string of plaintext, **x**, and the corresponding ci-
> phertext, **y**.

chosen plaintext attack
> The opponent has obtained temporary access to the encryption machinery.
> Hence he can choose a plaintext string, **x**, and construct the corresponding
> ciphertext string, **y**.

chosen ciphertext attack
> The opponent has obtained temporary access to the decryption machinery.
> Hence he can choose a ciphertext string, **y**, and construct the corresponding
> plaintext string, **x**.

TABLE 1.1
Probabilities of occurrence of the 26 letters

letter	probability	letter	probability
A	.082	N	.067
B	.015	O	.075
C	.028	P	.019
D	.043	Q	.001
E	.127	R	.060
F	.022	S	.063
G	.020	T	.091
H	.061	U	.028
I	.070	V	.010
J	.002	W	.023
K	.008	X	.001
L	.040	Y	.020
M	.024	Z	.001

In each case, the objective of the adversary is to determine the key that was used. We note that a chosen ciphertext attack is most relevant to public-key cryptosystems, which we discuss in the later chapters.

We first consider the weakest type of attack, namely a ciphertext-only attack. We also assume that the plaintext string is ordinary English text, without punctuation or "spaces." (This makes cryptanalysis more difficult than if punctuation and spaces were encrypted.)

Many techniques of cryptanalysis use statistical properties of the English language. Various people have estimated the relative frequencies of the 26 letters by compiling statistics from numerous novels, magazines, and newspapers. The estimates in Table 1.1 were obtained by Beker and Piper. On the basis of these probabilities, Beker and Piper partition the 26 letters into five groups as follows:

1. E, having probability about 0.120

2. T, A, O, I, N, S, H, R, each having probability between 0.06 and 0.09

3. D, L, each having probability around 0.04

4. $C, U, M, W, F, G, Y, P, B$, each having probability between 0.015 and 0.028

5. V, K, J, X, Q, Z, each having probability less than 0.01.

It is also useful to consider sequences of two or three consecutive letters, called *digrams* and *trigrams*, respectively. The 30 most common digrams are (in decreasing order) $TH, HE, IN, ER, AN, RE, ED, ON, ES, ST, EN, AT, TO, NT, HA, ND, OU, EA, NG, AS, OR, TI, IS, ET, IT, AR, TE, SE, HI$, and OF. The twelve most common trigrams are $THE, ING, AND, HER, ERE, ENT, THA, NTH, WAS, ETH, FOR$, and DTH.

TABLE 1.2
Frequency of occurrence of the 26 **ciphertext letters**

letter	frequency	letter	frequency
A	2	*N*	1
B	1	*O*	1
C	0	*P*	2
D	7	*Q*	0
E	5	*R*	8
F	4	*S*	3
G	0	*T*	0
H	5	*U*	2
I	0	*V*	4
J	0	*W*	0
K	5	*X*	2
L	2	*Y*	1
M	2	*Z*	0

1.2.1   Cryptanalysis of the Affine Cipher

As a simple illustration of how cryptanalysis can be performed using statistical data, let's look first at the *Affine Cipher*. Suppose Oscar has intercepted the following ciphertext:

Example 1.10   Ciphertext obtained from an *Affine Cipher*

```
FMXVEDKAPHFERBNDKRXRSREFMORUDSDKDVSHVUFEDK
APRKDLYEVLRHHRH
```

The frequency analysis of this ciphertext is given in Table 1.2.

There are only 57 characters of ciphertext, but this is usually sufficient to cryptanalyze an *Affine Cipher*. The most frequent ciphertext characters are: *R* (8 occurrences), *D* (7 occurrences), *E*, *H*, *K* (5 occurrences each), and *F*, *S*, *V* (4 occurrences each). As a first guess, we might hypothesize that *R* is the encryption of *e* and *D* is the encryption of *t*, since *e* and *t* are (respectively) the two most common letters. Expressed numerically, we have $e_K(4) = 17$ and $e_K(19) = 3$. Recall that $e_K(x) = ax + b$, where a and b are unknowns. So we get two linear equations in two unknowns:

$$4a + b = 17$$

$$19a + b = 3.$$

This system has the unique solution $a = 6$, $b = 19$ (in $\mathbb{Z}_{26}$). But this is an illegal key, since $\gcd(a, 26) = 2 > 1$. So our hypothesis must be incorrect.

Our next guess might be that R is the encryption of e and E is the encryption of t. Proceeding as above, we obtain $a = 13$, which is again illegal. So we try the next possibility, that R is the encryption of e and H is the encryption of t. This yields $a = 8$, again impossible. Continuing, we suppose that R is the encryption of e and K is the encryption of t. This produces $a = 3$, $b = 5$, which is at least a legal key. It remains to compute the decryption function corresponding to $K = (3, 5)$, and then to decrypt the ciphertext to see if we get a meaningful string of English, or nonsense. This will confirm the validity of $(3, 5)$.

If we perform these operations, we obtain $d_K(y) = 9y - 19$ and the given ciphertext decrypts to yield:

```
algorithmsarequitegeneraldefinitionsofarit
hmeticprocesses
```

We conclude that we have determined the correct key.   ⬜

1.2.2   Cryptanalysis of the Substitution Cipher

Here, we look at the more complicated situation, the *Substitution Cipher*. Consider the ciphertext in the following example:

Example 1.11   Ciphertext obtained from a *Substitution Cipher*

```
YIFQFMZRWQFYVECFMDZPCVMRZWNMDZVEJBTXCDDUMJ
NDIFEFMDZCDMQZKCEYFCJMYRNCWJCSZREXCHZUNMXZ
NZUCDRJXYYSMRTMEYIFZWDYVZVYFZUMRZCRWNZDZJJ
XZWGCHSMRNMDHNCMFQCHZJMXJZWIEJYUCFWDJNZDIR
```

The frequency analysis of this ciphertext is given in Table 1.3.

Since Z occurs significantly more often than any other ciphertext character, we might conjecture that $d_K(Z) = e$. The remaining ciphertext characters that occur at least ten times (each) are C, D, F, J, M, R, Y. We might expect that these letters are encryptions of (a subset of) t, a, o, i, n, s, h, r, but the frequencies really do not vary enough to tell us what the correspondence might be.

At this stage we might look at digrams, especially those of the form $-Z$ or $Z-$, since we conjecture that Z decrypts to e. We find that the most common digrams of this type are DZ and ZW (four times each); NZ and ZU (three times each); and $RZ, HZ, XZ, FZ, ZR, ZV, ZC, ZD$, and ZJ (twice each). Since ZW occurs four times and WZ not at all, and W occurs less often than many other

TABLE 1.3
Frequency of occurrence of the 26 ciphertext letters

letter	frequency	letter	frequency
A	0	N	9
B	1	O	0
C	15	P	1
D	13	Q	4
E	7	R	10
F	11	S	3
G	1	T	2
H	4	U	5
I	5	V	5
J	11	W	8
K	1	X	6
L	0	Y	10
M	16	Z	20

characters, we might guess that $d_K(W) = d$. Since DZ occurs four times and ZD occurs twice, we would think that $D_K(D) \in \{r, s, t\}$, but it is not clear which of the three possibilities is the correct one.

If we proceed on the assumption that $d_K(Z) = e$ and $d_K(W) = d$, we might look back at the ciphertext and notice that we have ZRW and RZW both occurring near the beginning of the ciphertext, and RW occurs again later on. Since R occurs frequently in the ciphertext and nd is a common digram, we might try $d_K(R) = n$ as the most likely possibility.

At this point, we have the following:

```
------end---------e----ned---e-----------
YIFQFMZRWQFYVECFMDZPCVMRZWNMDZVEJBTXCDDUMJ

--------e----e---------n--d---en----e----e
NDIFEFMDZCDMQZKCEYFCJMYRNCWJCSZREXCHZUNMXZ

-e---n------n------ed---e---e--ne-nd-e-e--
NZUCDRJXYYSMRTMEYIFZWDYVZVYFZUMRZCRWNZDZJJ

-ed-----n-----------e----ed-------d---e--n
XZWGCHSMRNMDHNCMFQCHZJMXJZWIEJYUCFWDJNZDIR
```

Our next step might be to try $d_K(N) = h$, since NZ is a common digram and ZN is not. If this is correct, then the segment of plaintext $ne - ndhe$ suggests that $d_K(C) = a$. Incorporating these guesses, we have:

```
------end-----a---e-a--nedh--e-------a-----
YIFQFMZRWQFYVECFMDZPCVMRZWNMDZVEJBTXCDDUMJ

h-------ea---e-a---a---nhad-a-en--a-e-h--e
NDIFEFMDZCDMQZKCEYFCJMYRNCWJCSZREXCHZUNMXZ

he-a-n------n------ed---e---e--neandhe-e--
NZUCDRJXYYSMRTMEYIFZWDYVZVYFZUMRZCRWNZDZJJ

-ed-a---nh---ha---a-e----ed-----a-d--he--n
XZWGCHSMRNMDHNCMFQCHZJMXJZWIEJYUCFWDJNZDIR
```

Now, we might consider M, the second most common ciphertext character. The ciphertext segment RNM, which we believe decrypts to $nh-$, suggests that $h-$ begins a word, so M probably represents a vowel. We have already accounted for a and e, so we expect that $d_K(M) = i$ or o. Since ai is a much more likely digram than ao, the ciphertext digram CM suggests that we try $d_K(M) = i$ first. Then we have:

```
-----iend-----a-i-e-a-inedhi-e-------a---i-
YIFQFMZRWQFYVECFMDZPCVMRZWNMDZVEJBTXCDDUMJ

h-----i-ea-i-e-a---a-i-nhad-a-en--a-e-hi-e
NDIFEFMDZCDMQZKCEYFCJMYRNCWJCSZREXCHZUNMXZ

he-a-n-----in-i----ed---e---e-ineandhe-e--
NZUCDRJXYYSMRTMEYIFZWDYVZVYFZUMRZCRWNZDZJJ

-ed-a--inhi--hai--a-e-i--ed-----a-d--he--n
XZWGCHSMRNMDHNCMFQCHZJMXJZWIEJYUCFWDJNZDIR
```

Next, we might try to determine which letter is encrypted to o. Since o is a common letter, we guess that the corresponding ciphertext letter is one of D, F, J, Y. Y seem to be the most likely possibility; otherwise, we would get long strings of vowels, namely aoi from CFM or CJM. Hence, let's suppose $d_E(Y) = o$.

The three most frequent remaining ciphertext letters are D, F, J, which we conjecture could decrypt to r, s, t in some order. Two occurrences of the trigram NMD suggest that $d_E(D) = s$, giving the trigram his in the plaintext (this is consistent with our earlier hypothesis that $d_E(D) \in \{r, s, t\}$). The segment $HNCMF$ could be an encryption of *chair*, which would give $d_E(F) = r$ (and $d_E(H) = c$) and so we would then have $d_E(J) = t$ by process of elimination.

Now, we have:

```
o-r-riend-ro--arise-a-inedhise--t---ass-it
YIFQFMZRWQFYVECFMDZPCVMRZWNMDZVEJBTXCDDUMJ

hs-r-riseasi-e-a-orationhadta-en--ace-hi-e
NDIFEFMDZCDMQZKCEYFCJMYRNCWJCSZREXCHZUNMXZ

he-asnt-oo-in-i-o-redso-e-ore-ineandhesett
NZUCDRJXYYSMRTMEYIFZWDYVZVYFZUMRZCRWNZDZJJ

-ed-ac-inhischair-aceti-ted--to-ardsthes-n
XZWGCHSMRNMDHNCMFQCHZJMXJZWIEJYUCFWDJNZDIR
```

It is now very easy to determine the plaintext and the key for Example 1.11. The complete decryption is the following:

> Our friend from Paris examined his empty glass with surprise, as if evaporation had taken place while he wasn't looking. I poured some more wine and he settled back in his chair, face tilted up towards the sun.[1]

□

1.2.3 Cryptanalysis of the Vigenère Cipher

In this section we describe some methods for cryptanalyzing the *Vigenère Cipher*. The first step is to determine the keyword length, which we denote by m. There are a couple of techniques that can be employed. The first of these is the so-called Kasiski test and the second uses the index of coincidence.

The *Kasiski test* was described by Friedrich Kasiski in 1863; however, it was apparently discovered earlier, around 1854, by Charles Babbage. It is based on the observation that two identical segments of plaintext will be encrypted to the same ciphertext whenever their occurrence in the plaintext is δ positions apart, where $\delta \equiv 0 \pmod{m}$. Conversely, if we observe two identical segments of ciphertext, each of length at least three, say, then there is a good chance that they correspond to identical segments of plaintext.

The Kasiski test works as follows. We search the ciphertext for pairs of identical segments of length at least three, and record the distance between the starting positions of the two segments. If we obtain several such distances, say $\delta_1, \delta_2, \ldots$, then we would conjecture that m divides all of the δ_i's, and hence m divides the greatest common divisor of the δ_i's.

---

[1] P. Mayle, *A Year in Provence*, A. Knopf, Inc., 1989.

Further evidence for the value of m can be obtained by the index of coincidence. This concept was defined by Wolfe Friedman in 1920, as follows.

Definition 1.7:   Suppose $\mathbf{x} = x_1 x_2 \cdots x_n$ is a string of n alphabetic characters. The *index of coincidence* of $\mathbf{x}$, denoted $I_c(\mathbf{x})$, is defined to be the probability that two random elements of $\mathbf{x}$ are identical.

Suppose we denote the frequencies of $A, B, C, \ldots, Z$ in $\mathbf{x}$ by $f_0, f_1, \ldots, f_{25}$ (respectively). We can choose two elements of $\mathbf{x}$ in $\binom{n}{2}$ ways.[2] For each i, $0 \leq i \leq 25$, there are $\binom{f_i}{2}$ ways of choosing both elements to be i. Hence, we have the formula

$$I_c(\mathbf{x}) = \frac{\sum_{i=0}^{25} \binom{f_i}{2}}{\binom{n}{2}} = \frac{\sum_{i=0}^{25} f_i(f_i - 1)}{n(n-1)}.$$

Suppose $\mathbf{x}$ is a string of English language text. Denote the expected probabilities of occurrence of the letters $A, B, \ldots, Z$ in Table 1.1 by $p_0, \ldots, p_{25}$, respectively. Then, we would expect that

$$I_c(\mathbf{x}) \approx \sum_{i=0}^{25} p_i{}^2 = 0.065,$$

since the probability that two random elements both are A is $p_0{}^2$, the probability that both are B is $p_1{}^2$, etc. The same reasoning applies if $\mathbf{x}$ is a ciphertext string obtained using any monoalphabetic cipher. In this case, the individual probabilities will be permuted, but the quantity $\sum p_i{}^2$ will be unchanged.

Now, suppose we start with a ciphertext string $\mathbf{y} = y_1 y_2 \cdots y_n$ that has been constructed by using a *Vigenère Cipher*. Define m substrings of $\mathbf{y}$, denoted $\mathbf{y}_1, \mathbf{y}_2, \ldots, \mathbf{y}_m$, by writing out the ciphertext, in columns, in a rectangular array of dimensions $m \times (n/m)$. The rows of this matrix are the substrings $\mathbf{y}_i$, $1 \leq i \leq m$. In other words, we have that

$$\mathbf{y}_1 = y_1 y_{m+1} y_{2m+1} \cdots,$$

$$\mathbf{y}_2 = y_2 y_{m+2} y_{2m+2} \cdots,$$

$$\vdots \quad \vdots \quad \vdots$$

$$\mathbf{y}_m = y_m y_{2m} y_{3m} \cdots.$$

If $\mathbf{y}_1, \mathbf{y}_2, \ldots, \mathbf{y}_m$ are constructed in this way, and m is indeed the keyword length, then each value $I_c(\mathbf{y}_i)$ should be roughly equal to 0.065. On the other

---

[2]The *binomial coefficient* $\binom{n}{k} = n!/(k!(n-k)!)$ denotes the number of ways of choosing a subset of k objects from a set of n objects.

hand, if m is not the keyword length, then the substrings $\mathbf{y}_i$ will look much more random, since they will have been obtained by shift encryption with different keys. Observe that a completely random string will have

$$I_c \approx 26 \left(\frac{1}{26} \right)^2 = \frac{1}{26} = 0.038.$$

The two values 0.065 and 0.038 are sufficiently far apart that we will often be able to determine the correct keyword length by this method (or confirm a guess that has already been made using the Kasiski test).

Let us illustrate these two techniques with an example.

Example 1.12   Ciphertext obtained from a *Vigenère Cipher*

```
CHREEVOAHMAERATBIAXXWTNXBEEOPHBSBQMQEQERBW
RVXUOAKXAOSXXWEAHBWGJMMQMNKGRFVGXWTRZXWIAK
LXFPSKAUTEMNDCMGTSXMXBTUIADNGMGPSRELXNJELX
VRVPRTULHDNQWTWDTYGBPHXTFALJHASVBFXNGLLCHR
ZBWELEKMSJIKNBHWRJGNMGJSGLXFEYPHAGNRBIEQJT
AMRVLCRREMNDGLXRRIMGNSNRWCHRQHAEYEVTAQEBBI
PEEWEVKAKOEWADREMXMTBHHCHRTKDNVRZCHRCLQOHP
WQAIIWXNRMGWOIIFKEE
```

First, let's try the Kasiski test. The ciphertext string CHR occurs in five places in the ciphertext, beginning at positions 1, 166, 236, 276 and 286. The distances from the first occurrence to the other four occurrences are (respectively) 165, 235, 275 and 285. The greatest common divisor of these four integers is 5, so that is very likely the keyword length.

Let's see if computation of indices of coincidence gives the same conclusion. With $m = 1$, the index of coincidence is 0.045. With $m = 2$, the two indices are 0.046 and 0.041. With $m = 3$, we get 0.043, 0.050, 0.047. With $m = 4$, we have indices 0.042, 0.039, 0.046, 0.040. Then, trying $m = 5$, we obtain the values 0.063, 0.068, 0.069, 0.061 and 0.072. This also provides strong evidence that the keyword length is five. ⬚

Assuming that we have determined the correct value of m, how do we determine the actual key, $K = (k_1, k_2, \ldots, k_m)$? We describe a simple and effective method now. Let $1 \leq i \leq m$, and let $f_0, \ldots, f_{25}$ denote the frequencies of $A, B, \ldots, Z$, respectively, in the string $\mathbf{y}_i$. Also, let $n' = n/m$ denote the length of the string $\mathbf{y}_i$. Then the probability distribution of the 26 letters in $\mathbf{y}_i$ is

$$\frac{f_0}{n'}, \ldots, \frac{f_{25}}{n'}.$$

Now, recall that the substring $\mathbf{y}_i$ is obtained by shift encryption of a subset of the plaintext elements using a shift k_i. Therefore, we would hope that the shifted probability distribution

$$\frac{f_{k_i}}{n'}, \ldots, \frac{f_{25+k_i}}{n'}$$

would be "close to" the ideal probability distribution $p_0, \ldots, p_{25}$ tabulated in Table 1.1, where subscripts in the above formula are evaluated modulo 26.

Suppose that $0 \leq g \leq 25$, and define the quantity

$$M_g = \sum_{i=0}^{25} \frac{p_i f_{i+g}}{n'}. \tag{1.1}$$

If $g = k_i$, then we would expect that

$$M_g \approx \sum_{i=0}^{25} {p_i}^2 = 0.065,$$

as in the consideration of the index of coincidence. If $g \neq k_i$, then M_g will usually be significantly smaller than 0.065 (see the exercises for a justification of this statement). Hopefully this technique will allow us to determine the correct value of k_i for each value of i, $1 \leq i \leq m$.

Let us illustrate by returning to Example 1.12.

Example 1.12   (Cont.)   We have hypothesized that the keyword length is 5. We now compute the values M_g as described above, for $1 \leq i \leq 5$. These values are tabulated in Table 1.4. For each i, we look for a value of M_g that is close to 0.065. These g's determine the shifts $k_1, \ldots, k_5$.

From the data in Table 1.4, we see that the key is likely to be $K = (9, 0, 13, 4, 19)$, and hence the keyword likely is $JANET$. This is correct, and the complete decryption of the ciphertext is the following:

> The almond tree was in tentative blossom. The days were longer, often ending with magnificent evenings of corrugated pink skies. The hunting season was over, with hounds and guns put away for six months. The vineyards were busy again as the well-organized farmers treated their vines and the more lackadaisical neighbors hurried to do the pruning they should have done in November.[3]

                                                                                    $\square$

1.2.4   Cryptanalysis of the Hill Cipher

The *Hill Cipher* can be difficult to break with a ciphertext-only attack, but it succumbs easily to a known plaintext attack. Let us first assume that the opponent

---

[3] P. Mayle, *A Year in Provence*, A. Knopf, Inc., 1989.

TABLE 1.4
Values of M_g

i	value of $M_g(\mathbf{y}_i)$								
1	.035	.031	.036	.037	.035	.039	.028	.028	.048
	.061	.039	.032	.040	.038	.038	.044	.036	.030
	.042	.043	.036	.033	.049	.043	.041	.036	
2	.069	.044	.032	.035	.044	.034	.036	.033	.030
	.031	.042	.045	.040	.045	.046	.042	.037	.032
	.034	.037	.032	.034	.043	.032	.026	.047	
3	.048	.029	.042	.043	.044	.034	.038	.035	.032
	.049	.035	.031	.035	.065	.035	.038	.036	.045
	.027	.035	.034	.034	.037	.035	.046	.040	
4	.045	.032	.033	.038	.060	.034	.034	.034	.050
	.033	.033	.043	.040	.033	.028	.036	.040	.044
	.037	.050	.034	.034	.039	.044	.038	.035	
5	.034	.031	.035	.044	.047	.037	.043	.038	.042
	.037	.033	.032	.035	.037	.036	.045	.032	.029
	.044	.072	.036	.027	.030	.048	.036	.037	

has determined the value of m being used. Suppose he has at least m distinct plaintext-ciphertext pairs, say

$$x_j = (x_{1,j}, x_{2,j}, \ldots, x_{m,j})$$

and

$$y_j = (y_{1,j}, y_{2,j}, \ldots, y_{m,j}),$$

for $1 \leq j \leq m$, such that $y_j = e_K(x_j)$, $1 \leq j \leq m$. If we define two $m \times m$ matrices $X = (x_{i,j})$ and $Y = (y_{i,j})$, then we have the matrix equation $Y = XK$, where the $m \times m$ matrix K is the unknown key. Provided that the matrix X is invertible, Oscar can compute $K = X^{-1}Y$ and thereby break the system. (If X is not invertible, then it will be necessary to try other sets of m plaintext-ciphertext pairs.)

Let's look at a simple example.

Example 1.13 Suppose the plaintext *friday* is encrypted using a *Hill Cipher* with $m = 2$, to give the ciphertext $PQCFKU$.

We have that $e_K(5, 17) = (15, 16)$, $e_K(8, 3) = (2, 5)$ and $e_K(0, 24) = (10, 20)$. From the first two plaintext-ciphertext pairs, we get the matrix equation

$$\begin{pmatrix} 15 & 16 \\ 2 & 5 \end{pmatrix} = \begin{pmatrix} 5 & 17 \\ 8 & 3 \end{pmatrix} K.$$

Using Corollary 1.4, it is easy to compute

$$\begin{pmatrix} 5 & 17 \\ 8 & 3 \end{pmatrix}^{-1} = \begin{pmatrix} 9 & 1 \\ 2 & 15 \end{pmatrix},$$

so

$$K = \begin{pmatrix} 9 & 1 \\ 2 & 15 \end{pmatrix} \begin{pmatrix} 15 & 16 \\ 2 & 5 \end{pmatrix} = \begin{pmatrix} 7 & 19 \\ 8 & 3 \end{pmatrix}.$$

This can be verified by using the third plaintext-ciphertext pair.   ☐

What would the opponent do if he does not know m? Assuming that m is not too big, he could simply try $m = 2, 3, \ldots$, until the key is found. If a guessed value of m is incorrect, then an $m \times m$ matrix found by using the algorithm described above will not agree with further plaintext-ciphertext pairs. In this way, the value of m can be determined if it is not known ahead of time.

1.2.5   Cryptanalysis of the LFSR Stream Cipher

Recall that the ciphertext is the sum modulo 2 of the plaintext and the keystream, i.e., $y_i = (x_i + z_i) \bmod 2$. The keystream is produced from an initial m-tuple, $(z_1, \ldots, z_m) = (k_1, \ldots, k_m)$, using the linear recurrence

$$z_{m+i} = \sum_{j=0}^{m-1} c_j z_{i+j} \bmod 2,$$

$i \geq 1$, where $c_0, \ldots, c_{m-1} \in \mathbb{Z}_2$.

Since all operations in this cryptosystem are linear, we might suspect that the cryptosystem is vulnerable to a known-plaintext attack, as is the case with the *Hill Cipher*. Suppose Oscar has a plaintext string $x_1 x_2 \cdots x_n$ and the corresponding ciphertext string $y_1 y_2 \cdots y_n$. Then he can compute the keystream bits $z_i = (x_i + y_i) \bmod 2$, $1 \leq i \leq n$. Let us also suppose that Oscar knows the value of m. Then Oscar needs only to compute $c_0, \ldots, c_{m-1}$ in order to be able to reconstruct the entire keystream. In other words, he needs to be able to determine the values of m unknowns.

Now, for any $i \geq 1$, we have

$$z_{m+i} = \sum_{j=0}^{m-1} c_j z_{i+j} \bmod 2,$$

which is a linear equation in the m unknowns. If $n \geq 2m$, then there are m linear equations in m unknowns, which can subsequently be solved.

The system of m linear equations can be written in matrix form as follows:

$$(z_{m+1}, z_{m+2}, \ldots, z_{2m}) = (c_0, c_1, \ldots, c_{m-1}) \begin{pmatrix} z_1 & z_2 & \cdots & z_m \\ z_2 & z_3 & \cdots & z_{m+1} \\ \vdots & \vdots & & \vdots \\ z_m & z_{m+1} & \cdots & z_{2m-1} \end{pmatrix}.$$

If the coefficient matrix has an inverse (modulo 2), we obtain the solution

$$(c_0, c_1, \ldots, c_{m-1}) = (z_{m+1}, z_{m+2}, \ldots, z_{2m}) \begin{pmatrix} z_1 & z_2 & \cdots & z_m \\ z_2 & z_3 & \cdots & z_{m+1} \\ \vdots & \vdots & & \vdots \\ z_m & z_{m+1} & \cdots & z_{2m-1} \end{pmatrix}^{-1}.$$

In fact, the matrix will have an inverse if m is the degree of the recurrence used to generate the keystream (see the exercises for a proof).

Let's illustrate with an example.

Example 1.14 Suppose Oscar obtains the ciphertext string

$$101101011110010$$

corresponding to the plaintext string

$$011001111111000.$$

Then he can compute the keystream bits:

$$110100100001010.$$

Suppose also that Oscar knows that the keystream was generated using a 5-stage LFSR. Then he would solve the following matrix equation, which is obtained from the first 10 keystream bits:

$$(0, 1, 0, 0, 0) = (c_0, c_1, c_2, c_3, c_4) \begin{pmatrix} 1 & 1 & 0 & 1 & 0 \\ 1 & 0 & 1 & 0 & 0 \\ 0 & 1 & 0 & 0 & 1 \\ 1 & 0 & 0 & 1 & 0 \\ 0 & 0 & 1 & 0 & 0 \end{pmatrix}.$$

It can be verified that

$$\begin{pmatrix} 1 & 1 & 0 & 1 & 0 \\ 1 & 0 & 1 & 0 & 0 \\ 0 & 1 & 0 & 0 & 1 \\ 1 & 0 & 0 & 1 & 0 \\ 0 & 0 & 1 & 0 & 0 \end{pmatrix}^{-1} = \begin{pmatrix} 0 & 1 & 0 & 0 & 1 \\ 1 & 0 & 0 & 1 & 0 \\ 0 & 0 & 0 & 0 & 1 \\ 0 & 1 & 0 & 1 & 1 \\ 1 & 0 & 1 & 1 & 0 \end{pmatrix},$$

by checking that the product of the two matrices, computed modulo 2, is the identity matrix. This yields

$$(c_0, c_1, c_2, c_3, c_4) = (0, 1, 0, 0, 0) \begin{pmatrix} 0 & 1 & 0 & 0 & 1 \\ 1 & 0 & 0 & 1 & 0 \\ 0 & 0 & 0 & 0 & 1 \\ 0 & 1 & 0 & 1 & 1 \\ 1 & 0 & 1 & 1 & 0 \end{pmatrix}$$

$$= (1, 0, 0, 1, 0).$$

Thus the recurrence used to generate the keystream is

$$z_{i+5} = (z_i + z_{i+3}) \bmod 2.$$

◻

1.3 Notes

Material on classical cryptography is covered in various textbooks and monographs, such as "Decrypted Secrets, Methods and Maxims of Cryptology" by Bauer [5]; "Cipher Systems, The Protection of Communications" by Beker and Piper [8]; "Cryptology" by Beutelspacher [20]; "Cryptography and Data Security" by Denning [63]; "Code Breaking, A History and Exploration" by Kippenhahn [116]; "Cryptography, A Primer" by Konheim [123]; and "Basic Methods of Cryptography" by van der Lubbe [136].

We have used the statistical data on frequency of English letters that is reported in Beker and Piper [8].

A good reference for elementary number theory is "Elementary Number Theory and its Applications" by Rosen [182]. Background in linear algebra can be found in "Elementary Linear Algebra" by Anton [3].

Two very enjoyable and readable books that provide interesting histories of cryptography are "The Codebreakers" by Kahn [109] and "The Code Book" by Singh [197].

Exercises

   1.1 Evaluate the following:
         (a) 7503 mod 81
         (b) (−7503) mod 81
         (c) 81 mod 7503

(d) (-81) mod 7503.

1.2 Suppose that $a, m > 0$, and $a \not\equiv 0 \pmod{m}$. Prove that

$$(-a) \bmod m = m - (a \bmod m).$$

1.3 Prove that $a \bmod m = b \bmod m$ if and only if $a \equiv b \pmod{m}$.

1.4 Prove that $a \bmod m = a - \lfloor \frac{a}{m} \rfloor m$, where $\lfloor x \rfloor = \max\{y \in \mathbb{Z} : y \leq x\}$.

1.5 Use exhaustive key search to decrypt the following ciphertext, which was encrypted using a *Shift Cipher*:

BEEAKFYDJXUQYHYJIQRYHTYJIQFBQDUYJIIKFUHCQD.

1.6 If an encryption function e_K is identical to the decryption function d_K, then the key K is said to be an *involutory key*. Find all the involutory keys in the *Shift Cipher* over $\mathbb{Z}_{26}$.

1.7 Determine the number of keys in an *Affine Cipher* over $\mathbb{Z}_m$ for $m = 30, 100$ and 1225.

1.8 List all the invertible elements in $\mathbb{Z}_m$ for $m = 28, 33$ and 35.

1.9 For $1 \leq a \leq 28$, determine a^{-1} mod 29 by trial and error.

1.10 Suppose that $K = (5, 21)$ is a key in an *Affine Cipher* over $\mathbb{Z}_{29}$.

    (a) Express the decryption function $d_K(y)$ in the form $d_K(y) = a'y + b'$, where $a', b' \in \mathbb{Z}_{29}$.

    (b) Prove that $d_K(e_K(x)) = x$ for all $x \in \mathbb{Z}_{29}$.

1.11   (a) Suppose that $K = (a, b)$ is a key in an *Affine Cipher* over $\mathbb{Z}_n$. Prove that K is an involutory key if and only if $a^{-1} \bmod n = a$ and $b(a + 1) \equiv 0 \pmod{n}$.

    (b) Determine all the involutory keys in the *Affine Cipher* over $\mathbb{Z}_{15}$.

    (c) Suppose that $n = pq$, where p and q are distinct odd primes. Prove that the number of involutory keys in the *Affine Cipher* over $\mathbb{Z}_n$ is $n + p + q + 1$.

1.12   (a) Let p be prime. Prove that the number of 2×2 matrices that are invertible over $\mathbb{Z}_p$ is $(p^2 - 1)(p^2 - p)$.

> **HINT**   Since p is prime, $\mathbb{Z}_p$ is a field. Use the fact that a matrix over a field is invertible if and only if its rows are linearly independent vectors (i.e., there does not exist a non-zero linear combination of the rows whose sum is the vector of all 0's).

    (b) For p prime and $m \geq 2$ an integer, find a formula for the number of $m \times m$ matrices that are invertible over $\mathbb{Z}_p$.

1.13 For $n = 6, 9$ and 26, how many 2×2 matrices are there that are invertible over $\mathbb{Z}_n$?

1.14   (a) Prove that $\det A \equiv \pm 1 \pmod{26}$ if A is a matrix over $\mathbb{Z}_{26}$ such that $A = A^{-1}$.

    (b) Use the formula given in Corollary 1.4 to determine the number of involutory keys in the *Hill Cipher* (over $\mathbb{Z}_{26}$) in the case $m = 2$.

1.15 Determine the inverses of the following matrices over $\mathbb{Z}_{26}$:

    (a) $\begin{pmatrix} 2 & 5 \\ 9 & 5 \end{pmatrix}$

    (b) $\begin{pmatrix} 1 & 11 & 12 \\ 4 & 23 & 2 \\ 17 & 15 & 9 \end{pmatrix}$

1.16   (a) Suppose that π is the following permutation of $\{1, \ldots, 8\}$:

x	1	2	3	4	5	6	7	8
$\pi(x)$	4	1	6	2	7	3	8	5

Compute the permutation π^{-1}.

(b) Decrypt the following ciphertext, for a *Permutation Cipher* with $m = 8$, which was encrypted using the key π:

ETEGENLMDNTNEOORDAHATECOESAHLRMI.

1.17   (a) Prove that a permutation π in the *Permutation Cipher* is an involutory key if and only if $\pi(i) = j$ implies $\pi(j) = i$, for all $i, j \in \{1, \ldots, m\}$.

(b) Determine the number of involutory keys in the *Permutation Cipher* for $m = 2, 3, 4, 5$ and 6.

1.18 Consider the following linear recurrence over $\mathbb{Z}_2$ of degree four:

$$z_{i+4} = (z_i + z_{i+1} + z_{i+2} + z_{i+3}) \bmod 2,$$

$i \geq 0$. For each of the 16 possible initialization vectors $(z_0, z_1, z_2, z_4) \in (\mathbb{Z}_2)^4$, determine the period of the resulting keystream.

1.19 Redo the preceding question, using the recurrence

$$z_{i+4} = (z_i + z_{i+3}) \bmod 2,$$

$i \geq 0$.

1.20 Suppose we construct a keystream in a synchronous stream cipher using the following method. Let $K \in \mathcal{K}$ be the key, let $\mathcal{L}$ be the keystream alphabet, and let Σ be a finite set of *states*. First, an *initial state* $\sigma_0 \in \Sigma$ is determined from K by some method. For all $i \geq 1$, the state σ_i is computed from the previous state σ_{i-1} according to the following rule:

$$\sigma_i = f(\sigma_{i-1}, K),$$

where $f : \Sigma \times \mathcal{K} \to \Sigma$. Also, for all $i \geq 1$, the keystream element z_i is computed using the following rule:

$$z_i = g(\sigma_i, K),$$

where $g : \Sigma \times \mathcal{K} \to \mathcal{L}$. Prove that any keystream produced by this method has period at most $|\Sigma|$.

1.21 Below are given four examples of ciphertext, one obtained from a *Substitution Cipher*, one from a *Vigenère Cipher*, one from an *Affine Cipher*, and one unspecified. In each case, the task is to determine the plaintext.

Give a clearly written description of the steps you followed to decrypt each ciphertext. This should include all statistical analysis and computations you performed.

The first two plaintexts were taken from "The Diary of Samuel Marchbanks," by Robertson Davies, Clarke Irwin, 1947; the fourth was taken from "Lake Wobegon Days," by Garrison Keillor, Viking Penguin, Inc., 1985.

(a) *Substitution Cipher*:

```
EMGLOSUDCGDNCUSWYSFHNSFCYKDPUMLWGYICOXYSIPJCK
QPKUGKMGOLICGINCGACKSNISACYKZSCKXECJCKSHYSXCG
OIDPKZCNKSHICGIWYGKKGKGOLDSILKGOIUSIGLEDSPWZU
GFZCCNDGYYSFUSZCNXEOJNCGYEOWEUPXEZGACGNFGLKNS
ACIGOIYCKXCJUCIUZCFZCCNDGYYSFEUEKUZCSOCFZCCNC
IACZEJNCSHFZEJZEGMXCYHCJUMGKUCY
```

HINT   *F* decrypts to *w*.

(b) *Vigenère Cipher*:

KCCPKBGUFDPHQTYAVINRRTMVGRKDNBVFDETDGILTXRGUD
DKOTFMBPVGEGLTGCKQRACQCWDNAWCRXIZAKFTLEWRPTYC
QKYVXCHKFTPONCQQRHJVAJUWETMCMSPKQDYHJVDAHCTRL
SVSKCGCZQQDZXGSFRLSWCWSJTBHAFSIASPRJAHKJRJUMV
GKMITZHFPDISPZLVLGWTFPLKKEBDPGCEBSHCTJRWXBAFS
PEZQNRWXCVYCGAONWDDKACKAWBBIKFTIOVKCGGHJVLNHI
FFSQESVYCLACNVRWBBIREPBBVFEXOSCDYGZWPFDTKFQIY
CWHJVLNHIQIBTKHJVNPIST

(c) *Affine Cipher*:

KQEREJEBCPPCJCRKIEACUZBKRVPKRBCIBQCARBJCVFCUP
KRIOFKPACUZQEPBKRXPEIIEABDKPBCPFCDCCAFIEABDKP
BCPFEQPKAZBKRHAIBKAPCCIBURCCDKDCCJCIDFUIXPAFF
ERBICZDFKABICBBENEFCUPJCVKABPCYDCCDPKBCOCPERK
IVKSCPICBRKIJPKABI

(d) unspecified cipher:

BNVSNSIHQCEELSSKKYERIFJKXUMBGYKAMQLJTYAVFBKVT
DVBPVVRJYYLAOKYMPQSCGDLFSRLLPROYGESEBUUALRWXM
MASAZLGLEDFJBZAVVPXWICGJXASCBYEHOSNMULKCEAHTQ
OKMFLEBKFXLRRFDTZXCIWBJSICBGAWDVYDHAVFJXZIBKC
GJIWEAHTTOEWTUHKRQVVRGZBXYIREMMASCSPBNLHJMBLR
FFJELHWEYLWISTFVVYFJCMHYUYRUFSFMGESIGRLWALSWM
NUHSIMYYITCCQPZSICEHBCCMZFEGVJYOCDEMMPGHVAAUM
ELCMOEHVLTIPSUYILVGFLMVWDVYDBTHFRAYISYSGKVSUU
HYHGGCKTMBLRX

1.22 (a) Suppose that $p_1, \ldots, p_n$ and $q_1, \ldots, q_n$ are both probability distributions, and $p_1 \geq \cdots \geq p_n$. Let $q_1', \ldots, q_n'$ be any permutation of $q_1, \ldots, q_n$. Prove that the quantity

$$\sum_{i=1}^{n} p_i q_i'$$

is maximized when $q_1' \geq \cdots \geq q_n'$.

(b) Explain why the expression in Equation (1.1) is likely to be maximized when $g = k_i$.

1.23 Suppose we are told that the plaintext

breathtaking

yields the ciphertext

UPOTENTOIFV

where the *Hill Cipher* is used (but m is not specified). Determine the encryption matrix.

1.24 An *Affine-Hill Cipher* is the following modification of a *Hill Cipher*: Let m be a positive integer, and define $\mathcal{P} = \mathcal{C} = (\mathbb{Z}_{26})^m$. In this cryptosystem, a key K consists of a pair (L, b), where L is an $m \times m$ invertible matrix over $\mathbb{Z}_{26}$, and $b \in (\mathbb{Z}_{26})^m$. For $x = (x_1, \ldots, x_m) \in \mathcal{P}$ and $K = (L, b) \in \mathcal{K}$, we compute $y = e_K(x) = (y_1, \ldots, y_m)$ by means of the formula $y = xL + b$. Hence, if

$L = (\ell_{i,j})$ and $b = (b_1, \ldots, b_m)$, then

$$(y_1, \ldots, y_m) = (x_1, \ldots, x_m) \begin{pmatrix} \ell_{1,1} & \ell_{1,2} & \cdots & \ell_{1,m} \\ \ell_{2,1} & \ell_{2,2} & \cdots & \ell_{2,m} \\ \vdots & \vdots & & \vdots \\ \ell_{m,1} & \ell_{m,2} & \cdots & \ell_{m,m} \end{pmatrix} + (b_1, \ldots, b_m).$$

Suppose Oscar has learned that the plaintext

adisplayedequation

is encrypted to give the ciphertext

DSRMSIOPLXLJBZULLM

and Oscar also knows that $m = 3$. Determine the key, showing all computations.

1.25 Here is how we might cryptanalyze the *Hill Cipher* using a ciphertext-only attack. Suppose that we know that $m = 2$. Break the ciphertext into blocks of length two letters (digrams). Each such digram is the encryption of a plaintext digram using the unknown encryption matrix. Pick out the most frequent ciphertext digram and assume it is the encryption of a common digram in the list following Table 1.1 (for example, TH or ST). For each such guess, proceed as in the known-plaintext attack, until the correct encryption matrix is found.

Here is a sample of ciphertext for you to decrypt using this method:

LMQETXYEAGTXCTUIEWNCTXLZEWUAISPZYVAPEWLMGQWYA
XFTCJMSQCADAGTXLMDXNXSNPJQSYVAPRIQSMHNOCVAXFV

1.26 We describe a special case of a *Permutation Cipher*. Let m, n be positive integers. Write out the plaintext, by rows, in $m \times n$ rectangles. Then form the ciphertext by taking the columns of these rectangles. For example, if $m = 4, n = 3$, then we would encrypt the plaintext "*cryptography*" by forming the following rectangle:

cryp
togr
aphy

The ciphertext would be "*CTAROPYGHPRY*."

(a) Describe how Bob would decrypt a ciphertext string (given values for m and n).

(b) Decrypt the following ciphertext, which was obtained by using this method of encryption:

MYAMRARUYIQTENCTORAHROYWDSOYEOUARRGDERNOGW

1.27 The purpose of this exercise is to prove the statement made in Section 1.2.5 that the $m \times m$ coefficient matrix is invertible. This is equivalent to saying that the rows of this matrix are linearly independent vectors over $\mathbb{Z}_2$.

As before, we suppose that the recurrence has the form

$$z_{m+i} = \sum_{j=0}^{m-1} c_j z_{i+j} \bmod 2.$$

$(z_1, \ldots, z_m)$ comprises the initialization vector. For $i \geq 1$, define

$$v_i = (z_i, \ldots, z_{i+m-1}).$$

Note that the coefficient matrix has the vectors $v_1, \ldots, v_m$ as its rows, so our objective is to prove that these m vectors are linearly independent.

Prove the following assertions:

(a) For any $i \geq 1$,

$$v_{m+i} = \sum_{j=0}^{m-1} c_j v_{i+j} \bmod 2.$$

(b) Choose h to be the minimum integer such that there exists a non-trivial linear combination of the vectors $v_1, \ldots, v_h$ which sums to the vector $(0, \ldots, 0)$ modulo 2. Then

$$v_h = \sum_{j=0}^{h-2} \alpha_j v_{j+1} \bmod 2,$$

and not all the α_j's are zero. Observe that $h \leq m + 1$, since any $m + 1$ vectors in an m-dimensional vector space are dependent.

(c) Prove that the keystream must satisfy the recurrence

$$z_{h-1+i} = \sum_{j=0}^{h-2} \alpha_j z_{j+i} \bmod 2$$

for any $i \geq 1$.

(d) Observe that if $h \leq m$, then the keystream satisfies a linear recurrence of degree less than m, a contradiction. Hence, $h = m + 1$, and the matrix must be invertible.

1.28 Decrypt the following ciphertext, obtained from the *Autokey Cipher*, by using exhaustive key search:

MALVVMAFBHBUQPTSOXALTGVWWRG

1.29 We describe a stream cipher that is a modification of the *Vigenère Cipher*. Given a keyword $(K_1, \ldots, K_m)$ of length m, construct a keystream by the rule $z_i = K_i$ $(1 \leq i \leq m)$, $z_{i+m} = (z_i + 1) \bmod 26$ $(i \geq 1)$. In other words, each time we use the keyword, we replace each letter by its successor modulo 26. For example, if $SUMMER$ is the keyword, we use $SUMMER$ to encrypt the first six letters, we use $TVNNFS$ for the next six letters, and so on.

(a) Describe how you can use the concept of index of coincidence to first determine the length of the keyword, and then actually find the keyword.

(b) Test your method by cryptanalyzing the following ciphertext:
IYMYSILONRFNCQXQJEDSHBUIBCJUZBOLFQYSCHATPEQGQ
JEJNGNXZWHHGWFSUKULJQACZKKJOAAHGKEMTAFGMKVRDO
PXNEHEKZNKFSKIFRQVHHOVXINPHMRTJPYWQGJWPUUVKFP
OAWPMRKKQZWLQDYAZDRMLPBJKJOBWIWPSEPVVQMBCRYVC
RUZAAOUMBCHDAGDIEMSZFZHALIGKEMJJFPCIWKRMLMPIN
AYOFIREAOLDTHITDVRMSE

The plaintext was taken from "The Codebreakers," by D. Kahn, Scribner, 1996.

1.30 We describe another stream cipher, which incorporates one of the ideas from the "Enigma" system used by Germany in World War II. Suppose that π is a fixed permutation of $\mathbb{Z}_{26}$. The key is an element $K \in \mathbb{Z}_{26}$. For all integers $i \geq 1$, the keystream element $z_i \in \mathbb{Z}_{26}$ is defined according to the rule $z_i = (K + i - 1) \bmod 26$. Encryption and decryption are performed using the permutations π and π^{-1}, respectively, as follows:

$$e_z(x) = \pi((x + z) \bmod 26)$$

and

$$d_z(y) = (\pi^{-1}(y) - z) \bmod 26,$$

where $z \in \mathbb{Z}_{26}$.

Suppose that π is the following permutation of $\mathbb{Z}_{26}$:

x	0	1	2	3	4	5	6	7	8	9	10	11	12
$\pi(x)$	23	13	24	0	7	15	14	6	25	16	22	1	19

x	13	14	15	16	17	18	19	20	21	22	23	24	25
$\pi(x)$	18	5	11	17	2	21	12	20	4	10	9	3	8

The following ciphertext has been encrypted using this stream cipher; use exhaustive key search to decrypt it:

WRTCNRLDSAFARWKXFTXCZRNHNYPDTZUUKMPLUSOXNEUDO
KLXRMCBKGRCCURR

2
Shannon's Theory

2.1 Introduction

In 1949, Claude Shannon published a paper entitled "Communication Theory of Secrecy Systems" in the *Bell Systems Technical Journal*. This paper had a great influence on the scientific study of cryptography. In this chapter, we discuss several of Shannon's ideas. First, however, we consider some of the various approaches to evaluating the security of a cryptosystem. We define some of the most useful criteria now.

computational security

> This measure concerns the computational effort required to break a cryptosystem. We might define a cryptosystem to be *computationally secure* if the best algorithm for breaking it requires at least N operations, where N is some specified, very large number. The problem is that no known practical cryptosystem can be proved to be secure under this definition. In practice, people often study the computational security of a cryptosystem with respect to certain specific types of attacks (e.g., an exhaustive key search). Of course, security against one specific type of attack does not guarantee security against some other type of attack.

provable security

> Another approach is to provide evidence of security by reducing the security of the cryptosystem to some well-studied problem that is thought to be difficult. For example, it may be able to prove a statement of the type "a given cryptosystem is secure if a given integer n cannot be factored." Cryptosystems of this type are sometimes termed *provably secure*, but it must be understood that this approach only provides a proof of security relative to some other problem, not an absolute proof of security. This is a similar

situation to proving that a problem is NP-complete: it proves that the given problem is at least as difficult as any other NP-complete problem, but it does not provide an absolute proof of the computational difficulty of the problem.

unconditional security

This measure concerns the security of cryptosystems when there is no bound placed on the amount of computation that Oscar is allowed to do. A cryptosystem is defined to be *unconditionally secure* if it cannot be broken, even with infinite computational resources.

When we discuss the security of a cryptosystem, we should also specify the type of attack that is being considered. For example, in Chapter 1, we saw that neither the *Shift Cipher*, the *Substitution Cipher* nor the *Vigenère Cipher* is computationally secure against a ciphertext-only attack (given a sufficient amount of ciphertext).

What we will do in Section 2.3 is to develop a theory of cryptosystems that are unconditionally secure against a ciphertext-only attack. This theory allows us to prove mathematically that certain cryptosystems are secure if the amount of ciphertext is sufficiently small. For example, it turns out that the *Shift Cipher* and the *Substitution Cipher* are both unconditionally secure if a single element of plaintext is encrypted with a given key. Similarly, the *Vigenère Cipher* with keyword length m is unconditionally secure if the key is used to encrypt only one element of plaintext (which consists of m alphabetic characters).

2.2   Elementary Probability Theory

The unconditional security of a cryptosystem obviously cannot be studied from the point of view of computational complexity because we allow computation time to be infinite. The appropriate framework in which to study unconditional security is probability theory. We need only elementary facts concerning probability; the main definitions are reviewed now. First, we define the idea of a random variable.

Definition 2.1:   A *discrete random variable*, say $\mathbf{X}$, consists of a finite set X and a *probability distribution* defined on X. The probability that the random variable $\mathbf{X}$ takes on the value x is denoted $\mathbf{Pr}[\mathbf{X} = x]$; sometimes we will abbreviate this to $\mathbf{Pr}[x]$ if the random variable $\mathbf{X}$ is fixed. It must be the case that $0 \leq \mathbf{Pr}[x]$ for all $x \in X$, and

$$\sum_{x \in X} \mathbf{Pr}[x] = 1.$$

As an example, we could consider a coin toss to be a random variable defined on the set {*heads, tails*}. The associated probability distribution would be $\mathbf{Pr}[heads] = \mathbf{Pr}[tails] = 1/2$.

Suppose we have random variable $\mathbf{X}$ defined on X, and $E \subseteq X$. The probability that $\mathbf{X}$ takes on a value in the subset E is computed to be

$$\mathbf{Pr}[x \in E] = \sum_{x \in E} \mathbf{Pr}[x]. \tag{2.1}$$

The subset E is often called an *event*.

Example 2.1 Suppose we consider a random throw of a pair of dice. This can be modeled by a random variable $\mathbf{Z}$ defined on the set

$$Z = \{1, 2, 3, 4, 5, 6\} \times \{1, 2, 3, 4, 5, 6\},$$

where $\mathbf{Pr}[(i, j)] = 1/36$ for all $(i, j) \in Z$. Let's consider the sum of the two dice. Each possible sum defines an event, and the probabilities of these events can be computed using equation (2.1). For example, suppose that we want to compute the probability that the sum is 4. This corresponds to the event

$$S_4 = \{(1, 3), (2, 2), (3, 1)\},$$

and therefore $\mathbf{Pr}[S_4] = 3/36 = 1/12$.

The probabilities of all the sums can be computed in a similar fashion. If we denote by S_j the event that the sum is j, then we obtain the following: $\mathbf{Pr}[S_2] = \mathbf{Pr}[S_{12}] = 1/36$, $\mathbf{Pr}[S_3] = \mathbf{Pr}[S_{11}] = 1/18$, $\mathbf{Pr}[S_4] = \mathbf{Pr}[S_{10}] = 1/12$, $\mathbf{Pr}[S_5] = \mathbf{Pr}[S_9] = 1/9$, $\mathbf{Pr}[S_6] = \mathbf{Pr}[S_8] = 5/36$, and $\mathbf{Pr}[S_7] = 1/6$. □

Since the events $S_2, \ldots, S_{12}$ partition the set S, it follows that we can consider the value of the sum of a pair of dice to be a random variable in its own right, which has the probability distribution computed in Example 2.1.

We next consider the concepts of joint and conditional probabilities.

Definition 2.2: Suppose $\mathbf{X}$ and $\mathbf{Y}$ are random variables defined on finite sets X and Y, respectively. The *joint probability* $\mathbf{Pr}[x, y]$ is the probability that $\mathbf{X}$ takes on the value x and $\mathbf{Y}$ takes on the value y. The *conditional probability* $\mathbf{Pr}[x|y]$ denotes the probability that $\mathbf{X}$ takes on the value x given that $\mathbf{Y}$ takes on the value y. The random variables $\mathbf{X}$ and $\mathbf{Y}$ are said to be *independent random variables* if $\mathbf{Pr}[x, y] = \mathbf{Pr}[x]\mathbf{Pr}[y]$ for all $x \in X$ and $y \in Y$.

Joint probability can be related to conditional probability by the formula

$$\mathbf{Pr}[x, y] = \mathbf{Pr}[x|y]\mathbf{Pr}[y].$$

Interchanging x and y, we have that

$$\mathbf{Pr}[x, y] = \mathbf{Pr}[y|x]\mathbf{Pr}[x].$$

From these two expressions, we immediately obtain the following result, which is known as Bayes' theorem.

THEOREM 2.1 *(Bayes' theorem)* *If* $\mathbf{Pr}[y] > 0$, *then*

$$\mathbf{Pr}[x|y] = \frac{\mathbf{Pr}[x]\mathbf{Pr}[y|x]}{\mathbf{Pr}[y]}.$$

COROLLARY 2.2 $\mathbf{X}$ *and* $\mathbf{Y}$ *are independent random variables if and only if* $\mathbf{Pr}[x|y] = \mathbf{Pr}[x]$ *for all* $x \in X$ *and* $y \in Y$.

Example 2.2 Suppose we consider a random throw of a pair of dice. Let $\mathbf{X}$ be the random variable defined on the set $X = \{2, \ldots, 12\}$, obtained by considering the sum of two dice, as in Example 2.1. Further, suppose that $\mathbf{Y}$ is a random variable which takes on the value D if the two dice are the same (i.e., if we throw "doubles"), and the value N, otherwise. Then we have that $\mathbf{Pr}[D] = 1/6$, $\mathbf{Pr}[N] = 5/6$.

It is straightforward to compute joint and conditional probabilities for these random variables. For example, the reader can check that $\mathbf{Pr}[D|4] = 1/3$ and $\mathbf{Pr}[4|D] = 1/6$, so

$$\mathbf{Pr}[D|4]\mathbf{Pr}[4] = \mathbf{Pr}[D]\mathbf{Pr}[4|D],$$

as stated by Bayes' theorem.   ⬚

2.3 Perfect Secrecy

Throughout this section, we assume that a cryptosystem $(\mathcal{P}, \mathcal{C}, \mathcal{K}, \mathcal{E}, \mathcal{D})$ is specified, and a particular key $K \in \mathcal{K}$ is used for only one encryption. Let us suppose that there is a probability distribution on the plaintext space, $\mathcal{P}$. Thus the plaintext element defines a random variable, denoted $\mathbf{x}$. We denote the *a priori* probability that plaintext x occurs by $\mathbf{Pr}[\mathbf{x} = x]$. We also assume that the key K is chosen (by Alice and Bob) using some fixed probability distribution (often a key is chosen at random, so all keys will be equiprobable, but this need not be the case). So the key also defines a random variable, which we denote by $\mathbf{K}$. Denote the probability that key K is chosen by $\mathbf{Pr}[\mathbf{K} = K]$. Recall that the key is chosen before Alice knows what the plaintext will be. Hence, we make the reasonable assumption that the key and the plaintext are independent random variables.

The two probability distributions on $\mathcal{P}$ and $\mathcal{K}$ induce a probability distribution on $\mathcal{C}$. Thus, we can also consider the ciphertext element to be a random variable, say $\mathbf{y}$. It is not hard to compute the probability $\mathbf{Pr}[\mathbf{y} = y]$ that y is the ciphertext that is transmitted. For a key $K \in \mathcal{K}$, define

$$C(K) = \{e_K(x) : x \in \mathcal{P}\}.$$

That is, $C(K)$ represents the set of possible ciphertexts if K is the key. Then, for every $y \in \mathcal{C}$, we have that

$$\mathbf{Pr}[\mathbf{y} = y] = \sum_{\{K : y \in C(K)\}} \mathbf{Pr}[\mathbf{K} = K]\mathbf{Pr}[\mathbf{x} = d_K(y)].$$

We also observe that, for any $y \in \mathcal{C}$ and $x \in \mathcal{P}$, we can compute the conditional probability $\mathbf{Pr}[\mathbf{y} = y | \mathbf{x} = x]$ (i.e., the probability that y is the ciphertext, given that x is the plaintext) to be

$$\mathbf{Pr}[\mathbf{y} = y | \mathbf{x} = x] = \sum_{\{K : x = d_K(y)\}} \mathbf{Pr}[\mathbf{K} = K].$$

It is now possible to compute the conditional probability $\mathbf{Pr}[\mathbf{x} = x | \mathbf{y} = y]$ (i.e., the probability that x is the plaintext, given that y is the ciphertext) using Bayes' theorem. The following formula is obtained:

$$\mathbf{Pr}[\mathbf{x} = x | \mathbf{y} = y] = \frac{\mathbf{Pr}[\mathbf{x} = x] \times \displaystyle\sum_{\{K : x = d_K(y)\}} \mathbf{Pr}[\mathbf{K} = K]}{\displaystyle\sum_{\{K : y \in C(K)\}} \mathbf{Pr}[\mathbf{K} = K]\mathbf{Pr}[\mathbf{x} = d_K(y)]}.$$

Observe that all these calculations can be performed by anyone who knows the probability distributions.

We present a toy example to illustrate the computation of these probability distributions.

Example 2.3 Let $\mathcal{P} = \{a, b\}$ with $\mathbf{Pr}[a] = 1/4, \mathbf{Pr}[b] = 3/4$. Let $\mathcal{K} = \{K_1, K_2, K_3\}$ with $\mathbf{Pr}[K_1] = 1/2, \mathbf{Pr}[K_2] = \mathbf{Pr}[K_3] = 1/4$. Let $\mathcal{C} = \{1, 2, 3, 4\}$, and suppose the encryption functions are defined to be $e_{K_1}(a) = 1, e_{K_1}(b) = 2$; $e_{K_2}(a) = 2, e_{K_2}(b) = 3$; and $e_{K_3}(a) = 3, e_{K_3}(b) = 4$. This cryptosystem can be represented by the following *encryption matrix*:

	a	b
K_1	1	2
K_2	2	3
K_3	3	4

We now compute the probability distribution on $\mathcal{C}$. We obtain the following:

$$\mathbf{Pr}[1] = \frac{1}{8}$$

$$\mathbf{Pr}[2] = \frac{3}{8} + \frac{1}{16} = \frac{7}{16}$$

$$\mathbf{Pr}[3] = \frac{3}{16} + \frac{1}{16} = \frac{1}{4}$$

$$\mathbf{Pr}[4] = \frac{3}{16}.$$

Now we can compute the conditional probability distributions on the plaintext, given that a certain ciphertext has been observed. We have:

$$\mathbf{Pr}[a|1] = 1 \qquad\qquad \mathbf{Pr}[b|1] = 0$$

$$\mathbf{Pr}[a|2] = \frac{1}{7} \qquad\qquad \mathbf{Pr}[b|2] = \frac{6}{7}$$

$$\mathbf{Pr}[a|3] = \frac{1}{4} \qquad\qquad \mathbf{Pr}[b|3] = \frac{3}{4}$$

$$\mathbf{Pr}[a|4] = 0 \qquad\qquad \mathbf{Pr}[b|4] = 1.$$

$\square$

We are now ready to define the concept of perfect secrecy. Informally, perfect secrecy means that Oscar can obtain no information about the plaintext by observing the ciphertext. This idea is made precise by formulating it in terms of the probability distributions we have defined, as follows.

Definition 2.3:   A cryptosystem has *perfect secrecy* if $\mathbf{Pr}[x|y] = \mathbf{Pr}[x]$ for all $x \in \mathcal{P}$, $y \in \mathcal{C}$. That is, the *a posteriori* probability that the plaintext is x, given that the ciphertext y is observed, is identical to the *a priori* probability that the plaintext is x.

In Example 2.3, the perfect secrecy property is satisfied for the ciphertext $y = 3$, but not for the other three ciphertexts.

We now prove that the *Shift Cipher* provides perfect secrecy. This seems quite obvious intuitively. For, if we are given any ciphertext element $y \in \mathbb{Z}_{26}$, then any plaintext element $x \in \mathbb{Z}_{26}$ is a possible decryption of y, depending on the value of the key. The following theorem gives the formal statement and proof using probability distributions.

THEOREM 2.3 *Suppose the* 26 *keys in the Shift Cipher are used with equal probability* $1/26$. *Then for any plaintext probability distribution, the Shift Cipher has perfect secrecy.*

PROOF   Recall that $\mathcal{P} = \mathcal{C} = \mathcal{K} = \mathbb{Z}_{26}$, and for $0 \leq K \leq 25$, the encryption rule e_K is defined as $e_K(x) = (x + K) \bmod 26$ ($x \in \mathbb{Z}_{26}$). First, we compute the probability distribution on $\mathcal{C}$. Let $y \in \mathbb{Z}_{26}$; then

$$\mathbf{Pr}[\mathbf{y} = y] = \sum_{K \in \mathbb{Z}_{26}} \mathbf{Pr}[\mathbf{K} = K]\mathbf{Pr}[\mathbf{x} = d_K(y)]$$

$$= \sum_{K \in \mathbb{Z}_{26}} \frac{1}{26}\mathbf{Pr}[\mathbf{x} = y - K]$$

$$= \frac{1}{26} \sum_{K \in \mathbb{Z}_{26}} \mathbf{Pr}[\mathbf{x} = y - K].$$

Now, for fixed y, the values $(y - K) \bmod 26$ comprise a permutation of $\mathbb{Z}_{26}$. Hence we have that

$$\sum_{K \in \mathbb{Z}_{26}} \mathbf{Pr}[\mathbf{x} = y - K] = \sum_{x \in \mathbb{Z}_{26}} \mathbf{Pr}[\mathbf{x} = x]$$

$$= 1.$$

Consequently,

$$\mathbf{Pr}[y] = \frac{1}{26}$$

for any $y \in \mathbb{Z}_{26}$.

Next, we have that

$$\mathbf{Pr}[y|x] = \mathbf{Pr}[\mathbf{K} = (y - x) \bmod 26]$$

$$= \frac{1}{26}$$

for every x, y. (This is true because, for every x, y, the unique key K such that $e_K(x) = y$ is $K = (y - x) \bmod 26$.) Now, using Bayes' theorem, it is trivial to compute

$$\mathbf{Pr}[x|y] = \frac{\mathbf{Pr}[x]\mathbf{Pr}[y|x]}{\mathbf{Pr}[y]}$$

$$= \frac{\mathbf{Pr}[x]\frac{1}{26}}{\frac{1}{26}}$$

$$= \mathbf{Pr}[x],$$

so we have perfect secrecy.   ∎

Hence, the *Shift Cipher* is "unbreakable" provided that a new random key is used to encrypt every plaintext character.

Let us next investigate perfect secrecy in general. First, we observe that, using Bayes' theorem, the condition that $\mathbf{Pr}[x|y] = \mathbf{Pr}[x]$ for all $x \in \mathcal{P}$, $y \in \mathcal{C}$ is equivalent to $\mathbf{Pr}[y|x] = \mathbf{Pr}[y]$ for all $x \in \mathcal{P}$, $y \in \mathcal{C}$. Now, let us make the reasonable assumption that $\mathbf{Pr}[y] > 0$ for all $y \in \mathcal{C}$ (if $\mathbf{Pr}[y] = 0$, then ciphertext y is never used and can be omitted from $\mathcal{C}$). Fix any $x \in \mathcal{P}$. For each $y \in \mathcal{C}$, we have $\mathbf{Pr}[y|x] = \mathbf{Pr}[y] > 0$. Hence, for each $y \in \mathcal{C}$, there must be at least one key K such that $e_K(x) = y$. It follows that $|\mathcal{K}| \geq |\mathcal{C}|$. In any cryptosystem, we must have $|\mathcal{C}| \geq |\mathcal{P}|$ since each encoding rule is injective. In the case of equality, where $|\mathcal{K}| = |\mathcal{C}| = |\mathcal{P}|$, we can give a nice characterization of when perfect secrecy can be obtained. This characterization is originally due to Shannon.

THEOREM 2.4 *Suppose* $(\mathcal{P}, \mathcal{C}, \mathcal{K}, \mathcal{E}, \mathcal{D})$ *is a cryptosystem where* $|\mathcal{K}| = |\mathcal{C}| = |\mathcal{P}|$. *Then the cryptosystem provides perfect secrecy if and only if every key is used with equal probability* $1/|\mathcal{K}|$, *and for every* $x \in \mathcal{P}$ *and every* $y \in \mathcal{C}$, *there is a unique key* K *such that* $e_K(x) = y$.

PROOF Suppose the given cryptosystem provides perfect secrecy. As observed above, for each $x \in \mathcal{P}$ and $y \in \mathcal{C}$, there must be at least one key K such that $e_K(x) = y$. So we have the inequalities:

$$|\mathcal{C}| = |\{e_K(x) : K \in \mathcal{K}\}|$$
$$\leq |\mathcal{K}|.$$

But we are assuming that $|\mathcal{C}| = |\mathcal{K}|$. Hence, it must be the case that

$$|\{e_K(x) : K \in \mathcal{K}\}| = |\mathcal{K}|.$$

That is, there do not exist two distinct keys K_1 and K_2 such that $e_{K_1}(x) = e_{K_2}(x) = y$. Hence, we have shown that for any $x \in \mathcal{P}$ and $y \in \mathcal{C}$, there is exactly one key K such that $e_K(x) = y$.

Denote $n = |\mathcal{K}|$. Let $\mathcal{P} = \{x_i : 1 \leq i \leq n\}$ and fix a ciphertext element $y \in \mathcal{C}$. We can name the keys $K_1, K_2, \ldots, K_n$, in such a way that $e_{K_i}(x_i) = y$, $1 \leq i \leq n$. Using Bayes' theorem, we have

$$\mathbf{Pr}[x_i|y] = \frac{\mathbf{Pr}[y|x_i]\mathbf{Pr}[x_i]}{\mathbf{Pr}[y]}$$
$$= \frac{\mathbf{Pr}[K = K_i]\mathbf{Pr}[x_i]}{\mathbf{Pr}[y]}.$$

Consider the perfect secrecy condition $\mathbf{Pr}[x_i|y] = \mathbf{Pr}[x_i]$. From this, it follows that $\mathbf{Pr}[K_i] = \mathbf{Pr}[y]$, for $1 \leq i \leq n$. This says that all the keys are used with equal probability (namely, $\mathbf{Pr}[y]$). But since the number of keys is $|\mathcal{K}|$, we must have that $\mathbf{Pr}[K] = 1/|\mathcal{K}|$ for every $K \in \mathcal{K}$.

Conversely, suppose the two hypothesized conditions are satisfied. Then the cryptosystem is easily seen to provide perfect secrecy for any plaintext probability distribution, in a manner similar to the proof of Theorem 2.3. We leave the details for the reader. ∎

One well-known realization of perfect secrecy is the *One-time Pad*, which was first described by Gilbert Vernam in 1917 for use in automatic encryption and decryption of telegraph messages. It is interesting that the *One-time Pad* was thought for many years to be an "unbreakable" cryptosystem, but there was no mathematical proof of this until Shannon developed the concept of perfect secrecy over 30 years later. The description of the *One-time Pad* is as follows.

Cryptosystem 2.1: *One-time Pad*

Let $n \geq 1$ be an integer, and take $\mathcal{P} = \mathcal{C} = \mathcal{K} = (\mathbb{Z}_2)^n$. For $K \in (\mathbb{Z}_2)^n$, define $e_K(x)$ to be the vector sum modulo 2 of K and x (or, equivalently, the exclusive-or of the two associated bitstrings). So, if $x = (x_1, \ldots, x_n)$ and $K = (K_1, \ldots, K_n)$, then

$$e_K(x) = (x_1 + K_1, \ldots, x_n + K_n) \bmod 2.$$

Decryption is identical to encryption. If $y = (y_1, \ldots, y_n)$, then

$$d_K(y) = (y_1 + K_1, \ldots, y_n + K_n) \bmod 2.$$

Using Theorem 2.4, it is easily seen that the *One-time Pad* provides perfect secrecy. The system is also attractive because of the ease of encryption and decryption. Vernam patented his idea in the hope that it would have widespread commercial use. Unfortunately, there are major disadvantages to unconditionally secure cryptosystems such as the *One-time Pad*. The fact that $|\mathcal{K}| \geq |\mathcal{P}|$ means that the amount of key that must be communicated securely is at least as big as the amount of plaintext. For example, in the case of the *One-time Pad*, we require n bits of key to encrypt n bits of plaintext. This would not be a major problem if the same key could be used to encrypt different messages; however, the security of unconditionally secure cryptosystems depends on the fact that each key is used for only one encryption. (This is the reason for the adjective "one-time" in the *One-time Pad*.)

For example, the *One-time Pad* is vulnerable to a known-plaintext attack, since K can be computed as the exclusive-or of the bitstrings x and $e_K(x)$. Hence, a new key needs to be generated and communicated over a secure channel for every message that is going to be sent. This creates severe key management problems, which has limited the use of the *One-time Pad* in commercial applications. However, the *One-time Pad* has been employed in military and diplomatic contexts, where unconditional security may be of great importance.

The historical development of cryptography has been to try to design cryptosystems where one key can be used to encrypt a relatively long string of plaintext (i.e., one key can be used to encrypt many messages) and still maintain some measure of computational security. One such cryptosystem of this type is the *Data Encryption Standard*, which we will discuss in the next chapter.

2.4 Entropy

In the previous section, we discussed the concept of perfect secrecy. We restricted our attention to the special situation where a key is used for only one encryption. We now want to look at what happens as more and more plaintexts are encrypted using the same key, and how likely a cryptanalyst will be able to carry out a successful ciphertext-only attack, given sufficient time.

The basic tool in studying this question is the idea of entropy, a concept from information theory introduced by Shannon in 1948. Entropy can be thought of as a mathematical measure of information or uncertainty, and is computed as a function of a probability distribution.

Suppose we have a discrete random variable $\mathbf{X}$ which takes values from a finite set X according to a specified probability distribution. What is the information gained by the outcome of an experiment which takes place according to this probability distribution? Equivalently, if the experiment has not (yet) taken place, what is the uncertainty about the outcome? This quantity is called the entropy of $\mathbf{X}$ and is denoted by $H(\mathbf{X})$.

These ideas may seem rather abstract, so let's look at a more concrete example. Suppose our random variable $\mathbf{X}$ represents the toss of a coin. As mentioned earlier, the associated probability distribution is $\mathbf{Pr}[heads] = \mathbf{Pr}[tails] = 1/2$. It would seem reasonable to say that the information, or entropy, of a coin toss is one bit, since we could encode *heads* by 1 and *tails* by 0, for example. In a similar fashion, the entropy of n independent coin tosses is n, since the n coin tosses can be encoded by a bit string of length n.

As a slightly more complicated example, suppose we have a random variable $\mathbf{X}$ that takes on three possible values x_1, x_2, x_3 with probabilities $1/2, 1/4, 1/4$ respectively. The most efficient "encoding" of the three possible outcomes is to encode x_1 as 0, to encode x_2 as 10 and to encode x_3 as 11. Then the average number of bits in an encoding of $\mathbf{X}$ is

$$\frac{1}{2} \times 1 + \frac{1}{4} \times 2 + \frac{1}{4} \times 2 = \frac{3}{2}.$$

The above examples suggest that an event which occurs with probability 2^{-n} could perhaps be encoded as a bit string of length n. More generally, we could plausibly imagine that an outcome occurring with probability p might be encoded by a bit string of length approximately $-\log_2 p$. Given an arbitrary probability distribution, taking on the values $p_1, p_2, \ldots, p_n$ for a random variable $\mathbf{X}$, we take the weighted average of the quantities $-\log_2 p_i$ to be our measure of information. This motivates the following formal definition.

Definition 2.4: Suppose $\mathbf{X}$ is a discrete random variable which takes on values from a finite set X. Then, the *entropy* of the random variable $\mathbf{X}$ is defined to be the quantity

$$H(\mathbf{X}) = -\sum_{x \in X} \mathbf{Pr}[x] \log_2 \mathbf{Pr}[x].$$

REMARK Observe that $\log_2 y$ is undefined if $y = 0$. Hence, entropy is sometimes defined to be the relevant sum over all the non-zero probabilities. However, since $\lim_{y \to 0} y \log_2 y = 0$, there is no real difficulty with allowing $\mathbf{Pr}[x] = 0$ for some x's.

Also, we note that the choice of two as the base of the logarithms is arbitrary: another base would only change the value of the entropy by a constant factor. ∎

Note that if $|X| = n$ and $\mathbf{Pr}[x] = 1/n$ for all $x \in X$, then $H(\mathbf{X}) = \log_2 n$. Also, it is easy to see that $H(\mathbf{X}) \geq 0$ for any random variable $\mathbf{X}$, and $H(\mathbf{X}) = 0$ if and only if $\mathbf{Pr}[x_0] = 1$ for some $x_0 \in X$ and $\mathbf{Pr}[x] = 0$ for all $x \neq x_0$.

Let us look at the entropy of the various components of a cryptosystem. We can think of the key as being a random variable $\mathbf{K}$ that takes on values in $\mathcal{K}$, and hence we can compute the entropy $H(\mathbf{K})$. Similarly, we can compute entropies $H(\mathbf{P})$ and $H(\mathbf{C})$ of random variables associated with the plaintext and ciphertext, respectively.

To illustrate, we compute the entropies of the cryptosystem of Example 2.3.

Example 2.3 *(Cont.)* We compute as follows:

$$\begin{aligned}
H(\mathbf{P}) &= -\frac{1}{4} \log_2 \frac{1}{4} - \frac{3}{4} \log_2 \frac{3}{4} \\
&= -\frac{1}{4}(-2) - \frac{3}{4}(\log_2 3 - 2) \\
&= 2 - \frac{3}{4}(\log_2 3) \\
&\approx 0.81.
\end{aligned}$$

Similar calculations yield $H(\mathbf{K}) = 1.5$ and $H(\mathbf{C}) \approx 1.85$. ▯

2.4.1 Huffman Encodings

In this section, we discuss briefly the connection between entropy and Huffman encodings. As the results in this section are not relevant to the cryptographic applications of entropy, it may be skipped without loss of continuity. However, this discussion may serve to further motivate the concept of entropy.

We introduced entropy in the context of encodings of random events which occur according to a specified probability distribution. We first make these ideas more precise. As before, $\mathbf{X}$ is a random variable which takes on values from a finite set X and p is the associated probability distribution.

An *encoding* of $\mathbf{X}$ is any mapping

$$f : X \to \{0,1\}^*,$$

where $\{0,1\}^*$ denotes the set of all finite strings of 0's and 1's. Given a finite list (or string) of events $x_1 \cdots x_n$, where each $x_i \in X$, we can extend the encoding f in an obvious way by defining

$$f(x_1 \cdots x_n) = f(x_1) \, \| \, \cdots \, \| \, f(x_n),$$

where $\|$ denotes concatenation. In this way, we can think of f as a mapping

$$f : X^* \to \{0,1\}^*.$$

Now, suppose a string $x_1 \ldots x_n$ is produced by a *memoryless source*, such that each x_i in the string occurs according to a specified probability distribution on X. This means that the probability of any string $x_1 \cdots x_n$ is computed to be

$$\mathbf{Pr}[x_1 \cdots x_n] = \mathbf{Pr}[x_1] \times \cdots \times \mathbf{Pr}[x_n].$$

REMARK   The string $x_1 \cdots x_n$ need not consist of distinct values, since the source is memoryless. As a simple example, consider a sequence of n tosses of a fair coin. If we encode "*heads*" as "1" and "*tails*" as "0," then every binary string of length n corresponds to a sequence of n coin tosses.   ∎

Now, given that we are going to encode strings using the mapping f, it is important that we are able to decode in an unambiguous fashion. Thus it should be the case that the encoding f is injective.

Example 2.4   Suppose $X = \{a, b, c, d\}$, and consider the following three encodings:

$$
\begin{array}{llllllll}
f(a) & = & 1 & f(b) & = & 10 & f(c) & = & 100 & f(d) & = & 1000 \\
g(a) & = & 0 & g(b) & = & 10 & g(c) & = & 110 & g(d) & = & 111 \\
h(a) & = & 0 & h(b) & = & 01 & h(c) & = & 10 & h(d) & = & 11
\end{array}
$$

It can be seen that f and g are injective encodings, but h is not. Any encoding using f can be decoded by starting at the end and working backwards: every time a 1 is encountered, it signals the beginning of the current element.

An encoding using g can be decoded by starting at the beginning and proceeding sequentially. At any point where we have a substring that is an encoding of a, b, c, or d, we decode it and chop off the substring. For example, given the string 10101110, we decode 10 to b, then 10 to b, then 111 to d, and finally 0 to a. So the decoded string is $bbda$.

To see that h is not injective, it suffices to give an example:

$$h(ac) = h(ba) = 010.$$

$\square$

From the point of view of ease of decoding, we would prefer the encoding g to f. This is because decoding can be done sequentially from beginning to end if g is used, so no memory is required. The property that allows the simple sequential decoding of g is called the prefix-free property. (An encoding g is a *prefix-free encoding* if there do not exist two elements $x, y \in X$, and a string $z \in \{0, 1\}^*$ such that $g(x) = g(y) \parallel z$.)

The discussion to this point has not involved entropy. Not surprisingly, entropy is related to the efficiency of an encoding. We will measure the efficiency of an encoding f as we did before: it is the weighted average length (denoted by $\ell(f)$) of an encoding of an element of $\mathbf{X}$. So we have the following definition:

$$\ell(f) = \sum_{x \in X} \mathbf{Pr}[x] |f(x)|,$$

where $|y|$ denotes the length of a string y.

Now, our fundamental problem is to find an injective encoding, f, that minimizes $\ell(f)$. There is a well-known algorithm, known as Huffman's algorithm, that accomplishes this goal. Moreover, the encoding f produced by Huffman's algorithm is prefix-free, and

$$H(\mathbf{X}) \le \ell(f) < H(\mathbf{X}) + 1.$$

Thus, the value of the entropy of $\mathbf{X}$ provides a close estimate to the average length of the optimal injective encoding.

We will not prove the results stated above, but we will give a short, informal description of Huffman's algorithm. Huffman's algorithm begins with the probability distribution on the set X, and the code of each element is initially empty. In each iteration, the two elements having lowest probability are combined into one element having as its probability the sum of the two smaller probabilities. The element having lowest probability is assigned the value "0" and the element having next lowest probability is assigned the value "1." When only one element remains, the coding for each $x \in X$ can be constructed by following the sequence of elements "backwards" from the final element to the initial element x.

This is easily illustrated with an example.

Example 2.5 Suppose $X = \{a, b, c, d, e\}$ has the following probability distribution: $\mathbf{Pr}[a] = .05$, $\mathbf{Pr}[b] = .10$, $\mathbf{Pr}[c] = .12$, $\mathbf{Pr}[d] = .13$ and $\mathbf{Pr}[e] = .60$. Huffman's algorithm would proceed as indicated in the following table:

a	b	c	d	e
.05	.10	.12	.13	.60
0	1			
	.15	.12	.13	.60
		0	1	
	.15		.25	.60
	0		1	
		.40		.60
		0		1
		1.0		

This leads to the following encodings:

x	$f(x)$
a	000
b	001
c	010
d	011
e	1

Thus, the average length encoding is

$$\ell(f) = .05 \times 3 + .10 \times 3 + .12 \times 3 + .13 \times 3 + .60 \times 1$$

$$= 1.8.$$

Compare this to the entropy:

$$H(\mathbf{X}) = .2161 + .3322 + .3671 + .3842 + .4422$$

$$= 1.7402.$$

It is seen that the average length encoding is very close to the entropy.   ∏

2.5 Properties of Entropy

In this section, we prove some fundamental results concerning entropy. First, we state a fundamental result, known as Jensen's inequality, that will be very useful to us. Jensen's inequality involves concave functions, which we now define.

Definition 2.5: A real-valued function f is a *concave function* on an interval I if

$$f\left(\frac{x+y}{2}\right) \geq \frac{f(x) + f(y)}{2}$$

for all $x, y \in I$. f is a *strictly concave function* on an interval I if

$$f\left(\frac{x+y}{2}\right) > \frac{f(x) + f(y)}{2}$$

for all $x, y \in I$, $x \neq y$.

Here is Jensen's inequality, which we state without proof.

THEOREM 2.5 *(Jensen's inequality)* *Suppose f is a continuous strictly concave function on the interval I,*

$$\sum_{i=1}^{n} a_i = 1,$$

and $a_i > 0$, $1 \leq i \leq n$. Then

$$\sum_{i=1}^{n} a_i f(x_i) \leq f\left(\sum_{i=1}^{n} a_i x_i\right),$$

where $x_i \in I$, $1 \leq i \leq n$. Further, equality occurs if and only if $x_1 = \cdots = x_n$.

We now proceed to derive several results on entropy. In the next theorem, we make use of the fact that the function $\log_2 x$ is strictly concave on the interval $(0, \infty)$. (In fact, this follows easily from elementary calculus since the second derivative of the logarithm function is negative on the interval $(0, \infty)$.)

THEOREM 2.6 *Suppose $\mathbf{X}$ is a random variable having a probability distribution which takes on the values $p_1, p_2, \ldots, p_n$, where $p_i > 0$, $1 \leq i \leq n$. Then $H(\mathbf{X}) \leq \log_2 n$, with equality if and only if $p_i = 1/n$, $1 \leq i \leq n$.*

PROOF Applying Jensen's inequality, we have the following:

$$H(\mathbf{X}) = -\sum_{i=1}^{n} p_i \log_2 p_i$$

$$= \sum_{i=1}^{n} p_i \log_2 \frac{1}{p_i}$$

$$\leq \log_2 \sum_{i=1}^{n} \left(p_i \times \frac{1}{p_i} \right)$$

$$= \log_2 n.$$

Further, equality occurs if and only if $p_i = 1/n, 1 \leq i \leq n$. ∎

THEOREM 2.7 $H(\mathbf{X}, \mathbf{Y}) \leq H(\mathbf{X}) + H(\mathbf{Y})$, *with equality if and only if* $\mathbf{X}$ *and* $\mathbf{Y}$ *are independent random variables.*

PROOF Suppose $\mathbf{X}$ takes on values x_i, $1 \leq i \leq m$, and $\mathbf{Y}$ takes on values y_j, $1 \leq j \leq n$. Denote $p_i = \mathbf{Pr}[(\mathbf{X} = x_i], 1 \leq i \leq m$, and $q_j = \mathbf{Pr}[\mathbf{Y} = y_j]$, $1 \leq j \leq n$. Denote $r_{ij} = \mathbf{Pr}[\mathbf{X} = x_i, \mathbf{Y} = y_j], 1 \leq i \leq m, 1 \leq j \leq n$ (this is the joint probability distribution).

Observe that

$$p_i = \sum_{j=1}^{n} r_{ij}$$

$(1 \leq i \leq m)$, and

$$q_j = \sum_{i=1}^{m} r_{ij}$$

$(1 \leq j \leq n)$. We compute as follows:

$$H(\mathbf{X}) + H(\mathbf{Y}) = -\left(\sum_{i=1}^{m} p_i \log_2 p_i + \sum_{j=1}^{n} q_j \log_2 q_j \right)$$

$$= -\left(\sum_{i=1}^{m} \sum_{j=1}^{n} r_{ij} \log_2 p_i + \sum_{j=1}^{n} \sum_{i=1}^{m} r_{ij} \log_2 q_j \right)$$

$$= -\sum_{i=1}^{m} \sum_{j=1}^{n} r_{ij} \log_2 p_i q_j.$$

On the other hand,

$$H(\mathbf{X}, \mathbf{Y}) = -\sum_{i=1}^{m} \sum_{j=1}^{n} r_{ij} \log_2 r_{ij}.$$

Combining, we obtain the following:

$$H(\mathbf{X}, \mathbf{Y}) - H(\mathbf{X}) - H(\mathbf{Y}) = \sum_{i=1}^{m} \sum_{j=1}^{n} r_{ij} \log_2 \frac{1}{r_{ij}} + \sum_{i=1}^{m} \sum_{j=1}^{n} r_{ij} \log_2 p_i q_j$$

$$= \sum_{i=1}^{m} \sum_{j=1}^{n} r_{ij} \log_2 \frac{p_i q_j}{r_{ij}}$$

$$\leq \log_2 \sum_{i=1}^{m} \sum_{j=1}^{n} p_i q_j$$

$$= \log_2 1$$

$$= 0.$$

(In the above computations, we apply Jensen's inequality, using the fact that the r_{ij}'s are positive real numbers that sum to 1.)

We can also say when equality occurs: it must be the case that there is a constant c such that $p_i q_j / r_{ij} = c$ for all i, j. Using the fact that

$$\sum_{j=1}^{n} \sum_{i=1}^{m} r_{ij} = \sum_{j=1}^{n} \sum_{i=1}^{m} p_i q_j = 1,$$

it follows that $c = 1$. Hence, equality occurs if and only if $r_{ij} = p_i q_j$, i.e., if and only if

$$\mathbf{Pr}[\mathbf{X} = x_i, \mathbf{Y} = y_j] = \mathbf{Pr}[\mathbf{X} = x_i]\mathbf{Pr}[\mathbf{Y} = y_j],$$

$1 \leq i \leq m, 1 \leq j \leq n$. But this says that $\mathbf{X}$ and $\mathbf{Y}$ are independent. ∎

We next define the idea of conditional entropy.

Definition 2.6: Suppose $\mathbf{X}$ and $\mathbf{Y}$ are two random variables. Then for any fixed value y of $\mathbf{Y}$, we get a (conditional) probability distribution on X; we denote the associated random variable by $\mathbf{X}|y$. Clearly,

$$H(\mathbf{X}|y) = -\sum_x \mathbf{Pr}[x|y] \log_2 \mathbf{Pr}[x|y].$$

We define the *conditional entropy*, denoted $H(\mathbf{X}|\mathbf{Y})$, to be the weighted average (with respect to the probabilities $\mathbf{Pr}[y]$) of the entropies $H(\mathbf{X}|y)$ over all possible values y. It is computed to be

$$H(\mathbf{X}|\mathbf{Y}) = -\sum_y \sum_x \mathbf{Pr}[y]\mathbf{Pr}[x|y] \log_2 \mathbf{Pr}[x|y].$$

The conditional entropy measures the average amount of information about $\mathbf{X}$ that is revealed by $\mathbf{Y}$.

The next two results are straightforward; we leave the proofs as exercises.

THEOREM 2.8 $H(\mathbf{X}, \mathbf{Y}) = H(\mathbf{Y}) + H(\mathbf{X}|\mathbf{Y})$.

COROLLARY 2.9 $H(\mathbf{X}|\mathbf{Y}) \leq H(\mathbf{X})$, *with equality if and only if* $\mathbf{X}$ *and* $\mathbf{Y}$ *are independent.*

2.6   Spurious Keys and Unicity Distance

In this section, we apply the entropy results we have proved to cryptosystems. First, we show a fundamental relationship exists among the entropies of the components of a cryptosystem. The conditional entropy $H(\mathbf{K}|\mathbf{C})$ is called the *key equivocation*, and is a measure of how much information about the key is revealed by the ciphertext.

THEOREM 2.10   *Let* $(\mathcal{P}, \mathcal{C}, \mathcal{K}, \mathcal{E}, \mathcal{D})$ *be a cryptosystem. Then*

$$H(\mathbf{K}|\mathbf{C}) = H(\mathbf{K}) + H(\mathbf{P}) - H(\mathbf{C}).$$

PROOF   First, observe that $H(\mathbf{K}, \mathbf{P}, \mathbf{C}) = H(\mathbf{C}|\mathbf{K}, \mathbf{P}) + H(\mathbf{K}, \mathbf{P})$. Now, the key and plaintext determine the ciphertext uniquely, since $y = e_K(x)$. This implies that $H(\mathbf{C}|\mathbf{K}, \mathbf{P}) = 0$. Hence, $H(\mathbf{K}, \mathbf{P}, \mathbf{C}) = H(\mathbf{K}, \mathbf{P})$. But $\mathbf{K}$ and $\mathbf{P}$ are independent, so $H(\mathbf{K}, \mathbf{P}) = H(\mathbf{K}) + H(\mathbf{P})$. Hence,

$$H(\mathbf{K}, \mathbf{P}, \mathbf{C}) = H(\mathbf{K}, \mathbf{P}) = H(\mathbf{K}) + H(\mathbf{P}).$$

In a similar fashion, since the key and ciphertext determine the plaintext uniquely (i.e., $x = d_K(y)$), we have that $H(\mathbf{P}|\mathbf{K}, \mathbf{C}) = 0$ and hence $H(\mathbf{K}, \mathbf{P}, \mathbf{C}) = H(\mathbf{K}, \mathbf{C})$.

Now, we compute as follows:

$$\begin{aligned} H(\mathbf{K}|\mathbf{C}) &= H(\mathbf{K}, \mathbf{C}) - H(\mathbf{C}) \\ &= H(\mathbf{K}, \mathbf{P}, \mathbf{C}) - H(\mathbf{C}) \\ &= H(\mathbf{K}) + H(\mathbf{P}) - H(\mathbf{C}), \end{aligned}$$

giving the desired formula.   ∎

Let us return to Example 2.3 to illustrate this result.

Example 2.3   (Cont.)   We have already computed $H(\mathbf{P}) \approx 0.81, H(\mathbf{K}) = 1.5$ and $H(\mathbf{C}) \approx 1.85$. Theorem 2.10 tells us that $H(\mathbf{K}|\mathbf{C}) \approx 1.5 + 0.81 - 1.85 \approx$

0.46. This can be verified directly by applying the definition of conditional entropy, as follows. First, we need to compute the probabilities $\mathbf{Pr}[\mathbf{K} = K_i | \mathbf{y} = j]$, $1 \le i \le 3, 1 \le j \le 4$. This can be done using Bayes' theorem, and the following values result:

$$\mathbf{Pr}[K_1|1] = 1 \qquad \mathbf{Pr}[K_2|1] = 0 \qquad \mathbf{Pr}[K_3|1] = 0$$

$$\mathbf{Pr}[K_1|2] = \frac{6}{7} \qquad \mathbf{Pr}[K_2|2] = \frac{1}{7} \qquad \mathbf{Pr}[K_3|2] = 0$$

$$\mathbf{Pr}[K_1|3] = 0 \qquad \mathbf{Pr}[K_2|3] = \frac{3}{4} \qquad \mathbf{Pr}[K_3|3] = \frac{1}{4}$$

$$\mathbf{Pr}[K_1|4] = 0 \qquad \mathbf{Pr}[K_2|4] = 0 \qquad \mathbf{Pr}[K_3|4] = 1.$$

Now we compute

$$H(\mathbf{K}|\mathbf{C}) = \frac{1}{8} \times 0 + \frac{7}{16} \times 0.59 + \frac{1}{4} \times 0.81 + \frac{3}{16} \times 0 = 0.46,$$

agreeing with the value predicted by Theorem 2.10. ▯

Suppose $(\mathcal{P}, \mathcal{C}, \mathcal{K}, \mathcal{E}, \mathcal{D})$ is the cryptosystem being used, and a string of plaintext

$$x_1 x_2 \cdots x_n$$

is encrypted with one key, producing a string of ciphertext

$$y_1 y_2 \cdots y_n.$$

Recall that the basic goal of the cryptanalyst is to determine the key. We are looking at ciphertext-only attacks, and we assume that Oscar has infinite computational resources. We also assume that Oscar knows that the plaintext is a "natural" language, such as English. In general, Oscar will be able to rule out certain keys, but many "possible" keys may remain, only one of which is the correct key. The remaining possible, but incorrect, keys are called *spurious keys*.

For example, suppose Oscar obtains the ciphertext string $WNAJW$, which has been obtained by encryption using a shift cipher. It is easy to see that there are only two "meaningful" plaintext strings, namely *river* and *arena*, corresponding respectively to the possible encryption keys F $(= 5)$ and W $(= 22)$. Of these two keys, one will be the correct key and the other will be spurious. (Actually, it is moderately difficult to find a ciphertext of length 5 for the *Shift Cipher* that has two meaningful decryptions; the reader might search for other examples.)

Our goal is to prove a bound on the expected number of spurious keys. First, we have to define what we mean by the entropy (per letter) of a natural language L, which we denote H_L. H_L should be a measure of the average information per letter in a "meaningful" string of plaintext. (Note that a random string of alphabetic characters would have entropy (per letter) equal to $\log_2 26 \approx 4.70$.) As a "first-order" approximation to H_L, we could take $H(\mathbf{P})$. In the case where L is the English language, we get $H(\mathbf{P}) \approx 4.19$ by using the probability distribution given in Table 1.1.

Of course, successive letters in a language are not independent, and correlations among successive letters reduce the entropy. For example, in English, the letter "Q" is always followed by the letter "U." For a "second-order" approximation, we would compute the entropy of the probability distribution of all digrams and then divide by 2. In general, define $\mathbf{P}^n$ to be the random variable that has as its probability distribution that of all n-grams of plaintext. We make use of the following definitions.

Definition 2.7:   Suppose L is a natural language. The *entropy* of L is defined to be the quantity

$$H_L = \lim_{n \to \infty} \frac{H(\mathbf{P}^n)}{n}$$

and the *redundancy* of L is defined to be

$$R_L = 1 - \frac{H_L}{\log_2 |\mathcal{P}|}.$$

REMARK H_L measures the entropy per letter of the language L. A random language would have entropy $\log_2 |\mathcal{P}|$. So the quantity R_L measures the fraction of "excess characters," which we think of as redundancy.   ∎

In the case of the English language, a tabulation of a large number of digrams and their frequencies would produce an estimate for $H(\mathbf{P}^2)$. $H(\mathbf{P}^2)/2 \approx 3.90$ is one estimate obtained in this way. One could continue, tabulating trigrams, etc. and thus obtain an estimate for H_L. In fact, various experiments have yielded the empirical result that $1.0 \le H_L \le 1.5$. That is, the average information content in English is something like one to one-and-a-half bits per letter!

Using 1.25 as our estimate of H_L gives a redundancy of about 0.75. This means that the English language is 75% redundant! (This is not to say that one can arbitrarily remove three out of every four letters from English text and hope to still be able to read it. What it does mean is that it is possible to find a Huffman encoding of n-grams, for a large enough value of n, which will compress English text to about one quarter of its original length.)

Given probability distributions on $\mathcal{K}$ and $\mathcal{P}^n$, we can define the induced probability distribution on $\mathcal{C}^n$, the set of n-grams of ciphertext (we already did this in the case $n = 1$). We have defined $\mathbf{P}^n$ to be a random variable representing an n-gram of plaintext. Similarly, define $\mathbf{C}^n$ to be a random variable representing an n-gram of ciphertext.

Given $\mathbf{y} \in \mathbf{C}^n$, define

$$K(\mathbf{y}) = \{K \in \mathcal{K} : \exists \mathbf{x} \in \mathcal{P}^n \text{ such that } \mathbf{Pr}[\mathbf{x}] > 0 \text{ and } e_K(\mathbf{x}) = \mathbf{y}\}.$$

That is, $K(\mathbf{y})$ is the set of keys K for which $\mathbf{y}$ is the encryption of a meaningful string of plaintext of length n, i.e., the set of "possible" keys, given that $\mathbf{y}$ is the ciphertext. If $\mathbf{y}$ is the observed string of ciphertext, then the number of spurious keys is $|K(\mathbf{y})| - 1$, since only one of the "possible" keys is the correct key. The average number of spurious keys (over all possible ciphertext strings of length n) is denoted by $\bar{s}_n$. Its value is computed to be

$$\bar{s}_n = \sum_{\mathbf{y} \in \mathcal{C}^n} \mathbf{Pr}[\mathbf{y}](|K(\mathbf{y})| - 1)$$

$$= \sum_{\mathbf{y} \in \mathcal{C}^n} \mathbf{Pr}[\mathbf{y}]|K(\mathbf{y})| - \sum_{\mathbf{y} \in \mathcal{C}^n} \mathbf{Pr}[\mathbf{y}]$$

$$= \sum_{\mathbf{y} \in \mathcal{C}^n} \mathbf{Pr}[\mathbf{y}]|K(\mathbf{y})| - 1.$$

From Theorem 2.10, we have that

$$H(\mathbf{K}|\mathbf{C}^n) = H(\mathbf{K}) + H(\mathbf{P}^n) - H(\mathbf{C}^n).$$

Also, we can use the estimate

$$H(\mathbf{P}^n) \approx nH_L = n(1 - R_L) \log_2 |\mathcal{P}|,$$

provided n is reasonably large. Certainly,

$$H(\mathbf{C}^n) \leq n \log_2 |\mathcal{C}|.$$

Then, if $|\mathcal{C}| = |\mathcal{P}|$, it follows that

$$H(\mathbf{K}|\mathbf{C}^n) \geq H(\mathbf{K}) - nR_L \log_2 |\mathcal{P}|. \tag{2.2}$$

Next, we relate the quantity $H(\mathbf{K}|\mathbf{C}^n)$ to the number of spurious keys, $\bar{s}_n$. We compute as follows:

$$H(\mathbf{K}|\mathbf{C}^n) = \sum_{\mathbf{y} \in \mathcal{C}^n} \mathbf{Pr}[\mathbf{y}]H(\mathbf{K}|\mathbf{y})$$

$$\leq \sum_{\mathbf{y} \in \mathcal{C}^n} \mathbf{Pr}[\mathbf{y}] \log_2 |K(\mathbf{y})|$$

$$\leq \log_2 \sum_{\mathbf{y} \in \mathcal{C}^n} \mathbf{Pr}[\mathbf{y}]|K(\mathbf{y})|$$

$$= \log_2(\bar{s}_n + 1),$$

where we apply Jensen's inequality (Theorem 2.5) with $f(x) = \log_2 x$. Thus we obtain the inequality

$$H(\mathbf{K}|\mathbf{C}^n) \leq \log_2(\overline{s}_n + 1). \tag{2.3}$$

Combining the two inequalities (2.2) and (2.3), we get that

$$\log_2(\overline{s}_n + 1) \geq H(\mathbf{K}) - nR_L \log_2 |\mathcal{P}|.$$

In the case where keys are chosen equiprobably (which maximizes $H(\mathbf{K})$), we have the following result.

THEOREM 2.11   *Suppose* $(\mathcal{P}, \mathcal{C}, \mathcal{K}, \mathcal{E}, \mathcal{D})$ *is a cryptosystem where* $|\mathcal{C}| = |\mathcal{P}|$ *and keys are chosen equiprobably. Let* R_L *denote the redundancy of the underlying language. Then given a string of ciphertext of length* n, *where* n *is sufficiently large, the expected number of spurious keys,* $\overline{s}_n$, *satisfies*

$$\overline{s}_n \geq \frac{|\mathcal{K}|}{|\mathcal{P}|^{nR_L}} - 1.$$

The quantity $|\mathcal{K}|/|\mathcal{P}|^{nR_L} - 1$ approaches 0 exponentially quickly as n increases. Also, note that the estimate may not be accurate for small values of n, especially since $H(\mathbf{P}^n)/n$ may not be a good estimate for H_L if n is small.

We have one more concept to define.

Definition 2.8:   The *unicity distance* of a cryptosystem is defined to be the value of n, denoted by n_0, at which the expected number of spurious keys becomes zero; i.e., the average amount of ciphertext required for an opponent to be able to uniquely compute the key, given enough computing time.

If we set $\overline{s}_n = 0$ in Theorem 2.11 and solve for n, we get an estimate for the unicity distance, namely

$$n_0 \approx \frac{\log_2 |\mathcal{K}|}{R_L \log_2 |\mathcal{P}|}.$$

As an example, consider the *Substitution Cipher*. In this cryptosystem, $|\mathcal{P}| = 26$ and $|\mathcal{K}| = 26!$. If we take $R_L = 0.75$, then we get an estimate for the unicity distance of

$$n_0 \approx 88.4/(0.75 \times 4.7) \approx 25.$$

This suggests that, given a ciphertext string of length at least 25, (usually) a unique decryption is possible.

2.7 Product Cryptosystems

Another innovation introduced by Shannon in his 1949 paper was the idea of combining cryptosystems by forming their "product." This idea has been of fundamental importance in the design of present-day cryptosystems such as the *Advanced Encryption Standard*, which we study in the next chapter.

For simplicity, we will confine our attention in this section to cryptosystems in which $\mathcal{C} = \mathcal{P}$: a cryptosystem of this type is called an *endomorphic cryptosystem*. Suppose $\mathbf{S}_1 = (\mathcal{P}, \mathcal{P}, \mathcal{K}_1, \mathcal{E}_1, \mathcal{D}_1)$ and $\mathbf{S}_2 = (\mathcal{P}, \mathcal{P}, \mathcal{K}_2, \mathcal{E}_2, \mathcal{D}_2)$ are two endomorphic cryptosystems which have the same plaintext (and ciphertext) spaces. Then the *product cryptosystem* of $\mathbf{S}_1$ and $\mathbf{S}_2$, denoted by $\mathbf{S}_1 \times \mathbf{S}_2$, is defined to be the cryptosystem

$$(\mathcal{P}, \mathcal{P}, \mathcal{K}_1 \times \mathcal{K}_2, \mathcal{E}, \mathcal{D}).$$

A key of the product cryptosystem has the form $K = (K_1, K_2)$, where $K_1 \in \mathcal{K}_1$ and $K_2 \in \mathcal{K}_2$. The encryption and decryption rules of the product cryptosystem are defined as follows: For each $K = (K_1, K_2)$, we have an encryption rule e_K defined by the formula

$$e_{(K_1, K_2)}(x) = e_{K_2}(e_{K_1}(x)),$$

and a decryption rule defined by the formula

$$d_{(K_1, K_2)}(y) = d_{K_1}(d_{K_2}(y)).$$

That is, we first encrypt x with e_{K_1}, and then "re-encrypt" the resulting ciphertext with e_{K_2}. Decrypting is similar, but it must be done in the reverse order:

$$\begin{aligned} d_{(K_1, K_2)}(e_{(K_1, K_2)}(x)) &= d_{(K_1, K_2)}(e_{K_2}(e_{K_1}(x))) \\ &= d_{K_1}(d_{K_2}(e_{K_2}(e_{K_1}(x)))) \\ &= d_{K_1}(e_{K_1}(x)) \\ &= x. \end{aligned}$$

Recall also that cryptosystems have probability distributions associated with their keyspaces. Thus we need to define the probability distribution for the keyspace $\mathcal{K}$ of the product cryptosystem. We do this in a very natural way:

$$\mathbf{Pr}[(K_1, K_2)] = \mathbf{Pr}[K_1] \times \mathbf{Pr}[K_2].$$

In other words, choose K_1 and K_2 independently, using the probability distributions defined on $\mathcal{K}_1$ and $\mathcal{K}_2$, respectively.

Here is a simple example to illustrate the definition of a product cryptosystem. Suppose we define the *Multiplicative Cipher* as follows.

Cryptosystem 2.2: *Multiplicative Cipher*

Let $\mathcal{P} = \mathcal{C} = \mathbb{Z}_{26}$ and let

$$\mathcal{K} = \{a \in \mathbb{Z}_{26} : \gcd(a, 26) = 1\}.$$

For $a \in \mathcal{K}$, define

$$e_a(x) = ax \bmod 26$$

and

$$d_a(y) = a^{-1}y \bmod 26$$

$(x, y \in \mathbb{Z}_{26})$.

Suppose **M** is the *Multiplicative Cipher* (with keys chosen equiprobably) and **S** is the *Shift Cipher* (with keys chosen equiprobably). Then it is very easy to see that $\mathbf{M} \times \mathbf{S}$ is nothing more than the *Affine Cipher* (again, with keys chosen equiprobably). It is slightly more difficult to show that $\mathbf{S} \times \mathbf{M}$ is also the *Affine Cipher* with equiprobable keys.

Let's prove these assertions. A key in the *Shift Cipher* is an element $K \in \mathbb{Z}_{26}$, and the corresponding encryption rule is $e_K(x) = (x + K) \bmod 26$. A key in the *Multiplicative Cipher* is an element $a \in \mathbb{Z}_{26}$ such that $\gcd(a, 26) = 1$; the corresponding encryption rule is $e_a(x) = ax \bmod 26$. Hence, a key in the product cipher $\mathbf{M} \times \mathbf{S}$ has the form (a, K), where

$$e_{(a,K)}(x) = (ax + K) \bmod 26.$$

But this is precisely the definition of a key in the *Affine Cipher*. Further, the probability of a key in the *Affine Cipher* is $1/312 = 1/12 \times 1/26$, which is the product of the probabilities of the keys a and K, respectively. Thus $\mathbf{M} \times \mathbf{S}$ is the *Affine Cipher*.

Now let's consider $\mathbf{S} \times \mathbf{M}$. A key in this cipher has the form (K, a), where

$$e_{(K,a)}(x) = a(x + K) \bmod 26 = (ax + aK) \bmod 26.$$

Thus the key (K, a) of the product cipher $\mathbf{S} \times \mathbf{M}$ is identical to the key (a, aK) of the *Affine Cipher*. It remains to show that each key of the *Affine Cipher* arises with the same probability $1/312$ in the product cipher $\mathbf{S} \times \mathbf{M}$. Observe that $aK = K_1$ if and only if $K = a^{-1}K_1$ (recall that $\gcd(a, 26) = 1$, so a has a multiplicative inverse modulo 26). In other words, the key (a, K_1) of the *Affine Cipher* is equivalent to the key $(a^{-1}K_1, a)$ of the product cipher $\mathbf{S} \times \mathbf{M}$. We thus have a bijection between the two key spaces. Since each key is equiprobable, we conclude that $\mathbf{S} \times \mathbf{M}$ is indeed the *Affine Cipher*.

We have shown that $\mathbf{M} \times \mathbf{S} = \mathbf{S} \times \mathbf{M}$. Thus we would say that the two cryptosystems **M** and **S** *commute*. But not all pairs of cryptosystems commute;

it is easy to find counterexamples. On the other hand, the product operation is always *associative*: $(S_1 \times S_2) \times S_3 = S_1 \times (S_2 \times S_3)$.

If we take the product of an (endomorphic) cryptosystem S with itself, we obtain the cryptosystem $S \times S$, which we denote by S^2. If we take the n-fold product, the resulting cryptosystem is denoted by S^n.

A cryptosystem S is defined to be an *idempotent cryptosystem* if $S^2 = S$. Many of the cryptosystems we studied in Chapter 1 are idempotent. For example, the *Shift, Substitution, Affine, Hill, Vigenère* and *Permutation Ciphers* are all idempotent. Of course, if a cryptosystem S is idempotent, then there is no point in using the product system S^2, as it requires an extra key but provides no more security.

If a cryptosystem is not idempotent, then there is a potential increase in security by iterating it several times. This idea is used in the *Data Encryption Standard*, which consists of 16 iterations. But, of course, this approach requires a non-idempotent cryptosystem to start with. One way in which simple non-idempotent cryptosystems can sometimes be constructed is to take the product of two different (simple) cryptosystems.

REMARK It is not hard to show that if S_1 and S_2 are both idempotent, and they commute, then $S_1 \times S_2$ will also be idempotent. This follows from the following algebraic manipulations:

$$(S_1 \times S_2) \times (S_1 \times S_2) = S_1 \times (S_2 \times S_1) \times S_2$$
$$= S_1 \times (S_1 \times S_2) \times S_2$$
$$= (S_1 \times S_1) \times (S_2 \times S_2)$$
$$= S_1 \times S_2.$$

(Note the use of the associative property in this proof.)

So, if S_1 and S_2 are both idempotent, and we want $S_1 \times S_2$ to be non-idempotent, then it is necessary that S_1 and S_2 not commute. ∎

Fortunately, many simple cryptosystems are suitable building blocks in this type of approach. Taking the product of substitution-type ciphers with permutation-type ciphers is a commonly used technique. We will see several realizations of this in the next chapter.

2.8   Notes

The idea of perfect secrecy and the use of entropy techniques in cryptography was pioneered by Shannon [191]. Product cryptosystems are also discussed in this paper. The concept of entropy was defined by Shannon in [190]. Good introductions to entropy, Huffman coding and related topics can be found in the books by Welsh [213] and Goldie and Pinch [93].

The results of Section 2.6 are due to Beauchemin and Brassard [6], who generalized earlier results of Shannon.

Exercises

2.1 Referring to Example 2.2, determine all the joint and conditional probabilities, $\Pr[x, y]$, $\Pr[x|y]$ and $\Pr[y|x]$, where $x \in \{2, \ldots, 12\}$ and $y \in \{D, N\}$.

2.2 Let n be a positive integer. A *Latin square* of order n is an $n \times n$ array L of the integers $1, \ldots, n$ such that every one of the n integers occurs exactly once in each row and each column of L. An example of a Latin square of order 3 is as follows:

1	2	3
3	1	2
2	3	1

Given any Latin square L of order n, we can define a related cryptosystem. Take $\mathcal{P} = \mathcal{C} = \mathcal{K} = \{1, \ldots, n\}$. For $1 \leq i \leq n$, the encryption rule e_i is defined to be $e_i(j) = L(i, j)$. (Hence each row of L gives rise to one encryption rule.)

Give a complete proof that this *Latin Square Cryptosystem* achieves perfect secrecy provided that every key is used with equal probability.

2.3 (a) Prove that the *Affine Cipher* achieves perfect secrecy if every key is used with equal probability $1/312$.

(b) More generally, suppose we are given a probability distribution on the set

$$\{a \in \mathbb{Z}_{26} : \gcd(a, 26) = 1\}.$$

Suppose that every key (a, b) for the *Affine Cipher* is used with probability $1/(26 \times \Pr[a])$. Prove that the *Affine Cipher* achieves perfect secrecy when this probability distribution is defined on the keyspace.

2.4 Suppose a cryptosystem achieves perfect secrecy for a particular plaintext probability distribution. Prove that perfect secrecy is maintained for any plaintext probability distribution.

2.5 Prove that if a cryptosystem has perfect secrecy and $|\mathcal{K}| = |\mathcal{C}| = |\mathcal{P}|$, then every ciphertext is equally probable.

2.6 Suppose that y and y' are two ciphertext elements (i.e., binary n-tuples) in the *One-time Pad* that were obtained by encrypting plaintext elements x and x', respectively, using the same key, K. Prove that $x + x' \equiv y + y' \pmod{2}$.

2.7 (a) Construct the encryption matrix (as defined in Example 2.3) for the *One-time Pad* with $n = 3$.

     (b) For any positive integer n, give a direct proof that the encryption matrix of a *One-time Pad* defined over $(\mathbb{Z}_2)^n$ is a Latin square of order n.

2.8 Suppose X is a set of cardinality n, where $2^k \leq n < 2^{k+1}$, and $\mathbf{Pr}[x] = 1/n$ for all $x \in X$.

     (a) Find a prefix-free encoding of X, say f, such that $\ell(f) = k + 2 - 2^{k+1}/n$.

        HINT   Encode $2^{k+1} - n$ elements of X as strings of length k, and encode the remaining elements as strings of length $k + 1$.

     (b) Illustrate your construction for $n = 6$. Compute $\ell(f)$ and $H(\mathbf{X})$ in this case.

2.9 Suppose $X = \{a, b, c, d, e\}$ has the following probability distribution: $\mathbf{Pr}[a] = .32$, $\mathbf{Pr}[b] = .23$, $\mathbf{Pr}[c] = .20$, $\mathbf{Pr}[d] = .15$ and $\mathbf{Pr}[e] = .10$. Use Huffman's algorithm to find the optimal prefix-free encoding of X. Compare the length of this encoding to $H(\mathbf{X})$.

2.10 Prove that $H(\mathbf{X}, \mathbf{Y}) = H(\mathbf{Y}) + H(\mathbf{X}|\mathbf{Y})$. Then show as a corollary that $H(\mathbf{X}|\mathbf{Y}) \leq H(\mathbf{X})$, with equality if and only if $\mathbf{X}$ and $\mathbf{Y}$ are independent.

2.11 Prove that a cryptosystem has perfect secrecy if and only if $H(\mathbf{P}|\mathbf{C}) = H(\mathbf{P})$.

2.12 Prove that, in any cryptosystem, $H(\mathbf{K}|\mathbf{C}) \geq H(\mathbf{P}|\mathbf{C})$. (Intuitively, this result says that, given a ciphertext, the opponent's uncertainty about the key is at least as great as his uncertainty about the plaintext.)

2.13 Consider a cryptosystem in which $\mathcal{P} = \{a, b, c\}$, $\mathcal{K} = \{K_1, K_2, K_3\}$ and $\mathcal{C} = \{1, 2, 3, 4\}$. Suppose the encryption matrix is as follows:

	a	b	c
K_1	1	2	3
K_2	2	3	4
K_3	3	4	1

Given that keys are chosen equiprobably, and the plaintext probability distribution is $\mathbf{Pr}[a] = 1/2$, $\mathbf{Pr}[b] = 1/3$, $\mathbf{Pr}[c] = 1/6$, compute $H(\mathbf{P})$, $H(\mathbf{C})$, $H(\mathbf{K})$, $H(\mathbf{K}|\mathbf{C})$ and $H(\mathbf{P}|\mathbf{C})$.

2.14 Compute $H(\mathbf{K}|\mathbf{C})$ and $H(\mathbf{K}|\mathbf{P}, \mathbf{C})$ for the *Affine Cipher*.

2.15 Consider a *Vigenère Cipher* with keyword length m. Show that the unicity distance is $1/R_L$, where R_L is the redundancy of the underlying language. (This result is interpreted as follows. If n_0 denotes the number of alphabetic characters being encrypted, then the "length" of the plaintext is n_0/m, since each plaintext element consists of m alphabetic characters. So, a unicity distance of $1/R_L$ corresponds to a plaintext consisting of m/R_L alphabetic characters.)

2.16 Show that the unicity distance of the *Hill Cipher* (with an $m \times m$ encryption matrix) is less than m/R_L. (Note that the number of alphabetic characters in a plaintext of this length is m^2/R_L.)

2.17 A *Substitution Cipher* over a plaintext space of size n has $|\mathcal{K}| = n!$. Stirling's formula gives the following estimate for $n!$:

$$n! \approx \sqrt{2\pi n}\left(\frac{n}{e}\right)^n.$$

     (a) Using Stirling's formula, derive an estimate of the unicity distance of the *Substitution Cipher*.

     (b) Let $m \geq 1$ be an integer. The m-gram *Substitution Cipher* is the *Substitution Cipher* where the plaintext (and ciphertext) spaces consist of all 26^m m-grams. Estimate the unicity distance of the m-gram *Substitution Cipher* if $R_L = 0.75$.

2.18 Prove that the *Shift Cipher* is idempotent.

2.19 Suppose $\mathbf{S}_1$ is the *Shift Cipher* (with equiprobable keys, as usual) and $\mathbf{S}_2$ is the *Shift Cipher* where keys are chosen with respect to some probability distribution $p_{\mathcal{K}}$ (which need not be equiprobable). Prove that $\mathbf{S}_1 \times \mathbf{S}_2 = \mathbf{S}_1$.

2.20 Suppose $\mathbf{S}_1$ and $\mathbf{S}_2$ are *Vigenère Ciphers* with keyword lengths m_1, m_2 respectively, where $m_1 > m_2$.

   (a) If $m_2 \mid m_1$, then show that $\mathbf{S}_2 \times \mathbf{S}_1 = \mathbf{S}_1$.

   (b) One might try to generalize the previous result by conjecturing that $\mathbf{S}_2 \times \mathbf{S}_1 = \mathbf{S}_3$, where $\mathbf{S}_3$ is the *Vigenère Cipher* with keyword length $\operatorname{lcm}(m_1, m_2)$. Prove that this conjecture is false.

   HINT   If $m_1 \not\equiv 0 \pmod{m_2}$, then the number of keys in the product cryptosystem $\mathbf{S}_2 \times \mathbf{S}_1$ is less than the number of keys in $\mathbf{S}_3$.

3

Block Ciphers and the Advanced Encryption Standard

3.1 Introduction

Most modern-day block ciphers are product ciphers (product ciphers were introduced in Section 2.7). Product ciphers frequently incorporate a sequence of permutation and substitution operations. A commonly used design is that of an *iterated cipher*. Here is a description of a typical iterated cipher: The cipher requires the specification of a *round function* and a *key schedule*, and the encryption of a plaintext will proceed through Nr similar *rounds*.

Let K be a random binary key of some specified length. K is used to construct Nr *round keys* (also called *subkeys*), which are denoted $K^1, \ldots, K^{Nr}$. The list of round keys, $(K^1, \ldots, K^{Nr})$, is the key schedule. The key schedule is constructed from K using a fixed, public algorithm.

The round function, say g, takes two inputs: a round key (K^r) and a current *state* (which we denote w^{r-1}). The next state is defined as $w^r = g(w^{r-1}, K^r)$. The initial state, w^0, is defined to be the plaintext, x. The ciphertext, y, is defined to be the state after all Nr rounds have been performed. Therefore, the encryption operation is carried out as follows:

$$w^0 \leftarrow x$$
$$w^1 \leftarrow g(w^0, K^1)$$
$$w^2 \leftarrow g(w^1, K^2)$$
$$\vdots \quad \vdots \quad \vdots$$
$$w^{Nr-1} \leftarrow g(w^{Nr-2}, K^{Nr-1})$$
$$w^{Nr} \leftarrow g(w^{Nr-1}, K^{Nr})$$
$$y \leftarrow w^{Nr}.$$

In order for decryption to be possible, the function g must have the property that it is injective (i.e., one-to-one) if its second argument is fixed. This is equivalent to saying that there exists a function g^{-1} with the property that

$$g^{-1}(g(w, y), y) = w$$

for all w and y. Then decryption can be accomplished as follows:

$$w^{\mathrm{Nr}} \leftarrow y$$
$$w^{\mathrm{Nr}-1} \leftarrow g^{-1}(w^{\mathrm{Nr}}, K^{\mathrm{Nr}})$$
$$\vdots \quad \vdots \quad \vdots$$
$$w^1 \leftarrow g^{-1}(w^2, K^2)$$
$$w^0 \leftarrow g^{-1}(w^1, K^1)$$
$$x \leftarrow w^0.$$

In Section 3.2, we describe a simple type of iterated cipher, the substitution-permutation network, which illustrates many of the main principles used in the design of practical block ciphers. Linear and differential attacks on substitution-permutation networks are described in Sections 3.3 and 3.4, respectively. In Section 3.5, we discuss Feistel-type ciphers and the *Data Encryption Standard*. In Section 3.6, we present the *Advanced Encryption Standard*. Finally, modes of operation of block ciphers are the topic of Section 3.7.

3.2 Substitution-Permutation Networks

We begin by defining a *substitution-permutation network*, or *SPN*. (An SPN is a special type of iterated cipher with a couple of small changes that we will indicate.) Suppose that ℓ and m are positive integers. A plaintext and ciphertext will both be binary vectors of length ℓm (i.e., ℓm is the *block length* of the cipher). An SPN is built from two components, which are denoted π_S and π_P.

$$\pi_S : \{0, 1\}^\ell \to \{0, 1\}^\ell$$

is a permutation, and

$$\pi_P : \{1, \ldots, \ell m\} \to \{1, \ldots, \ell m\}$$

is also a permutation. The permutation π_S is called an *S-box* (the letter "S" denotes "substitution"). It is used to replace ℓ bits with a different set of ℓ bits. π_P, on the other hand, is used to permute ℓm bits.

Given an ℓm-bit binary string, say $x = (x_1, \ldots, x_{\ell m})$, we can regard x as the concatenation of m ℓ-bit substrings, which we denote $x_{(1)}, \ldots, x_{(m)}$. Thus

$$x = x_{(1)} \, \| \, \cdots \, \| \, x_{(m)}$$

and for $1 \leq i \leq m$, we have that

$$x_{(i)} = \left(x_{(i-1)\ell+1}, \ldots, x_{i\ell} \right).$$

The SPN will consist of Nr rounds. In each round (except for the last round, which is slightly different), we will perform m substitutions using π_S, followed by a permutation using π_P. Prior to each substitution operation, we will incorporate round key bits via a simple exclusive-or operation. We now present an SPN, based on π_S and π_P, as Cryptosystem 3.1.

Cryptosystem 3.1: *Substitution-Permutation Network*

Let ℓ, m and Nr be positive integers, let $\pi_S : \{0,1\}^\ell \to \{0,1\}^\ell$ be a permutation, and let $\pi_P : \{1, \ldots, \ell m\} \to \{1, \ldots, \ell m\}$ be a permutation. Let $\mathcal{P} = \mathcal{C} = \{0,1\}^{\ell m}$, and let $\mathcal{K} \subseteq (\{0,1\}^{\ell m})^{\mathrm{Nr}+1}$ consist of all possible key schedules that could be derived from an initial key K using the key scheduling algorithm. For a key schedule $(K^1, \ldots, K^{\mathrm{Nr}+1})$, we encrypt the plaintext x using Algorithm 3.1.

Algorithm 3.1: $\mathrm{SPN}(x, \pi_S, \pi_P, (K^1, \ldots, K^{\mathrm{Nr}+1}))$

$w^0 \leftarrow x$
for $r \leftarrow 1$ **to** $\mathrm{Nr} - 1$

$\mathbf{do} \begin{cases} u^r \leftarrow w^{r-1} \oplus K^r \\ \mathbf{for}\ i \leftarrow 1\ \mathbf{to}\ m \\ \quad \mathbf{do}\ v^r_{(i)} \leftarrow \pi_S(u^r_{(i)}) \\ w^r \leftarrow (v^r_{\pi_P(1)}, \ldots, v^r_{\pi_P(\ell m)}) \end{cases}$

$u^{\mathrm{Nr}} \leftarrow w^{\mathrm{Nr}-1} \oplus K^{\mathrm{Nr}}$
for $i \leftarrow 1$ **to** m
$\quad \mathbf{do}\ v^{\mathrm{Nr}}_{(i)} \leftarrow \pi_S(u^{\mathrm{Nr}}_{(i)})$
$y \leftarrow v^{\mathrm{Nr}} \oplus K^{\mathrm{Nr}+1}$
output (y)

In Algorithm 3.1, u^r is the input to the S-boxes in round r, and v^r is the output of the S-boxes in round r. w^r is obtained from v^r by applying the permutation π_P, and then u^{r+1} is constructed from v^r by x-or-ing with the round key K^{r+1} (this is called *round key mixing*). In the last round, the permutation π_P is not applied. As a consequence, the encryption algorithm can also be used for decryption, if appropriate modifications are made to the key schedule and the S-boxes are replaced by their inverses (see the Exercises).

Notice that the very first and last operations performed in this SPN are x-ors with subkeys. This is called *whitening*, and is regarded as a useful way to prevent an attacker from even beginning to carry out an encryption or decryption operation if the key is not known.

We illustrate the above general description with a particular SPN.

Example 3.1   Suppose that $\ell = m = \text{Nr} = 4$. Let π_S be defined as follows, where the input (i.e., z) and the output (i.e., $\pi_S(z)$) are written in hexadecimal notation, $(0 \leftrightarrow (0,0,0,0), 1 \leftrightarrow (0,0,0,1), \ldots, 9 \leftrightarrow (1,0,0,1), A \leftrightarrow (1,0,1,0), \ldots, F \leftrightarrow (1,1,1,1))$:

z	0	1	2	3	4	5	6	7	8	9	A	B	C	D	E	F
$\pi_S(z)$	E	4	D	1	2	F	B	8	3	A	6	C	5	9	0	7

Further, let π_P be defined as follows:

z	1	2	3	4	5	6	7	8	9	10	11	12	13	14	15	16
$\pi_P(z)$	1	5	9	13	2	6	10	14	3	7	11	15	4	8	12	16

See Figure 3.1 for a pictorial representation of this particular SPN. (In this diagram, we have named the S-boxes S_i^r $(1 \leq i \leq 4, 1 \leq r \leq 4)$ for ease of later reference. All 16 S-boxes incorporate the same substitution function based on π_S.)

In order to complete the description of the SPN, we need to specify a key scheduling algorithm. Here is a simple possibility: suppose that we begin with a 32-bit key $K = (k_1, \ldots, k_{32}) \in \{0,1\}^{32}$. For $1 \leq r \leq 5$, define K^r to consist of 16 consecutive bits of K, beginning with k_{4r-3}. (This is not a very secure way to define a key schedule; we have just chosen something easy for purposes of illustration.)

Now let's work out a sample encryption using this SPN. We represent all data in binary notation. Suppose the key is

$$K = 0011\ 1010\ 1001\ 0100\ 1101\ 0110\ 0011\ 1111.$$

Then the round keys are as follows:

$$K^1 = 0011\ 1010\ 1001\ 0100$$

$$K^2 = 1010\ 1001\ 0100\ 1101$$

$$K^3 = 1001\ 0100\ 1101\ 0110$$

$$K^4 = 0100\ 1101\ 0110\ 0011$$

$$K^5 = 1101\ 0110\ 0011\ 1111.$$

Suppose that the plaintext is

$$x = 0010\ 0110\ 1011\ 0111.$$

Then the encryption of x proceeds as follows:

$$w^0 = 0010\ 0110\ 1011\ 0111$$

$$K^1 = 0011\ 1010\ 1001\ 0100$$

$$u^1 = 0001\ 1100\ 0010\ 0011$$

$$v^1 = 0100\ 0101\ 1101\ 0001$$

$$w^1 = 0010\ 1110\ 0000\ 0111$$

$$K^2 = 1010\ 1001\ 0100\ 1101$$

$$u^2 = 1000\ 0111\ 0100\ 1010$$

$$v^2 = 0011\ 1000\ 0010\ 0110$$

$$w^2 = 0100\ 0001\ 1011\ 1000$$

$$K^3 = 1001\ 0100\ 1101\ 0110$$

$$u^3 = 1101\ 0101\ 0110\ 1110$$

$$v^3 = 1001\ 1111\ 1011\ 0000$$

$$w^3 = 1110\ 0100\ 0110\ 1110$$

$$K^4 = 0100\ 1101\ 0110\ 0011$$

$$u^4 = 1010\ 1001\ 0000\ 1101$$

$$v^4 = 0110\ 1010\ 1110\ 1001$$

$$K^5 = 1101\ 0110\ 0011\ 1111, \quad \text{and}$$

$$y = 1011\ 1100\ 1101\ 0110$$

is the ciphertext. $\square$

SPNs have several attractive features. First, the design is simple and very efficient, in both hardware and software. In software, an S-box is usually implemented in the form of a look-up table. Observe that the memory requirement of the S-box $\pi_S : \{0,1\}^\ell \to \{0,1\}^\ell$ is $\ell 2^\ell$ bits, since we have to store 2^ℓ values, each of which needs ℓ bits of storage. Hardware implementations, in particular, necessitate the use of relatively small S-boxes.

In Example 3.1, we used four identical S-boxes in each round. The memory requirement of the S-box is 2^6 bits. If we instead used one S-box which mapped 16 bits to 16 bits, the memory requirement would be increased to 2^{20} bits, which

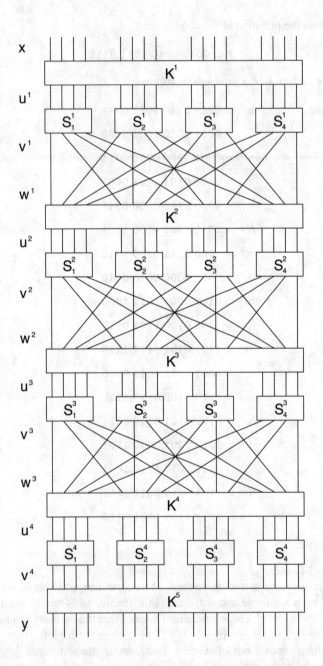

FIGURE 3.1
A substitution-permutation network

would be prohibitively high for some applications. The S-box used in the *Advanced Encryption Standard* (to be discussed in Section 3.6) maps eight bits to eight bits.

The SPN in Example 3.1 is not secure, if for no other reason than the key length (32 bits) is small enough that an exhaustive key search is feasible. However, "larger" SPNs can be designed that are secure against all known attacks. A practical, secure SPN would have a larger key size and block length than Example 3.1, would most likely use larger S-boxes, and would have more rounds. *Rijndael*, which was chosen to be the *Advanced Encryption Standard*, is an example of an SPN that is similar to Example 3.1 in many respects. *Rijndael* has a minimum key size of 128 bits, a block length of 128, a minimum of 10 rounds; and its S-box maps eight bits to eight bits (see Section 3.6 for a complete description).

Many variations of SPNs are possible. One common modification would be to use more than one S-box. In Example, 3.1, we could use four different S-boxes in each round if we so desired, instead of using the same S-box four times. This feature can be found in the *Data Encryption Standard*, which employs eight different S-boxes in each round (see Section 3.5.1). Another popular design strategy is to include an invertible linear transformation in each round, either as a replacement for, or in addition to, the permutation operation. This is done in the *Advanced Encryption Standard* (see Section 3.6.1).

3.3 Linear Cryptanalysis

We begin by informally describing the strategy behind linear cryptanalysis. The idea can be applied, in principle, to any iterated cipher. Suppose that it is possible to find a probabilistic linear relationship between a subset of plaintext bits and a subset of state bits immediately preceding the substitutions performed in the last round. In other words, there exists a subset of bits whose exclusive-or behaves in a non-random fashion (it takes on the value 0, say, with probability bounded away from $1/2$). Now assume that an attacker has a large number of plaintext-ciphertext pairs, all of which are encrypted using the same unknown key K (i.e., we consider a known-plaintext attack). For each of the plaintext-ciphertext pairs, we will begin to decrypt the ciphertext, using all possible candidate keys for the last round of the cipher. For each candidate key, we compute the values of the relevant state bits involved in the linear relationship, and determine if the above-mentioned linear relationship holds. Whenever it does, we increment a counter corresponding to the particular candidate key. At the end of this process, we hope that the candidate key that has a frequency count that is furthest from $1/2$ times the number of pairs contains the correct values for these key bits.

We will illustrate the above description with a detailed example later in this section. First, we need to establish some results from probability theory to provide a (non-rigorous) justification for the techniques involved in the attack.

3.3.1   The Piling-up Lemma

We use terminology and concepts introduced in Section 2.2. Suppose that $\mathbf{X_1}, \mathbf{X_2}, \ldots$ are independent random variables taking on values from the set $\{0, 1\}$. Let $p_1, p_2, \ldots$ be real numbers such that $0 \leq p_i \leq 1$ for all i, and suppose that

$$\mathbf{Pr}[\mathbf{X_i} = 0] = p_i,$$

$i = 1, 2, \ldots$. Hence,

$$\mathbf{Pr}[\mathbf{X_i} = 1] = 1 - p_i,$$

$i = 1, 2, \ldots$.

Suppose that $i \neq j$. The independence of $\mathbf{X_i}$ and $\mathbf{X_j}$ implies that

$$\mathbf{Pr}[\mathbf{X_i} = 0, \mathbf{X_j} = 0] = p_i p_j$$

$$\mathbf{Pr}[\mathbf{X_i} = 0, \mathbf{X_j} = 1] = p_i(1 - p_j)$$

$$\mathbf{Pr}[\mathbf{X_i} = 1, \mathbf{X_j} = 0] = (1 - p_i)p_j, \quad \text{and}$$

$$\mathbf{Pr}[\mathbf{X_i} = 1, \mathbf{X_j} = 1] = (1 - p_i)(1 - p_j).$$

Now consider the discrete random variable $\mathbf{X_i} \oplus \mathbf{X_j}$ (this is the same thing as $\mathbf{X_i} + \mathbf{X_j} \bmod 2$). It is easy to see that $\mathbf{X_i} \oplus \mathbf{X_j}$ has the following probability distribution:

$$\mathbf{Pr}[\mathbf{X_i} \oplus \mathbf{X_j} = 0] = p_i p_j + (1 - p_i)(1 - p_j)$$

$$\mathbf{Pr}[\mathbf{X_i} \oplus \mathbf{X_j} = 1] = p_i(1 - p_j) + (1 - p_i)p_j.$$

It is often convenient to express a probability distribution of a random variable taking on the values 0 and 1 in terms of a quantity called the bias of the distribution. The *bias* of $\mathbf{X_i}$ is defined to be the quantity

$$\epsilon_i = p_i - \frac{1}{2}.$$

Observe the following facts:

$$-\frac{1}{2} \leq \epsilon_i \leq \frac{1}{2},$$

$$\mathbf{Pr}[\mathbf{X_i} = 0] = \frac{1}{2} + \epsilon_i, \text{ and}$$

$$\mathbf{Pr}[\mathbf{X_i} = 1] = \frac{1}{2} - \epsilon_i,$$

$i = 1, 2, \ldots$. For $i_1 < i_2 < \cdots < i_k$, let $\epsilon_{i_1, i_2, \ldots, i_k}$ denote the bias of the random variable $\mathbf{X_{i_1}} \oplus \cdots \oplus \mathbf{X_{i_k}}$. It is not hard to verify that $\epsilon_{i_1, i_2} = 2\epsilon_{i_1}\epsilon_{i_2}$. In general, the following result holds, which is known as the "piling-up lemma."

LEMMA 3.1 (*Piling-up lemma*) *Let $\epsilon_{i_1, i_2, \ldots, i_k}$ denote the bias of the random variable $\mathbf{X_{i_1}} \oplus \cdots \oplus \mathbf{X_{i_k}}$. Then*

$$\epsilon_{i_1, i_2, \ldots, i_k} = 2^{k-1} \prod_{j=1}^{k} \epsilon_{i_j}.$$

PROOF The proof is by induction on k. Clearly the result is true when $k = 1$. As an induction hypothesis, assume that the result is true for $k = \ell$, for some positive integer ℓ.

Consider $k = \ell + 1$. We want to determine the bias of $\mathbf{X_{i_1}} \oplus \cdots \oplus \mathbf{X_{i_{\ell+1}}}$. By induction, the bias of $\mathbf{X_{i_1}} \oplus \cdots \oplus \mathbf{X_{i_\ell}}$ is $2^{\ell-1} \prod_{j=1}^{\ell} \epsilon_{i_j}$. Therefore, we have that

$$\mathbf{Pr}[\mathbf{X_{i_1}} \oplus \cdots \oplus \mathbf{X_{i_\ell}} = 0] = \frac{1}{2} + 2^{\ell-1} \prod_{j=1}^{\ell} \epsilon_{i_j}, \quad \text{and}$$

$$\mathbf{Pr}[\mathbf{X_{i_1}} \oplus \cdots \oplus \mathbf{X_{i_\ell}} = 1] = \frac{1}{2} - 2^{\ell-1} \prod_{j=1}^{\ell} \epsilon_{i_j}.$$

From this, it is easy to see that

$$\mathbf{Pr}[\mathbf{X_{i_1}} \oplus \cdots \oplus \mathbf{X_{i_{\ell+1}}} = 0]$$

$$= \left(\frac{1}{2} + 2^{\ell-1} \prod_{j=1}^{\ell} \epsilon_{i_j} \right) \left(\frac{1}{2} + \epsilon_{i_{\ell+1}} \right) + \left(\frac{1}{2} - 2^{\ell-1} \prod_{j=1}^{\ell} \epsilon_{i_j} \right) \left(\frac{1}{2} - \epsilon_{i_{\ell+1}} \right)$$

$$= \frac{1}{2} + 2^{\ell} \prod_{j=1}^{\ell+1} \epsilon_{i_j},$$

as desired. By induction, the proof is complete. ∎

COROLLARY 3.2 *Let $\epsilon_{i_1, i_2, \ldots, i_k}$ denote the bias of the random variable $\mathbf{X_{i_1}} \oplus \cdots \oplus \mathbf{X_{i_k}}$. Suppose that $\epsilon_{i_j} = 0$ for some j. Then $\epsilon_{i_1, i_2, \ldots, i_k} = 0$.*

It is important to realize that Lemma 3.1 holds, in general, only when the relevant random variables are independent. We illustrate this by considering an example. Suppose that $\epsilon_1 = \epsilon_2 = \epsilon_3 = 1/4$. Applying Lemma 3.1, we see that $\epsilon_{1,2} = \epsilon_{2,3} = \epsilon_{1,3} = 1/8$. Now, consider the random variable $\mathbf{X_1} \oplus \mathbf{X_3}$. It is clear that

$$\mathbf{X_1} \oplus \mathbf{X_3} = (\mathbf{X_1} \oplus \mathbf{X_2}) \oplus (\mathbf{X_2} \oplus \mathbf{X_3}).$$

If the two random variables $\mathbf{X_1} \oplus \mathbf{X_2}$ and $\mathbf{X_2} \oplus \mathbf{X_3}$ were independent, then Lemma 3.1 would say that $\epsilon_{1,3} = 2(1/8)^2 = 1/32$. However, we already know that this is not the case: $\epsilon_{1,3} = 1/8$. Lemma 3.1 does not yield the correct value of $\epsilon_{1,3}$ because $\mathbf{X_1} \oplus \mathbf{X_2}$ and $\mathbf{X_2} \oplus \mathbf{X_3}$ are not independent.

3.3.2   Linear Approximations of S-boxes

Consider an S-box $\pi_S : \{0, 1\}^m \rightarrow \{0, 1\}^n$. (We do not assume that π_S is a permutation, or even that $m = n$.) Let us write an input m-tuple as $X = (x_1, \ldots, x_m)$. This m-tuple is chosen uniformly at random from $\{0, 1\}^m$, which means that each co-ordinate x_i defines a random variable $\mathbf{X_i}$ taking on values 0 and 1, having bias $\epsilon_i = 0$. Further, these m random variables are independent.

Now write an output n-tuple as $Y = (y_1, \ldots, y_n)$. Each co-ordinate y_j defines a random variable $\mathbf{Y_j}$ taking on values 0 and 1. These n random variables are, in general, not independent from each other or from the $\mathbf{X_i}$'s. In fact, it is not hard to see that the following formula holds:

$$\mathbf{Pr}[\mathbf{X_1} = x_1, \ldots, \mathbf{X_m} = x_m, \mathbf{Y_1} = y_1, \ldots, \mathbf{Y_n} = y_n] = 0$$

if $(y_1, \ldots, y_n) \neq \pi_S(x_1, \ldots, x_m)$; and

$$\mathbf{Pr}[\mathbf{X_1} = x_1, \ldots, \mathbf{X_m} = x_m, \mathbf{Y_1} = y_1, \ldots, \mathbf{Y_n} = y_n] = 2^{-m}$$

if $(y_1, \ldots, y_n) = \pi_S(x_1, \ldots, x_m)$. (The last formula holds because

$$\mathbf{Pr}[\mathbf{X_1} = x_1, \ldots, \mathbf{X_m} = x_m] = 2^{-m}$$

and

$$\mathbf{Pr}[\mathbf{Y_1} = y_1, \ldots, \mathbf{Y_n} = y_n | \mathbf{X_1} = x_1, \ldots, \mathbf{X_m} = x_m] = 1$$

if $(y_1, \ldots, y_n) = \pi_S(x_1, \ldots, x_m)$.)

It is now relatively straightforward to compute the bias of a random variable of the form

$$\mathbf{X_{i_1}} \oplus \cdots \oplus \mathbf{X_{i_\kappa}} \oplus \mathbf{Y_{j_1}} \oplus \cdots \oplus \mathbf{Y_{j_\ell}}$$

using the formulas stated above. (A linear cryptanalytic attack can potentially be mounted when a random variable of this form has a bias that is bounded away from zero.)

Let's consider a small example.

Example 3.2   We use the S-box from Example 3.1, which is defined by a permutation $\pi_S : \{0, 1\}^4 \rightarrow \{0, 1\}^4$. We record the possible values taken on by the eight random variables $\mathbf{X_1}, \ldots, \mathbf{X_4}, \mathbf{Y_1}, \ldots, \mathbf{Y_4}$ in the rows of the following table:

X_1	X_2	X_3	X_4	Y_1	Y_2	Y_3	Y_4
0	0	0	0	1	1	1	0
0	0	0	1	0	1	0	0
0	0	1	0	1	1	0	1
0	0	1	1	0	0	0	1
0	1	0	0	0	0	1	0
0	1	0	1	1	1	1	1
0	1	1	0	1	0	1	1
0	1	1	1	1	0	0	0
1	0	0	0	0	0	1	1
1	0	0	1	1	0	1	0
1	0	1	0	0	1	1	0
1	0	1	1	1	1	0	0
1	1	0	0	0	1	0	1
1	1	0	1	1	0	0	1
1	1	1	0	0	0	0	0
1	1	1	1	0	1	1	1

Now, consider the random variable $X_1 \oplus X_4 \oplus Y_2$. The probability that this random variable takes on the value 0 can be determined by counting the number of rows in the above table in which $X_1 \oplus X_4 \oplus Y_2 = 0$, and then dividing by 16 ($16 = 2^4$ is the total number of rows in the table). It is seen that

$$\Pr[X_1 \oplus X_4 \oplus Y_2 = 0] = \frac{1}{2}$$

(and therefore

$$\Pr[X_1 \oplus X_4 \oplus Y_2 = 1] = \frac{1}{2},$$

as well.) Hence, the bias of this random variable is 0. $\quad\square$

If we instead analyzed the random variable $X_3 \oplus X_4 \oplus Y_1 \oplus Y_4$, we would find that the bias is $-3/8$. (We suggest that the reader verify this computation.) Indeed, it is not difficult to compute the biases of all $2^8 = 256$ possible random variables of this form.

We record this information using the following notation. We represent each of the relevant random variables in the form

$$\left(\bigoplus_{i=1}^{4} a_i X_i \right) \oplus \left(\bigoplus_{i=1}^{4} b_i Y_i \right),$$

where $a_i \in \{0, 1\}$, $b_i \in \{0, 1\}$, $i = 1, 2, 3, 4$. Then, in order to have a compact notation, we treat each of the binary vectors (a_1, a_2, a_3, a_4) and (b_1, b_2, b_3, b_4) as a hexadecimal digit (these are called the *input sum* and *output sum*, respectively).

a	0	1	2	3	4	5	6	7	8	9	A	B	C	D	E	F
0	16	8	8	8	8	8	8	8	8	8	8	8	8	8	8	8
1	8	8	6	6	8	8	6	14	10	10	8	8	10	10	8	8
2	8	8	6	6	8	8	6	6	8	8	10	10	8	8	2	10
3	8	8	8	8	8	8	8	8	10	2	6	6	10	10	6	6
4	8	10	8	6	6	4	6	8	8	6	8	10	10	4	10	8
5	8	6	6	8	6	8	12	10	6	8	4	10	8	6	6	8
6	8	10	6	12	10	8	8	10	8	6	10	12	6	8	8	6
7	8	6	8	10	10	4	10	8	6	8	10	8	12	10	8	10
8	8	8	8	8	8	8	8	8	6	10	10	6	10	6	6	2
9	8	8	6	6	8	8	6	6	4	8	6	10	8	12	10	6
A	8	12	6	10	4	8	10	6	10	10	8	8	10	10	8	8
B	8	12	8	4	12	8	12	8	8	8	8	8	8	8	8	8
C	8	6	12	6	6	8	10	8	10	8	10	12	8	10	8	6
D	8	10	10	8	6	12	8	10	4	6	10	8	10	8	8	10
E	8	10	10	8	6	4	8	10	6	8	8	6	4	10	6	8
F	8	6	4	6	6	8	10	8	8	6	12	6	6	8	10	8

(Column header b spans columns 0–F.)

FIGURE 3.2
Linear approximation table: values of $N_L(a, b)$

In this way, each of the 256 random variables is named by a (unique) pair of hexadecimal digits, representing the input and output sum.

As an example, consider the random variable $\mathbf{X_1} \oplus \mathbf{X_4} \oplus \mathbf{Y_2}$. The input sum is $(1, 0, 0, 1)$, which is 9 in hexadecimal; the output sum is $(0, 1, 0, 0)$, which is 4 in hexadecimal.

For a random variable having (hexadecimal) input sum a and output sum b (where $a = (a_1, a_2, a_3, a_4)$ and $b = (b_1, b_2, b_3, b_4)$, in binary), let $N_L(a, b)$ denote the number of binary eight-tuples $(x_1, x_2, x_3, x_4, y_1, y_2, y_3, y_4)$ such that

$$(y_1, y_2, y_3, y_4) = \pi_S(x_1, x_2, x_3, x_4)$$

and

$$\left(\bigoplus_{i=1}^{4} a_i x_i \right) \oplus \left(\bigoplus_{i=1}^{4} b_i y_i \right) = 0.$$

The bias of the random variable having input sum a and output sum b is computed as $\epsilon(a, b) = (N_L(a, b) - 8)/16$.

We computed $N_L(9, 4) = 8$, and hence $\epsilon(9, 4) = 0$, in Example 3.2. The table of all values N_L is called the *linear approximation table*; see Figure 3.2.

3.3.3 A Linear Attack on an SPN

Linear cryptanalysis requires finding a set of linear approximations of S-boxes that can be used to derive a linear approximation of the entire SPN (excluding the last round). We will illustrate the procedure using the SPN from Example 3.1. The diagram in Figure 3.3 illustrates the structure of the approximation we will use. This diagram can be interpreted as follows: Lines with arrows correspond to random variables which will be involved in linear approximations. The labeled S-boxes are the ones used in these approximations (they are called the *active S-boxes* in the approximation).

This approximation incorporates four active S-boxes:

- In S_2^1, the random variable $T_1 = U_5^1 \oplus U_7^1 \oplus U_8^1 \oplus V_6^1$ has bias $1/4$
- In S_2^2, the random variable $T_2 = U_6^2 \oplus V_6^2 \oplus V_8^2$ has bias $-1/4$
- In S_2^3, the random variable $T_3 = U_6^3 \oplus V_6^3 \oplus V_8^3$ has bias $-1/4$
- In S_4^3, the random variable $T_4 = U_{14}^3 \oplus V_{14}^3 \oplus V_{16}^3$ has bias $-1/4$

The four random variables T_1, T_2, T_3, T_4 have biases that are high in absolute value. Further, we will see that their exclusive-or will lead to cancellations of "intermediate" random variables.

If we make the assumption that these four random variables are independent, then we can compute the bias of their x-or using the piling-up lemma (Lemma 3.1). (The random variables are in fact not independent, which means that we cannot provide a mathematical justification of this approximation. Nevertheless, the approximation seems to work in practice, as we shall demonstrate.) We therefore hypothesize that the random variable

$$T_1 \oplus T_2 \oplus T_3 \oplus T_4$$

has bias equal to $2^3(1/4)(-1/4)^3 = -1/32$.

Now, the random variables T_1, T_2, T_3 and T_4 have the property that their x-or can be expressed in terms of plaintext bits, bits of u^4 (the input to the last round of S-boxes) and key bits. This can be done as follows: First, we have the following relations, which can be easily verified by inspecting Figure 3.3:

$$T_1 = U_5^1 \oplus U_7^1 \oplus U_8^1 \oplus V_6^1 = X_5 \oplus K_5^1 \oplus X_7 \oplus K_7^1 \oplus X_8 \oplus K_8^1 \oplus V_6^1$$

$$T_2 = U_6^2 \oplus V_6^2 \oplus V_8^2 \qquad = V_6^1 \oplus K_6^2 \oplus V_6^2 \oplus V_8^2$$

$$T_3 = U_6^3 \oplus V_6^3 \oplus V_8^3 \qquad = V_6^2 \oplus K_6^3 \oplus V_6^3 \oplus V_8^3$$

$$T_4 = U_{14}^3 \oplus V_{14}^3 \oplus V_{16}^3 \qquad = V_8^2 \oplus K_{14}^3 \oplus V_{14}^3 \oplus V_{16}^3.$$

If we compute the x-or of the random variables on the right sides of the above equations, we see that the random variable

$$X_5 \oplus X_7 \oplus X_8 \oplus V_6^3 \oplus V_8^3 \oplus V_{14}^3 \oplus V_{16}^3$$

$$\oplus K_5^1 \oplus K_7^1 \oplus K_8^1 \oplus K_6^2 \oplus K_6^3 \oplus K_{14}^3 \quad (3.1)$$

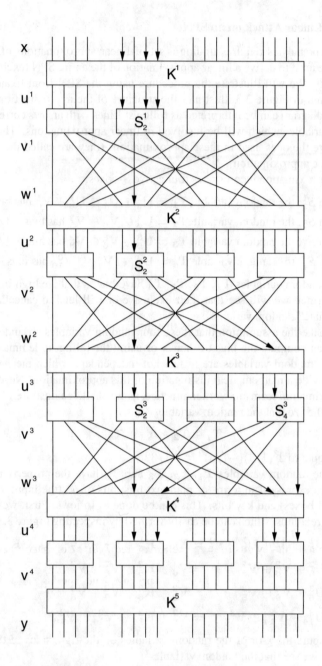

FIGURE 3.3
A linear approximation of a substitution-permutation network

has bias equal to $-1/32$. The next step is to replace the terms $\mathbf{V}_i^3$ in the above formula by expressions involving $\mathbf{U}_i^4$ and further key bits:

$$\mathbf{V}_6^3 = \mathbf{U}_6^4 \oplus \mathbf{K}_6^4$$

$$\mathbf{V}_8^3 = \mathbf{U}_{14}^4 \oplus \mathbf{K}_{14}^4$$

$$\mathbf{V}_{14}^3 = \mathbf{U}_8^4 \oplus \mathbf{K}_8^4$$

$$\mathbf{V}_{16}^3 = \mathbf{U}_{16}^4 \oplus \mathbf{K}_{16}^4$$

Now we substitute these four expressions into (3.1), to get the following:

$$\mathbf{X}_5 \oplus \mathbf{X}_7 \oplus \mathbf{X}_8 \oplus \mathbf{U}_6^4 \oplus \mathbf{U}_8^4 \oplus \mathbf{U}_{14}^4 \oplus \mathbf{U}_{16}^4$$

$$\oplus \, \mathbf{K}_5^1 \oplus \mathbf{K}_7^1 \oplus \mathbf{K}_8^1 \oplus \mathbf{K}_6^2 \oplus \mathbf{K}_6^3 \oplus \mathbf{K}_{14}^3 \oplus \mathbf{K}_6^4 \oplus \mathbf{K}_8^4 \oplus \mathbf{K}_{14}^4 \oplus \mathbf{K}_{16}^4 \quad (3.2)$$

This expression only involves plaintext bits, bits of u^4 and key bits. Suppose that the key bits in (3.2) are fixed. Then the random variable

$$\mathbf{K}_5^1 \oplus \mathbf{K}_7^1 \oplus \mathbf{K}_8^1 \oplus \mathbf{K}_6^2 \oplus \mathbf{K}_6^3 \oplus \mathbf{K}_{14}^3 \oplus \mathbf{K}_6^4 \oplus \mathbf{K}_8^4 \oplus \mathbf{K}_{14}^4 \oplus \mathbf{K}_{16}^4$$

has the (fixed) value 0 or 1. It follows that the random variable

$$\mathbf{X}_5 \oplus \mathbf{X}_7 \oplus \mathbf{X}_8 \oplus \mathbf{U}_6^4 \oplus \mathbf{U}_8^4 \oplus \mathbf{U}_{14}^4 \oplus \mathbf{U}_{16}^4 \quad (3.3)$$

has bias equal to $\pm 1/32$, where the sign of this bias depends on the values of unknown key bits. Note that the random variable (3.3) involves only plaintext bits and bits of u^4. The fact that (3.3) has bias bounded away from 0 allows us to carry out the linear attack mentioned at the beginning of Section 3.3.

Suppose that we have T plaintext-ciphertext pairs, all of which use the same unknown key, K. (It will turn out that we need $T \approx 8000$ in order for the attack to succeed.) Denote this set of T pairs by $\mathcal{T}$. The attack will allow us to obtain the eight key bits in $K_{(2)}^5$ and $K_{(4)}^5$, namely,

$$K_5^5, K_6^5, K_7^5, K_8^5, K_{13}^5, K_{14}^5, K_{15}^5, \text{ and } K_{16}^5.$$

These are the eight key bits that are exclusive-ored with the output of the S-boxes S_4^2 and S_4^4. Notice that there are $2^8 = 256$ possibilities for this list of eight key bits. We will refer to a binary 8-tuple (comprising values for these eight key bits) as a *candidate subkey*.

For each $(x, y) \in \mathcal{T}$ and for each candidate subkey, it is possible to compute a partial decryption of y and obtain the resulting value for $u_{(2)}^4$ and $u_{(4)}^4$. Then we compute the value

$$x_5 \oplus x_7 \oplus x_8 \oplus u_6^4 \oplus u_8^4 \oplus u_{14}^4 \oplus u_{16}^4 \quad (3.4)$$

taken on by the random variable (3.3). We maintain an array of counters indexed by the 256 possible candidate subkeys, and increment the counter corresponding to a particular subkey whenever (3.4) has the value 0. (This array is initialized to have all values equal to 0.)

At the end of this counting process, we expect that most counters will have a value close to $T/2$, but the counter for the correct candidate subkey will have a value that is close to $T/2 \pm T/32$. This will (hopefully) allow us to identify eight subkey bits.

The algorithm for this particular linear attack is presented as Algorithm 3.2. In this algorithm, the variables L_1 and L_2 take on hexadecimal values. The set $\mathcal{T}$ is the set of T plaintext-ciphertext pairs used in the attack. π_S^{-1} is the permutation corresponding to the inverse of the S-box; this is used to partially decrypt the ciphertexts. The output, $maxkey$, contains the "most likely" eight subkey bits identified in the attack.

Algorithm 3.2: LINEARATTACK$(\mathcal{T}, T, \pi_S^{-1})$

> **for** $(L_1, L_2) \leftarrow (0, 0)$ **to** (F, F)
> $\quad$ **do** $Count[L_1, L_2] \leftarrow 0$
> **for each** $(x, y) \in \mathcal{T}$
> $\quad$ **do** $\begin{cases} \textbf{for } (L_1, L_2) \leftarrow (0, 0) \textbf{ to } (F, F) \\ \quad \textbf{do} \begin{cases} v_{(2)}^4 \leftarrow L_1 \oplus y_{(2)} \\ v_{(4)}^4 \leftarrow L_2 \oplus y_{(4)} \\ u_{(2)}^4 \leftarrow \pi_S^{-1}(v_{(2)}^4) \\ u_{(4)}^4 \leftarrow \pi_S^{-1}(v_{(4)}^4) \\ z \leftarrow x_5 \oplus x_7 \oplus x_8 \oplus u_6^4 \oplus u_8^4 \oplus u_{14}^4 \oplus u_{16}^4 \\ \textbf{if } z = 0 \\ \quad \textbf{then } Count[L_1, L_2] \leftarrow Count[L_1, L_2] + 1 \end{cases} \end{cases}$
> $max \leftarrow -1$
> **for** $(L_1, L_2) \leftarrow (0, 0)$ **to** (F, F)
> $\quad$ **do** $\begin{cases} Count[L_1, L_2] \leftarrow |Count[L_1, L_2] - T/2| \\ \textbf{if } Count[L_1, L_2] > max \\ \quad \textbf{then } \begin{cases} max \leftarrow Count[L_1, L_2] \\ maxkey \leftarrow (L_1, L_2) \end{cases} \end{cases}$
> **output** $(maxkey)$

In general, it is suggested that a linear attack based on a linear approximation having bias equal to ϵ will be successful if the number of plaintext-ciphertext pairs, which we denote by T, is approximately $c\,\epsilon^{-2}$, for some "small" constant c. We implemented the attack described in Algorithm 3.2, and found that the attack was usually successful if we took $T = 8000$. Note that $T = 8000$ corresponds to $c \approx 8$, because $\epsilon^{-2} = 1024$.

3.4 Differential Cryptanalysis

Differential cryptanalysis is similar to linear cryptanalysis in many respects. The main difference from linear cryptanalysis is that differential cryptanalysis involves comparing the x-or of two inputs to the x-or of the corresponding two outputs. In general, we will be looking at inputs x and x^* (which are assumed to be binary strings) having a specified (fixed) x-or value denoted by $x' = x \oplus x^*$. Throughout this section, we will use prime markings ($'$) to indicate the x-or of two bitstrings.

Differential cryptanalysis is a chosen-plaintext attack. We assume that an attacker has a large number of tuples (x, x^*, y, y^*), where the x-or value $x' = x \oplus x^*$ is fixed. The plaintext elements (i.e., x and x^*) are encrypted using the same unknown key, K, yielding the ciphertexts y and y^*, respectively. For each of these tuples, we will begin to decrypt the ciphertexts y and y^*, using all possible candidate keys for the last round of the cipher. For each candidate key, we compute the values of certain state bits, and determine if their x-or has a certain value (namely, the most likely value for the given input x-or). Whenever it does, we increment a counter corresponding to the particular candidate key. At the end of this process, we hope that the candidate key that has the highest frequency count contains the correct values for these key bits. (As we did with linear cryptanalysis, we will illustrate the attack with a particular example.)

Definition 3.1: Let $\pi_S : \{0, 1\}^m \to \{0, 1\}^n$ be an S-box. Consider an (ordered) pair of bitstrings of length m, say (x, x^*). We say that the *input x-or* of the S-box is $x \oplus x^*$ and the *output x-or* is $\pi_S(x) \oplus \pi_S(x^*)$. Note that the output x-or is a bitstring of length n.

For any $x' \in \{0, 1\}^m$, define the set $\Delta(x')$ to consist of all the ordered pairs (x, x^*) having input x-or equal to x'.

It is easy to see that any set $\Delta(x')$ contains 2^m pairs, and that

$$\Delta(x') = \{(x, x \oplus x') : x \in \{0, 1\}^m\}.$$

For each pair in $\Delta(x')$, we can compute the output x-or of the S-box. Then we can tabulate the resulting distribution of output x-ors. There are 2^m output x-ors, which are distributed among 2^n possible values. A non-uniform output distribution will be the basis for a successful differential attack.

Example 3.3 We again use the S-box from Example 3.1. Suppose we consider input x-or $x' = 1011$. Then

$$\Delta(1011) = \{(0000, 1011), (0001, 1010), \ldots, (1111, 0100)\}.$$

For each ordered pair in the set $\Delta(1011)$, we compute output x-or of π_S. In each row of the following table, we have $x \oplus x^* = 1011$, $y = \pi_S(x)$, $y^* = \pi_S(x^*)$ and $y' = y \oplus y^*$:

x	x^*	y	y^*	y'
0000	1011	1110	1100	0010
0001	1010	0100	0110	0010
0010	1001	1101	1010	0111
0011	1000	0001	0011	0010
0100	1111	0010	0111	0101
0101	1110	1111	0000	1111
0110	1101	1011	1001	0010
0111	1100	1000	0101	1101
1000	0011	0011	0001	0010
1001	0010	1010	1101	0111
1010	0001	0110	0100	0010
1011	0000	1100	1110	0010
1100	0111	0101	1000	1101
1101	0110	1001	1011	0010
1110	0101	0000	1111	1111
1111	0100	0111	0010	0101

Looking at the last column of the above table, we obtain the following distribution of output x-ors:

0000	0001	0010	0011	0100	0101	0110	0111
0	0	8	0	0	2	0	2

1000	1001	1010	1011	1100	1101	1110	1111
0	0	0	0	0	2	0	2

$\square$

In Example 3.3, only five of the 16 possible output x-ors actually occur. This particular example has a very non-uniform distribution.

We can carry out computations, as was done in Example 3.3, for any possible input x-or. It will be convenient to have some notation to describe the distributions of the output x-ors, so we state the following definition. For a bitstring x' of length m and a bitstring y' of length n, define

$$N_D(x', y') = |\{(x, x^*) \in \Delta(x') : \pi_S(x) \oplus \pi_S(x^*) = y'\}|.$$

In other words, $N_D(x', y')$ counts the number of pairs with input x-or equal to x' which also have output x-or equal to y' (for a given S-box). All the values $N_D(a', b')$ for the S-box from Example 3.1 are tabulated in Figure 3.4 (a' and b'

a'	0	1	2	3	4	5	6	7	8	9	A	B	C	D	E	F
0	16	0	0	0	0	0	0	0	0	0	0	0	0	0	0	0
1	0	0	0	2	0	0	0	2	0	2	4	0	4	2	0	0
2	0	0	0	2	0	6	2	2	0	2	0	0	0	0	2	0
3	0	0	2	0	2	0	0	0	0	4	2	0	2	0	0	4
4	0	0	0	2	0	0	6	0	0	2	0	4	2	0	0	0
5	0	4	0	0	0	2	2	0	0	0	4	0	2	0	0	2
6	0	0	0	4	0	4	0	0	0	0	0	0	2	2	2	2
7	0	0	2	2	2	0	2	0	0	2	2	0	0	0	0	4
8	0	0	0	0	0	0	2	2	0	0	0	4	0	4	2	2
9	0	2	0	0	2	0	0	4	2	0	2	2	2	0	0	0
A	0	2	2	0	0	0	0	0	6	0	0	2	0	0	4	0
B	0	0	8	0	0	2	0	2	0	0	0	0	0	2	0	2
C	0	2	0	0	2	2	2	0	0	0	0	2	0	6	0	0
D	0	4	0	0	0	0	0	4	2	0	2	0	2	0	2	0
E	0	0	2	4	2	0	0	0	6	0	0	0	0	0	2	0
F	0	2	0	0	6	0	0	0	0	4	0	2	0	0	2	0

FIGURE 3.4
Difference distribution table: values of $N_D(a', b')$

are the hexadecimal representations of the input and output x-ors, respectively).
Observe that the distribution computed in Example 3.3 corresponds to row "*B*"
in the table in Figure 3.4.

Recall that the input to the ith S-box in round r of the SPN from Example 3.1
is denoted $u_{(i)}^r$, and

$$u_{(i)}^r = w_{(i)}^{r-1} \oplus K_{(i)}^r.$$

An input x-or is computed as

$$u_{(i)}^r \oplus (u_{(i)}^r)^* = (w_{(i)}^{r-1} \oplus K_{(i)}^r) \oplus ((w_{(i)}^{r-1})^* \oplus K_{(i)}^r)$$
$$= w_{(i)}^{r-1} \oplus (w_{(i)}^{r-1})^*$$

Therefore, this input x-or does not depend on the subkey bits used in round r; it
is equal to the (permuted) output x-or of round $r - 1$. (However, the output x-or
of round r certainly does depend on the subkey bits in round r.)

Let a' denote an input x-or and let b' denote an output x-or. The pair (a', b')
is called a *differential*. Each entry in the difference distribution table gives rise
to an *x-or propagation ratio* (or more simply, a *propagation ratio*) for the corresponding differential. The propagation ratio $R_p(a', b')$ for the differential (a', b')
is defined as follows:

$$R_p(a', b') = \frac{N_D(a', b')}{2^m}.$$

$R_p(a', b')$ can be interpreted as a conditional probability:

$$\mathbf{Pr}[\text{output x-or} = b' \mid \text{input x-or} = a'] = R_p(a', b').$$

Suppose we find propagation ratios for differentials in consecutive rounds of the SPN, such that the input x-or of a differential in any round is the same as the (permuted) output x-ors of the differentials in the previous round. Then these differentials can be combined to form a *differential trail*. We make the assumption that the various propagation ratios in a differential trail are independent (an assumption which may not be mathematically valid, in fact). This assumption allows us to multiply the propagation ratios of the differentials in order to obtain the propagation ratio of the differential trail.

We illustrate this process by returning to the SPN from Example 3.1. A particular differential trail is shown in Figure 3.5. Arrows are used to highlight the "1" bits in the input and output x-ors of the differentials that are used in the differential trail.

The differential attack arising from Figure 3.5 uses the following propagation ratios of differentials, all of which can be verified from Figure 3.4:

- In S_2^1, $R_p(1011, 0010) = 1/2$
- In S_3^2, $R_p(0100, 0110) = 3/8$
- In S_2^3, $R_p(0010, 0101) = 3/8$
- In S_3^3, $R_p(0010, 0101) = 3/8$

These differentials can be combined to form a differential trail. We therefore obtain a propagation ratio for a differential trail of the first three rounds of the SPN:

$$R_p(0000\ 1011\ 0000\ 0000, 0000\ 0101\ 0101\ 0000) = \frac{1}{2} \times \left(\frac{3}{8}\right)^3 = \frac{27}{1024}.$$

In other words,

$$x' = 0000\ 1011\ 0000\ 0000 \Rightarrow (v^3)' = 0000\ 0101\ 0101\ 0000$$

with probability $27/1024$. However,

$$(v^3)' = 0000\ 0101\ 0101\ 0000 \Leftrightarrow (u^4)' = 0000\ 0110\ 0000\ 0110.$$

Hence, it follows that

$$x' = 0000\ 1011\ 0000\ 0000 \Rightarrow (u^4)' = 0000\ 0110\ 0000\ 0110$$

with probability $27/1024$. Note that $(u^4)'$ is the x-or of two inputs to the last round of S-boxes.

Now we can present an algorithm, for this particular example, based on the informal description at the beginning of this section; see Algorithm 3.3. The input

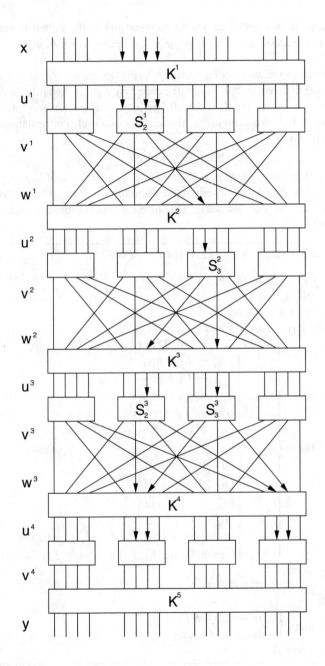

FIGURE 3.5
A differential trail for a substitution-permutation network

and output of this algorithm are similar to linear attack; the main difference is that $\mathcal{T}$ is a set of tuples of the form (x, x^*, y, y^*), where x' is fixed, in the differential attack.

Algorithm 3.3 makes use of a certain *filtering operation*. Tuples (x, x^*, y, y^*) for which the differential holds are often called *right pairs*, and it is the right pairs that allow us to determine the relevant key bits. (Tuples that are not right pairs basically constitute "random noise" that provides no useful information.)

A right pair has

$$(u_{(1)}^4)' = (u_{(3)}^4)' = 0000.$$

Hence, it follows that a right pair must have $y_{(1)} = (y_{(1)})^*$ and $y_{(3)} = (y_{(3)})^*$. If a tuple (x, x^*, y, y^*) does not satisfy these conditions, then we know that it is not a right pair, and we can discard it. This filtering process increases the efficiency of the attack.

Algorithm 3.3: DIFFERENTIALATTACK$(\mathcal{T}, T, \pi_S^{-1})$

for $(L_1, L_2) \leftarrow (0, 0)$ **to** (F, F)
  do $Count[L_1, L_2] \leftarrow 0$
for each $(x, y, x^*, y^*) \in \mathcal{T}$
do $\begin{cases} \textbf{if } (y_{(1)} = (y_{(1)})^*) \textbf{ and } (y_{(3)} = (y_{(3)})^*) \\ \textbf{then } \begin{cases} \textbf{for } (L_1, L_2) \leftarrow (0, 0) \textbf{ to } (F, F) \\ \textbf{do } \begin{cases} v_{(2)}^4 \leftarrow L_1 \oplus y_{(2)} \\ v_{(4)}^4 \leftarrow L_2 \oplus y_{(4)} \\ u_{(2)}^4 \leftarrow \pi_S^{-1}(v_{(2)}^4) \\ u_{(4)}^4 \leftarrow \pi_S^{-1}(v_{(4)}^4) \\ (v_{(2)}^4)^* \leftarrow L_1 \oplus (y_{(2)})^* \\ (v_{(4)}^4)^* \leftarrow L_2 \oplus (y_{(4)})^* \\ (u_{(2)}^4)^* \leftarrow \pi_S^{-1}((v_{(2)}^4)^*) \\ (u_{(4)}^4)^* \leftarrow \pi_S^{-1}((v_{(4)}^4)^*) \\ (u_{(2)}^4)' \leftarrow u_{(2)}^4 \oplus (u_{(2)}^4)^* \\ (u_{(4)}^4)' \leftarrow u_{(4)}^4 \oplus (u_{(4)}^4)^* \\ \textbf{if } ((u_{(2)}^4)' = 0110) \textbf{ and } ((u_{(4)}^4)' = 0110) \\ \quad \textbf{then } Count[L_1, L_2] \leftarrow Count[L_1, L_2] + 1 \end{cases} \end{cases} \end{cases}$

$max \leftarrow -1$
for $(L_1, L_2) \leftarrow (0, 0)$ **to** (F, F)
do $\begin{cases} \textbf{if } Count[L_1, L_2] > max \\ \textbf{then } \begin{cases} max \leftarrow Count[L_1, L_2] \\ maxkey \leftarrow (L_1, L_2) \end{cases} \end{cases}$
output $(maxkey)$

A differential attack based on a differential trail having propagation ratio equal to ϵ will often be successful if the number of tuples (x, x^*, y, y^*), which we denote by T, is approximately $c\epsilon^{-1}$, for a "small" constant c. We implemented the attack described in Algorithm 3.3, and found that the attack was often successful if we took T between 50 and 100. In this example, $\epsilon^{-1} \approx 38$.

3.5 The Data Encryption Standard

On May 15, 1973, the National Bureau of Standards (now the *National Institute of Standards and Technology*, or *NIST*) published a solicitation for cryptosystems in the Federal Register. This led ultimately to the adoption of the *Data Encryption Standard*, or *DES*, which became the most widely used cryptosystem in the world. *DES* was developed at IBM, as a modification of an earlier system known as *Lucifer*. *DES* was first published in the Federal Register of March 17, 1975. After a considerable amount of public discussion, *DES* was adopted as a standard for "unclassified" applications on January 15, 1977. It was initially expected that DES would only be used as a standard for 10–15 years; however, it proved to be much more durable. *DES* was reviewed approximately every five years after its adoption. Its last renewal was in January 1999; by that time, development of a replacement, the *Advanced Encryption Standard*, had already begun (see Section 3.6).

3.5.1 Description of DES

A complete description of the *Data Encryption Standard* is given in the *Federal Information Processing Standards* (*FIPS*) Publication 46, dated January 15, 1977. *DES* is a special type of iterated cipher called a *Feistel cipher*. We describe the basic form of a Feistel cipher now, using the terminology from Section 3.1. In a Feistel cipher, each state u^i is divided into two halves of equal length, say L^i and R^i. The round function g has the following form: $g(L^{i-1}, R^{i-1}, K^i) = (L^i, R^i)$, where

$$L^i = R^{i-1}$$
$$R^i = L^{i-1} \oplus f(R^{i-1}, K^i).$$

We observe that the function f does not need to satisfy any type of injectivity property. This is because a Feistel-type round function is always invertible, given the round key:

$$L^{i-1} = R^i \oplus f(L^i, K^i)$$
$$R^{i-1} = L^i.$$

DES is a 16-round Feistel cipher having block length 64: it encrypts a plaintext bitstring x (of length 64) using a 56-bit key, K, obtaining a ciphertext bitstring (of

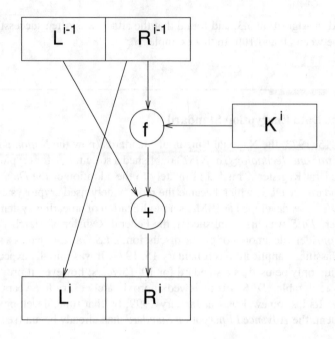

FIGURE 3.6
One round of DES encryption

length 64). Prior to the 16 rounds of encryption, there is a fixed *initial permutation* IP that is applied to the plaintext. We denote

$$IP(x) = L^0 R^0.$$

After the 16 rounds of encryption, the inverse permutation IP^{-1} is applied to the bitstring $R^{16} L^{16}$, yielding the ciphertext y. That is,

$$y = IP^{-1}(R^{16} L^{16})$$

(note that L^{16} and R^{16} are swapped before IP^{-1} is applied). The application of IP and IP^{-1} has no cryptographic significance, and is often ignored when the security of *DES* is discussed. One round of *DES* encryption is depicted in Figure 3.6.

Each L^i and R^i is 32 bits in length. The function

$$f : \{0, 1\}^{32} \times \{0, 1\}^{48} \to \{0, 1\}^{32}$$

takes as input a 32-bit string (the right half of the current state) and a round key. The key schedule, $(K^1, K^2, \ldots, K^{16})$, consists of 48-bit round keys that are derived from the 56-bit key, K. Each K^i is a permuted selection of bits from K.

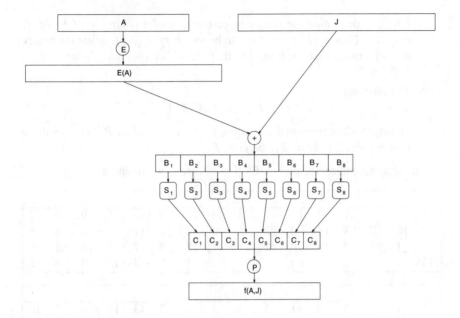

FIGURE 3.7
The DES f function

The f function is shown in Figure 3.7. Basically, it consists of a substitution (using an S-box) followed by a (fixed) permutation, denoted P. Suppose we denote the first argument of f by A, and the second argument by J. Then, in order to compute $f(A, J)$, the following steps are executed.

1. A is "expanded" to a bitstring of length 48 according to a fixed *expansion function* E. $E(A)$ consists of the 32 bits from A, permuted in a certain way, with 16 of the bits appearing twice.

2. Compute $E(A) \oplus J$ and write the result as the concatenation of eight 6-bit strings $B = B_1 B_2 B_3 B_4 B_5 B_6 B_7 B_8$.

3. The next step uses eight S-boxes, denoted $S_1, \ldots, S_8$. Each S-box

$$S_i : \{0, 1\}^6 \rightarrow \{0, 1\}^4$$

maps six bits to four bits, and is traditionally depicted as a 4×16 array whose entries come from the integers $0, \ldots, 15$. Given a bitstring of length six, say

$$B_j = b_1 b_2 b_3 b_4 b_5 b_6,$$

we compute $S_j(B_j)$ as follows. The two bits $b_1 b_6$ determine the binary representation of a row r of S_j (where $0 \leq r \leq 3$), and the four bits

$b_2b_3b_4b_5$ determine the binary representation of a column c of S_j ($0 \leq c \leq 15$). Then $S_j(B_j)$ is defined to be the entry $S_j(r, c)$, written in binary as a bitstring of length four. In this fashion, we compute $C_j = S_j(B_j)$, $1 \leq j \leq 8$.

4. The bitstring

$$C = C_1C_2C_3C_4C_5C_6C_7C_8$$

of length 32 is permuted according to the permutation P. The resulting bitstring $\mathsf{P}(C)$ is defined to be $f(A, J)$.

For future reference, the eight *DES* S-boxes are now presented:

S_1															
14	4	13	1	2	15	11	8	3	10	6	12	5	9	0	7
0	15	7	4	14	2	13	1	10	6	12	11	9	5	3	8
4	1	14	8	13	6	2	11	15	12	9	7	3	10	5	0
15	12	8	2	4	9	1	7	5	11	3	14	10	0	6	13

S_2															
15	1	8	14	6	11	3	4	9	7	2	13	12	0	5	10
3	13	4	7	15	2	8	14	12	0	1	10	6	9	11	5
0	14	7	11	10	4	13	1	5	8	12	6	9	3	2	15
13	8	10	1	3	15	4	2	11	6	7	12	0	5	14	9

S_3															
10	0	9	14	6	3	15	5	1	13	12	7	11	4	2	8
13	7	0	9	3	4	6	10	2	8	5	14	12	11	15	1
13	6	4	9	8	15	3	0	11	1	2	12	5	10	14	7
1	10	13	0	6	9	8	7	4	15	14	3	11	5	2	12

S_4															
7	13	14	3	0	6	9	10	1	2	8	5	11	12	4	15
13	8	11	5	6	15	0	3	4	7	2	12	1	10	14	9
10	6	9	0	12	11	7	13	15	1	3	14	5	2	8	4
3	15	0	6	10	1	13	8	9	4	5	11	12	7	2	14

S_5															
2	12	4	1	7	10	11	6	8	5	3	15	13	0	14	9
14	11	2	12	4	7	13	1	5	0	15	10	3	9	8	6
4	2	1	11	10	13	7	8	15	9	12	5	6	3	0	14
11	8	12	7	1	14	2	13	6	15	0	9	10	4	5	3

S_6															
12	1	10	15	9	2	6	8	0	13	3	4	14	7	5	11
10	15	4	2	7	12	9	5	6	1	13	14	0	11	3	8
9	14	15	5	2	8	12	3	7	0	4	10	1	13	11	6
4	3	2	12	9	5	15	10	11	14	1	7	6	0	8	13

S_7															
4	11	2	14	15	0	8	13	3	12	9	7	5	10	6	1
13	0	11	7	4	9	1	10	14	3	5	12	2	15	8	6
1	4	11	13	12	3	7	14	10	15	6	8	0	5	9	2
6	11	13	8	1	4	10	7	9	5	0	15	14	2	3	12

S_8															
13	2	8	4	6	15	11	1	10	9	3	14	5	0	12	7
1	15	13	8	10	3	7	4	12	5	6	11	0	14	9	2
7	11	4	1	9	12	14	2	0	6	10	13	15	3	5	8
2	1	14	7	4	10	8	13	15	12	9	0	3	5	6	11

Example 3.4   We show how to compute a sample output of an S-box, using the traditional presentation described above. Consider the S-box S_1, and suppose that the input is the binary 6-tuple 101000. The first and last bits are 10, which is the binary representation of the integer 2. The middle four bits are 0100, which is the binary representation of the integer 4. Row 2 of S_1 is the third row (because the rows are numbered $0, 1, 2, 3$); similarly, column 4 is the fifth column. The entry in row 2, column 4 of S_1 is 13, which is 1101 in binary. Therefore 1101 is the output of the S-box S_1, given the input 101000.   □

The *DES* S-boxes are not permutations, of course, because the number of possible inputs (64) exceeds the number of possible outputs (16). However, it can be verified that each row of each of the eight S-boxes is a permutation of the integers $0, \ldots, 15$. This property is one of several design criteria that were required of the S-boxes in order to prevent certain types of cryptanalytic attacks.

The expansion function E is specified by the following table:

E bit-selection table					
32	1	2	3	4	5
4	5	6	7	8	9
8	9	10	11	12	13
12	13	14	15	16	17
16	17	18	19	20	21
20	21	22	23	24	25
24	25	26	27	28	29
28	29	30	31	32	1

Given a bitstring of length 32, say $A = (a_1, a_2, \ldots, a_{32})$, $E(A)$ is the following bitstring of length 48:

$$E(A) = (a_{32}, a_1, a_2, a_3, a_4, a_5, a_4, \ldots, a_{31}, a_{32}, a_1).$$

The permutation P is as follows:

P			
16	7	20	21
29	12	28	17
1	15	23	26
5	18	31	10
2	8	24	14
32	27	3	9
19	13	30	6
22	11	4	25

Denote the bitstring $C = (c_1, c_2, \ldots, c_{32})$. Then the permuted bitstring $P(C)$ is as follows:

$$P(C) = (c_{16}, c_7, c_{20}, c_{21}, c_{29}, \ldots, c_{11}, c_4, c_{25}).$$

3.5.2 Analysis of DES

When *DES* was proposed as a standard, there was considerable criticism. One objection to *DES* concerned the S-boxes. All computations in *DES*, with the sole exception of the S-boxes, are linear, i.e., computing the exclusive-or of two outputs is the same as forming the exclusive-or of two inputs and then computing the output. The S-boxes, being the non-linear components of the cryptosystem, are vital to its security. (We saw in Chapter 1 how linear cryptosystems, such as the *Hill Cipher*, could easily be cryptanalyzed by a known plaintext attack.) At the time that *DES* was proposed, several people suggested that its S-boxes might contain hidden "trapdoors" which would allow the National Security Agency to easily decrypt messages while claiming falsely that *DES* is "secure." It is, of course, impossible to disprove such a speculation, but no evidence ever came to light that indicated that trapdoors in *DES* do, in fact, exist.

Actually, it was eventually revealed that the *DES* S-boxes were designed to prevent certain types of attacks. When Biham and Shamir invented the technique of differential cryptanalysis (which we discussed in Section 3.4) in the early 1990's, it was acknowledged that the purpose of certain unpublished design criteria of the S-boxes was to make differential cryptanalysis of *DES* infeasible. Differential cryptanalysis was known to IBM researchers at the time that *DES* was being developed, but it was kept secret for almost 20 years, until Biham and Shamir independently discovered the attack.

The most pertinent criticism of *DES* is that the size of the keyspace, 2^{56}, is too small to be really secure. The IBM *Lucifer* cryptosystem, a predecessor of *DES*, had a 128-bit key. The original proposal for *DES* had a 64-bit key, but this was later reduced to a 56-bit key. IBM claimed that the reason for this reduction was that it was necessary to include eight parity-check bits in the key, meaning that 64 bits of storage could only contain a 56-bit key.

Even in the 1970's, it was argued that a special-purpose machine could be built to carry out a known plaintext attack, which would essentially perform an exhaustive search for the key. That is, given a 64-bit plaintext x and corresponding ciphertext y, every possible key would be tested until a key K is found such that $e_K(x) = y$ (note that there may be more than one such key K). As early as 1977, Diffie and Hellman suggested that one could build a VLSI chip which could test 10^6 keys per second. A machine with 10^6 chips could search the entire key space in about a day. They estimated that such a machine could be built, at that time, for about $20,000,000.

Later, at the CRYPTO '93 Rump Session, Michael Wiener gave a very detailed design of a *DES* key search machine. The machine is based on a key search chip which is pipelined, so that 16 encryptions take place simultaneously. This chip would test 5×10^7 keys per second, and could have been built using 1993 technology for $10.50 per chip. A frame consisting of 5760 chips could be built for $100,000. This would allow a *DES* key to be found in about 1.5 days on average. A machine using ten frames would cost $1,000,000, but would reduce the average search time to about 3.5 hours.

Wiener's machine was never built, but a key search machine costing $250,000 was built in 1998 by the Electronic Frontier Foundation. This computer, called "DES Cracker," contained 1536 chips and could search 88 billion keys per second. It won RSA Laboratory's "DES Challenge II-2" by successfully finding a *DES* key in 56 hours in July 1998. In January 1999, RSA Laboratory's "DES Challenge III" was solved by the DES Cracker working in conjunction with a worldwide network (of 100,000 computers) known as `distributed.net`. This co-operative effort found a *DES* key in 22 hours, 15 minutes, testing over 245 billion keys per second.

Other than exhaustive key search, the two most important cryptanalytic attacks on *DES* are differential cryptanalysis and linear cryptanalysis. (For SPNs, these attacks were described in Sections 3.4 and 3.3, respectively.) In the case of *DES*, linear cryptanalysis is the more efficient of the two attacks, and an actual implementation of linear cryptanalysis was carried out in 1994 by its inventor, Matsui. This linear cryptanalysis of *DES* is a known-plaintext attack using 2^{43} plaintext-ciphertext pairs, all of which are encrypted using the same (unknown) key. It took 40 days to generate the 2^{43} pairs, and it took 10 days to actually find the key. This cryptanalysis did not have a practical impact on the security of *DES*, however, due to the extremely large number of plaintext-ciphertext pairs that are required to mount the attack: it is unlikely in practice that an adversary will be able to accumulate such a large number of plaintext-ciphertext pairs that are all encrypted using the same key.

3.6   The Advanced Encryption Standard

On January 2, 1997, NIST began the process of choosing a replacement for *DES*. The replacement would be called the *Advanced Encryption Standard*, or *AES*. A formal call for algorithms was made on September 12, 1997. It was required that the *AES* have a block length of 128 bits, and support key lengths of 128, 192 and 256 bits. It was also necessary that the *AES* should be available worldwide on a royalty-free basis.

Submissions were due on June 15, 1998. Of the 21 submitted cryptosystems, 15 met all the necessary criteria and were accepted as *AES* candidates. NIST announced the 15 *AES* candidates at the "First AES Candidate Conference" on August 20, 1998. A "Second AES Candidate Conference" was held in March 1999. Then, in August 1999, five of the candidates were chosen by NIST as finalists: *MARS*, *RC6*, *Rijndael*, *Serpent* and *Twofish*.

The "Third AES Candidate Conference" was held in April 2000. On October 2, 2000, *Rijndael* was selected to be the *Advanced Encryption Standard*. On February 28, 2001, NIST announced that a draft Federal Information Processing Standard for the *AES* was available for public review and comment. *AES* was adopted as a standard on November 26, 2001, and it was published as FIPS 197 in the Federal Register on December 4, 2001.

The selection process for the *AES* was notable for its openness and its international flavor. The three candidate conferences, as well as official solicitations for public comments, provided ample opportunity for feedback and public discussion and analysis of the candidates, and the process was viewed very favorably by everyone involved. The "international" aspect of *AES* is demonstrated by the variety of countries represented by the authors of the 15 candidate ciphers: Australia, Belgium, Canada, Costa Rica, France, Germany, Israel, Japan, Korea, Norway, the United Kingdom and the USA. *Rijndael*, which was ultimately selected as the *AES*, was invented by two Belgian researchers, Daemen and Rijmen. Another interesting departure from past practice was that the "Second AES Candidate Conference" was held outside the U.S., in Rome, Italy.

AES candidates were evaluated for their suitability according to three main criteria:

- security
- cost
- algorithm and implementation characteristics

Security of the proposed algorithm was absolutely essential, and any algorithm found not to be secure would not be considered further. "Cost" refers to the computational efficiency (speed and memory requirements) of various types of implementations, including software, hardware and smart cards. Algorithm and implementation characteristics include flexibility and algorithm simplicity, among other factors.

In the end, the five finalists were all felt to be secure. *Rijndael* was selected because its combination of security, performance, efficiency, implementability and flexibility was judged to be superior to the other finalists.

3.6.1 Description of AES

As mentioned above, the *AES* has block length 128, and there are three allowable key lengths, namely 128 bits, 192 bits and 256 bits. *AES* is an iterated cipher; the number of rounds, which we denote by Nr, depends on the key length. Nr = 10 if the key length is 128 bits; Nr = 12 if the key length is 192 bits; and Nr = 14 if the key length is 256 bits.

We first give a high-level description of *AES*. The algorithm proceeds as follows:

1. Given a plaintext x, initialize State to be x and perform an operation ADD-ROUNDKEY, which x-ors the RoundKey with State.

2. For each of the first Nr − 1 rounds, perform a substitution operation called SUBBYTES on State using an S-box; perform a permutation SHIFTROWS on State; perform an operation MIXCOLUMNS on State; and perform ADDROUNDKEY.

3. Perform SUBBYTES; perform SHIFTROWS; and perform ADDROUND-KEY.

4. Define the ciphertext y to be State.

From this high-level description, we can see that the structure of the *AES* is very similar in many respects to the SPN discussed in Section 3.2. In every round of both these cryptosystems, we have subkey mixing, a substitution step and a permutation step. Both ciphers also include whitening. *AES* is "larger," and it also includes an additional linear transformation (MIXCOLUMNS) in each round.

We now give precise descriptions of all the operations used in the *AES*; describe the structure of State; and discuss the construction of the key schedule. All operations in *AES* are byte-oriented operations, and all variables used are considered to be formed from an appropriate number of bytes. The plaintext x consists of 16 bytes, denoted $x_0, \ldots, x_{15}$. State is represented as a four by four array of bytes, as follows:

$s_{0,0}$	$s_{0,1}$	$s_{0,2}$	$s_{0,3}$
$s_{1,0}$	$s_{1,1}$	$s_{1,2}$	$s_{1,3}$
$s_{2,0}$	$s_{2,1}$	$s_{2,2}$	$s_{2,3}$
$s_{3,0}$	$s_{3,1}$	$s_{3,2}$	$s_{3,3}$

Initially, State is defined to consist of the 16 bytes of the plaintext x, as follows:

$s_{0,0}$	$s_{0,1}$	$s_{0,2}$	$s_{0,3}$
$s_{1,0}$	$s_{1,1}$	$s_{1,2}$	$s_{1,3}$
$s_{2,0}$	$s_{2,1}$	$s_{2,2}$	$s_{2,3}$
$s_{3,0}$	$s_{3,1}$	$s_{3,2}$	$s_{3,3}$

$\leftarrow$

x_0	x_4	x_8	x_{12}
x_1	x_5	x_9	x_{13}
x_2	x_6	x_{10}	x_{14}
x_3	x_7	x_{11}	x_{15}

We will often use hexadecimal notation to represent the contents of a byte. Each byte therefore consists of two hexadecimal digits.

The operation SUBBYTES performs a substitution on each byte of State independently, using an S-box, say π_S, which is a permutation of $\{0, 1\}^8$. To present this π_S, we represent bytes in hexadecimal notation. π_S is depicted as a 16 by 16 array, where the rows and columns are indexed by hexadecimal digits. The entry in row X and column Y is $\pi_S(XY)$. The array representation of π_S is presented in Figure 3.8.

In contrast to the S-boxes in *DES*, which are apparently "random" substitutions, the *AES* S-box can be defined algebraically. The algebraic formulation of the *AES* S-box involves operations in a finite field (finite fields are discussed in detail in Section 6.4). We include the following description for the benefit of readers who are already familiar with finite fields (other readers may want to skip this description, or read Section 6.4 first): The permutation π_S incorporates operations in the finite field

$$\mathbb{F}_{2^8} = \mathbb{Z}_2[x]/(x^8 + x^4 + x^3 + x + 1).$$

Let FIELDINV denote the multiplicative inverse of a field element; let BINARY-TOFIELD convert a byte to a field element; and let FIELDTOBINARY perform the inverse conversion. This conversion is done in the obvious way: the field element

$$\sum_{i=0}^{7} a_i x^i$$

corresponds to the byte

$$a_7 a_6 a_5 a_4 a_3 a_2 a_1 a_0,$$

where $a_i \in \mathbb{Z}_2$ for $0 \leq i \leq 7$. Then the permutation π_S is defined according to Algorithm 3.4. In this algorithm, the eight input bits $a_7 a_6 a_5 a_4 a_3 a_2 a_1 a_0$ are replaced by the eight output bits $b_7 b_6 b_5 b_4 b_3 b_2 b_1 b_0$.

Algorithm 3.4: SUBBYTES$(a_7a_6a_5a_4a_3a_2a_1a_0)$

external FIELDINV, BINARYTOFIELD, FIELDTOBINARY
$z \leftarrow$ BINARYTOFIELD$(a_7a_6a_5a_4a_3a_2a_1a_0)$
if $z \neq 0$
  then $z \leftarrow$ FIELDINV(z)
$(a_7a_6a_5a_4a_3a_2a_1a_0) \leftarrow$ FIELDTOBINARY(z)
$(c_7c_6c_5c_4c_3c_2c_1c_0) \leftarrow (01100011)$
comment: In the following loop, all subscripts are to be reduced modulo 8

for $i \leftarrow 0$ **to** 7
  do $b_i \leftarrow (a_i + a_{i+4} + a_{i+5} + a_{i+6} + a_{i+7} + c_i) \bmod 2$
return $(b_7b_6b_5b_4b_3b_2b_1b_0)$

Example 3.5 We do a small example to illustrate Algorithm 3.4, where we also include the conversions to hexadecimal. Suppose we begin with (hexadecimal) 53. In binary, this is

$$01010011,$$

which represents the field element

$$x^6 + x^4 + x^3 + 1.$$

The multiplicative inverse (in the field $\mathbb{F}_{2^8}$) can be shown to be

$$x^7 + x^6 + x^3 + x.$$

Therefore, in binary notation, we have

$$(a_7a_6a_5a_4a_3a_2a_1a_0) = (11001010).$$

Next, we compute

$$b_0 = a_0 + a_4 + a_5 + a_6 + a_7 + c_0 \bmod 2$$
$$= 0 + 0 + 0 + 1 + 1 + 1 \bmod 2$$
$$= 1$$
$$b_1 = a_1 + a_5 + a_6 + a_7 + a_0 + c_1 \bmod 2$$
$$= 1 + 0 + 1 + 1 + 0 + 1 \bmod 2$$
$$= 0,$$

etc. The result is that

$$(b_7b_6b_5b_4b_3b_2b_1b_0) = (11101101).$$

								Y								
X	0	1	2	3	4	5	6	7	8	9	A	B	C	D	E	F
0	63	7C	77	7B	F2	6B	6F	C5	30	01	67	2B	FE	D7	AB	76
1	CA	82	C9	7D	FA	59	47	F0	AD	D4	A2	AF	9C	A4	72	C0
2	B7	FD	93	26	36	3F	F7	CC	34	A5	E5	F1	71	D8	31	15
3	04	C7	23	C3	18	96	05	9A	07	12	80	E2	EB	27	B2	75
4	09	83	2C	1A	1B	6E	5A	A0	52	3B	D6	B3	29	E3	2F	84
5	53	D1	00	ED	20	FC	B1	5B	6A	CB	BE	39	4A	4C	58	CF
6	D0	EF	AA	FB	43	4D	33	85	45	F9	02	7F	50	3C	9F	A8
7	51	A3	40	8F	92	9D	38	F5	BC	B6	DA	21	10	FF	F3	D2
8	CD	0C	13	EC	5F	97	44	17	C4	A7	7E	3D	64	5D	19	73
9	60	81	4F	DC	22	2A	90	88	46	EE	B8	14	DE	5E	0B	DB
A	E0	32	3A	0A	49	06	24	5C	C2	D3	AC	62	91	95	E4	79
B	E7	C8	37	6D	8D	D5	4E	A9	6C	56	F4	EA	65	7A	AE	08
C	BA	78	25	2E	1C	A6	B4	C6	E8	DD	74	1F	4B	BD	8B	8A
D	70	3E	B5	66	48	03	F6	0E	61	35	57	B9	86	C1	1D	9E
E	E1	F8	98	11	69	D9	8E	94	9B	1E	87	E9	CE	55	28	DF
F	8C	A1	89	0D	BF	E6	42	68	41	99	2D	0F	B0	54	BB	16

FIGURE 3.8
The AES S-box

In hexadecimal notation, 11101101 is ED.

This computation can be checked by verifying that the entry in row 5 and column 3 of Figure 3.8 is ED.   ⬜

The operation SHIFTROWS acts on State as shown in the following diagram:

$s_{0,0}$	$s_{0,1}$	$s_{0,2}$	$s_{0,3}$
$s_{1,0}$	$s_{1,1}$	$s_{1,2}$	$s_{1,3}$
$s_{2,0}$	$s_{2,1}$	$s_{2,2}$	$s_{2,3}$
$s_{3,0}$	$s_{3,1}$	$s_{3,2}$	$s_{3,3}$

$\leftarrow$

$s_{0,0}$	$s_{0,1}$	$s_{0,2}$	$s_{0,3}$
$s_{1,1}$	$s_{1,2}$	$s_{1,3}$	$s_{1,0}$
$s_{2,2}$	$s_{2,3}$	$s_{2,0}$	$s_{2,1}$
$s_{3,3}$	$s_{3,0}$	$s_{3,1}$	$s_{3,2}$

The operation MIXCOLUMNS is carried out on each of the four columns of State; it is presented as Algorithm 3.5. Each column of State is replaced by a new column which is formed by multiplying that column by a certain matrix of elements of the field $\mathbb{F}_{2^8}$. (Here "multiplication" means multiplication in the field $\mathbb{F}_{2^8}$. We assume that the external procedure FIELDMULT takes as input two field elements, and computes their product in the field.) Field addition is just componentwise addition modulo 2 (i.e., the x-or of the corresponding bitstrings). This operation is denoted by "$\oplus$" in Algorithm 3.5.

Algorithm 3.5: MixColumn(c)

external FIELDMULT, BINARYTOFIELD, FIELDTOBINARY
for $i \leftarrow 0$ **to** 3
  do $t_i \leftarrow$ BINARYTOFIELD($s_{i,c}$)
$u_0 \leftarrow$ FIELDMULT(x, t_0) $\oplus$ FIELDMULT($x + 1, t_1$) $\oplus t_2 \oplus t_3$
$u_1 \leftarrow$ FIELDMULT(x, t_1) $\oplus$ FIELDMULT($x + 1, t_2$) $\oplus t_3 \oplus t_0$
$u_2 \leftarrow$ FIELDMULT(x, t_2) $\oplus$ FIELDMULT($x + 1, t_3$) $\oplus t_0 \oplus t_1$
$u_3 \leftarrow$ FIELDMULT(x, t_3) $\oplus$ FIELDMULT($x + 1, t_0$) $\oplus t_1 \oplus t_2$
for $i \leftarrow 0$ **to** 3
  do $s_{i,c} \leftarrow$ FIELDTOBINARY(u_i)

It remains to discuss the key schedule for the *AES*. We describe how to construct the key schedule for the 10-round version of *AES*, which uses a 128-bit key (key schedules for 12- and 14-round versions are similar to 10-round *AES*, but there are some minor differences in the key scheduling algorithm). We need 11 round keys, each of which consists of 16 bytes. The key scheduling algorithm is word-oriented (a *word* consists of 4 bytes, or, equivalently, 32 bits). Therefore each round key is comprised of four words. The concatenation of the round keys is called the *expanded key*, which consists of 44 words. It is denoted $w[0], \ldots, w[43]$, where each $w[i]$ is a word. The expanded key is constructed using the operation KEYEXPANSION, which is presented as Algorithm 3.6.

Algorithm 3.6: KEYEXPANSION(key)

external ROTWORD, SUBWORD
$RCon[1] \leftarrow$ 01000000
$RCon[2] \leftarrow$ 02000000
$RCon[3] \leftarrow$ 04000000
$RCon[4] \leftarrow$ 08000000
$RCon[5] \leftarrow$ 10000000
$RCon[6] \leftarrow$ 20000000
$RCon[7] \leftarrow$ 40000000
$RCon[8] \leftarrow$ 80000000
$RCon[9] \leftarrow$ 1B000000
$RCon[10] \leftarrow$ 36000000
for $i \leftarrow 0$ **to** 3
  do $w[i] \leftarrow (key[4i], key[4i + 1], key[4i + 2], key[4i + 3])$
for $i \leftarrow 4$ **to** 43
  do $\begin{cases} temp \leftarrow w[i - 1] \\ \textbf{if } i \equiv 0 \ (\text{mod } 4) \\ \quad \textbf{then } temp \leftarrow \text{SUBWORD}(\text{ROTWORD}(temp)) \oplus Rcon[i/4] \\ w[i] \leftarrow w[i - 4] \oplus temp \end{cases}$
return $(w[0], \ldots, w[43])$

The input to this algorithm is the 128-bit key, key, which is treated as an array of bytes, $key[0], \dots, key[15]$; and the output is the array of words, w, that was introduced above.

KEYEXPANSION incorporates two other operations, which are named ROT-WORD and SUBWORD. ROTWORD(B_0, B_1, B_2, B_3) performs a cyclic shift of the four bytes B_0, B_1, B_2, B_3, i.e.,

$$\text{ROTWORD}(B_0, B_1, B_2, B_3) = (B_1, B_2, B_3, B_0).$$

SUBWORD(B_0, B_1, B_2, B_3) applies the *AES* S-box to each of the four bytes B_0, B_1, B_2, B_3, i.e.,

$$\text{SUBWORD}(B_0, B_1, B_2, B_3) = (B_0', B_1', B_2', B_3')$$

where $B_i' = \text{SUBBYTES}(B_i)$, $i = 0, 1, 2, 3$. $RCon$ is an array of 10 words, denoted $RCon[1], \dots, RCon[10]$. These are constants that are defined in hexadecimal notation at the beginning of Algorithm 3.6.

We have now described all the operations required to perform an encryption operation in the *AES*. In order to decrypt, it is necessary to perform all operations in the reverse order, and use the key schedule in reverse order. Further the operations SHIFTROWS, SUBBYTES and MIXCOLUMNS must be replaced by their inverse operations (the operation ADDROUNDKEY is its own inverse). It is also possible to construct an "equivalent inverse cipher" which performs *AES* decryption by doing a sequence of (inverse) operations in the same order as is done for *AES* encryption. It is suggested that this can lead to implementation efficiencies.

3.6.2 Analysis of AES

Obviously, the *AES* is secure against all known attacks. Various aspects of its design incorporate specific features that help provide security against specific attacks. For example, the use of the finite field inversion operation in the construction of the S-box yields linear approximation and difference distribution tables in which the entries are close to uniform. This provides security against differential and linear attacks. As well, the linear transformation, MIXCOLUMNS, makes it impossible to find differential and linear attacks that involve "few" active S-boxes (the designers refer to this feature as the *wide trail strategy*). There are apparently no known attacks on *AES* that are faster than exhaustive search. The "best" attacks on *AES* apply to variants of the cipher in which the number of rounds is reduced, and are not effective for 10-round *AES*.

3.7 Modes of Operation

Four *modes of operation* were developed for *DES*. They were standardized in FIPS Publication 81 in December 1980. These modes of operation can be used for any block cipher. At the present time, modes of operation are being developed for *AES*. The *AES* modes will likely include the previous *DES* modes of operation, and may include additional, new modes.

We briefly discuss the four *DES* modes of operation now. In order to adapt these modes to *AES*, the only changes necessary would be to change the block length from 64 to 128. Here are the four modes of operation for *DES*:

- *electronic codebook mode* (ECB mode),
- *cipher feedback mode* (CFB mode),
- *cipher block chaining mode* (CBC mode), and
- *output feedback mode* (OFB mode).

ECB mode corresponds to the naive use of a block cipher: given a sequence $x_1 x_2 \cdots$ of 64-bit plaintext blocks, each x_i is encrypted with the same key K, producing a string of ciphertext blocks, $y_1 y_2 \cdots$.

In CBC mode, each ciphertext block y_i is x-ored with the next plaintext block, x_{i+1}, before being encrypted with the key K. More formally, we start with a 64-bit *initialization vector*, denoted by IV, and define $y_0 = \text{IV}$. Then we construct $y_1, y_2, \ldots$, using the rule

$$y_i = e_K(y_{i-1} \oplus x_i),$$

$i \geq 1$. The use of CBC mode is depicted in Figure 3.9.

In OFB and CFB modes, a keystream is generated, which is then x-ored with the plaintext (i.e., it operates as a stream cipher, cf. Section 1.1.7). OFB mode is actually a synchronous stream cipher: the keystream is produced by repeatedly encrypting a 64-bit initialization vector, IV. We define $z_0 = \text{IV}$, and then compute the keystream $z_1 z_2 \cdots$ using the rule

$$z_i = e_K(z_{i-1}),$$

for all $i \geq 1$. The plaintext sequence $x_1 x_2 \cdots$ is then encrypted by computing

$$y_i = x_i \oplus z_i,$$

for all $i \geq 1$.

In CFB mode, we start with $y_0 = \text{IV}$ (a 64-bit initialization vector) and we produce the keystream element z_i by encrypting the previous ciphertext block. That is,

$$z_i = e_K(y_{i-1}),$$

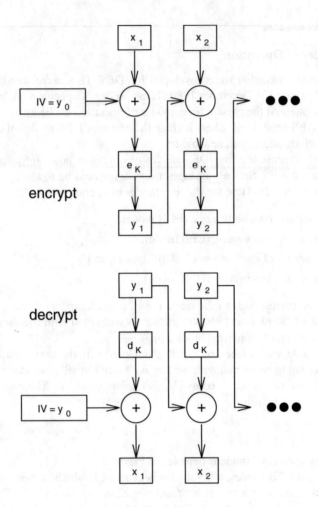

FIGURE 3.9
CBC mode

for all $i \geq 1$. As in OFB mode, we define

$$y_i = x_i \oplus z_i,$$

for all $i \geq 1$. The use of CFB mode is depicted in Figure 3.10 (note that the encryption function e_K is used for both encryption and decryption in CFB and OFB modes).

There are also variations of OFB and CFB mode called k-bit feedback modes ($1 \leq k \leq 64$). We have described the 64-bit feedback modes here. 1-bit and 8-bit feedback modes are often used in practice for encrypting data one bit (or byte) at a time.

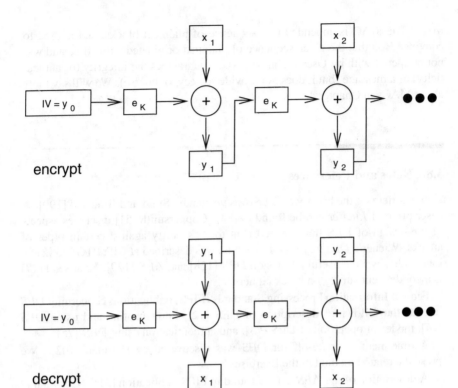

FIGURE 3.10
CFB mode

The four modes of operation have different advantages and disadvantages. One obvious weakness of ECB mode is that the encryption of identical plaintext blocks yields identical ciphertext blocks. This is a serious weakness if the underlying message blocks are chosen from a "low entropy" plaintext space. For example, if a plaintext block always consists of 64 0's or 64 1's, then ECB encryption is essentially useless.

In ECB and OFB modes, changing one 64-bit plaintext block, x_i, causes the corresponding ciphertext block, y_i, to be altered, but other ciphertext blocks are not affected. In some situations this might be a desirable property, for example, if the communication channel is not very reliable. OFB mode is often used to encrypt satellite transmissions.

On the other hand, if a plaintext block x_i is changed in CBC and CFB modes, then y_i and all subsequent ciphertext blocks will be affected. This property means that CBC and CFB modes are useful for purposes of authentication. More specifically, these modes can be used to produce a *message authentication code*, or

MAC. The MAC is appended to a sequence of plaintext blocks, and is used to convince Bob that the given sequence of plaintext originated with Alice and was not tampered with by Oscar. Thus the MAC guarantees the integrity (or authenticity) of a message (but it does not provide secrecy, of course). We will say more about MACs in Chapter 4.

3.8 Notes and References

A nice article on the history of *DES* was written by Smid and Branstad [199]. A description of *Lucifer* can be found in [83]. Coppersmith [51] discusses aspects of the design of DES that are pertinent to its security against certain types of attacks. Wiener's *DES* key search machine was described at CRYPTO '93 [215]. Landau has written useful articles on *DES* [128] and *AES* [129]. Knudsen [117] is a good recent survey on block ciphers.

Federal Information Processing Standards (FIPS) publications concerning *DES* include the following: description of *DES* [72]; implementing and using *DES* [74]; modes of operation of *DES* [73]; and authentication using *DES* [75].

A time-memory trade-off for *DES* was discovered by Hellman [102]. We present a related method in the Exercises.

A description of the *AES* can be found in FIPS publication [81]. Nechvatal *et al.* [153] is a detailed report on the development of the *AES*. *Rijndael* is described in [54]; its predecessor, known as *Square*, is presented in [53]. Daemon and Rijmen have also written a monograph [55] explaining *Rijndael* and the design strategies they incorporated into its design. Information on attacks on reduced-round variants of *Rijndael* can be found in Ferguson *et al.* [84]. A relatively simple algebraic representation of *Rijndael* is given in Ferguson, Schroeppel and Whiting [85].

The technique of differential cryptanalysis was developed by Biham and Shamir [22] (see also [24] and their book on the differential cryptanalysis of *DES* [23]). Linear cryptanalysis was developed by Matsui [137, 138]. Some works developing the theoretical basis for these attacks are Lai, Massey and Murphy [126] and Nyberg [156]. The results of recent experiments on the effectiveness of linear cryptanalysis of *DES* can be found in Junod [108].

Our treatment of differential and linear cryptanalysis is based on the excellent tutorial by Heys [104]; we have also used the differential and linear attacks on SPNs that are described in [104]. General design principles for SPNs that are resistant to linear and differential cryptanalysis are presented by Heys and Tavares [105]. Recent results on the security of *Rijndael* against linear cryptanalysis have been given by Keliher, Meijer and Tavares [112, 113].

Nyberg [155] suggests the use of field inverses to define S-boxes (the technique which was later used in *Rijndael*). Chabaud and Vaudenay [45] also study the design of S-boxes that are resistant to differential and linear cryptanalysis.

Exercises

3.1 Let y be the output of Algorithm 3.1 on input x, where π_S and π_P are defined as in Example 3.1. In other words,

$$y = \text{SPN}\left(x, \pi_S, \pi_P, (K^1, \ldots, K^{Nr+1})\right),$$

where $(K^1, \ldots, K^{Nr+1})$ is the key schedule. Find a substitution π_{S^*} and a permutation π_{P^*} such that

$$x = \text{SPN}\left(y, \pi_{S^*}, \pi_{P^*}, (K^{Nr+1}, \ldots, K^1)\right).$$

3.2 Prove that decryption in a Feistel cipher can be done by applying the encryption algorithm to the ciphertext, with the key schedule reversed.

3.3 Let $DES(x, K)$ represent the encryption of plaintext x with key K using the *DES* cryptosystem. Suppose $y = DES(x, K)$ and $y' = DES(c(x), c(K))$, where $c(\cdot)$ denotes the bitwise complement of its argument. Prove that $y' = c(y)$ (i.e., if we complement the plaintext and the key, then the ciphertext is also complemented). Note that this can be proved using only the "high-level" description of *DES* — the actual structure of S-boxes and other components of the system are irrelevant.

3.4 Before the *AES* was developed, it was suggested to increase the security of *DES* by using the product cipher *DES* × *DES*, as discussed in Section 2.7. This product cipher uses two 56-bit keys.

This exercise considers known-plaintext attacks on product ciphers. In general, suppose that we take the product of any endomorphic cipher $\mathbf{S} = (\mathcal{P}, \mathcal{P}, \mathcal{K}, \mathcal{E}, \mathcal{D})$ with itself. Further, suppose that $\mathcal{K} = \{0, 1\}^n$ and $\mathcal{P} = \{0, 1\}^m$.

Now, assume we have several plaintext-ciphertext pairs for the product cipher $\mathbf{S}^2$, say $(x_1, y_1), \ldots, (x_\ell, y_\ell)$, all of which are obtained using the same unknown key, (K_1, K_2).

(a) Prove that $e_{K_1}(x_i) = d_{K_2}(y_i)$ for all i, $1 \leq i \leq \ell$. Give a heuristic argument that the expected number of keys (K_1, K_2) such that $e_{K_1}(x_i) = d_{K_2}(y_i)$ for all i, $1 \leq i \leq \ell$, is roughly $2^{2n-\ell m}$.

(b) Assume that $\ell \geq 2n/m$. A time-memory trade-off can be used to compute the unknown key (K_1, K_2). We compute two lists, each containing 2^n items, where each item comtains an ℓ-tuple of elements of $\mathcal{P}$ as well as an element of $\mathcal{K}$. If the two lists are sorted, then a common ℓ-tuple can be identified by means of a linear search through each of the two lists. Show that this algorithm requires $2^{n+m+1}\ell + 2^{2n+1}$ bits of memory and $\ell 2^{n+1}$ encryptions and/or decryptions.

(c) Show that the memory requirement of the attack can be reduced by a factor of 2^t if the total number of encryptions is increased by a factor of 2^t.

HINT Break the problem up into 2^{2t} subcases, each of which is specified by simultaneaouly fixing t bits of K_1 and t bits of K_2.

3.5 Suppose that we have the following 128-bit *AES* key, given in hexadecimal nota-
     tion:

$$\text{2B7E151628AED2A6ABF7158809CF4F3C}$$

Construct the complete key schedule arising from this key.

3.6 Compute the encryption of the following plaintext (given in hexadecimal notation)
     using the 10-round *AES*:

$$\text{3243F6A8885A308D313198A2E0370734}$$

Use the 128-bit key from the previous exercise.

3.7 Suppose a sequence of plaintext blocks, $x_1 \cdots x_n$, yields the ciphertext sequence
     $y_1 \cdots y_n$. Suppose that one ciphertext block, say y_i, is transmitted incorrectly (i.e.,
     some 1's are changed to 0's and vice versa). Show that the number of plaintext
     blocks that will be decrypted incorrectly is equal to one if ECB or OFB modes are
     used for encryption; and equal to two if CBC or CFB modes are used.

3.8 The purpose of this question is to investigate a time-memory trade-off for a chosen
     plaintext attack on a certain type of cipher. Suppose we have a cryptosystem in
     which $\mathcal{P} = \mathcal{C} = \mathcal{K}$, which attains perfect secrecy. Then it must be the case that
     $e_K(x) = e_{K_1}(x)$ implies $K = K_1$. Denote $\mathcal{P} = Y = \{y_1, \ldots, y_N\}$. Let x be a
     fixed plaintext. Define the function $g : Y \to Y$ by the rule $g(y) = e_y(x)$. Define
     a directed graph G having vertex set Y, in which the edge set consists of all the
     directed edges of the form $(y_i, g(y_i))$, $1 \le i \le N$.

Algorithm 3.7: TIME-MEMORY TRADE-OFF (x)

$y_0 \leftarrow y$
$backup \leftarrow$ **false**
while $g(y) \ne y_0$
$$\textbf{do} \begin{cases} \textbf{if } y = z_j \text{ for some } j \textbf{ and not } backup \\ \quad \textbf{then } \begin{cases} y \leftarrow g^{-T}(z_j) \\ backup \leftarrow \textbf{true} \end{cases} \\ \quad \textbf{else } \begin{cases} y \leftarrow g(y) \\ K \leftarrow y \end{cases} \end{cases}$$

(a) Prove that G consists of the union of disjoint directed cycles.

(b) Let T be a desired time parameter. Suppose we have a set of elements
     $Z = \{z_1, \ldots, z_m\} \subseteq Y$ such that, for every element $y_i \in Y$, either y_i is
     contained in a cycle of length at most T, or there exists an element $z_j \ne y_i$
     such that the distance from y_i to z_j (in G) is at most T. Prove that there
     exists such a set Z such that

$$|Z| \le \frac{2N}{T},$$

     so $|Z|$ is $O(N/T)$.

(c) For each $z_j \in Z$, define $g^{-T}(z_j)$ to be the element y_i such that $g^T(y_i) = z_j$,
     where g^T is the function that consists of T iterations of g. Construct a table
     X consisting of the ordered pairs $(z_j, g^{-T}(z_j))$, sorted with respect to their
     first coordinates.

A pseudo-code description of an algorithm to find K, given $y = e_K(x)$, is presented. Prove that this algorithm finds K in at most T steps. (Hence the time-memory trade-off is $O(N)$.)

(d) Describe a pseudo-code algorithm to construct the desired set Z in time $O(NT)$ without using an array of size N.

3.9 Suppose that X_1, X_2 and X_3 are independent discrete random variables defined on the set $\{0, 1\}$. Let ϵ_i denote the bias of X_i, for $i = 1, 2, 3$. Prove that $X_1 \oplus X_2$ and $X_2 \oplus X_3$ are independent if and only if $\epsilon_1 = 0$, $\epsilon_3 = 0$ or $\epsilon_2 = \pm 1/2$.

3.10 For the each of eight *DES* S-boxes, compute the bias of the random variable

$$X_2 \oplus Y_1 \oplus Y_2 \oplus Y_3 \oplus Y_4.$$

(Note that these biases are all relatively large in absolute value.)

3.11 The *DES* S-box S_4 has some unusual properties:

(a) Prove that the second row of S_4 can be obtained from the first row by means of the following mapping:

$$(y_1, y_2, y_3, y_4) \mapsto (y_2, y_1, y_4, y_3) \oplus (0, 1, 1, 0),$$

where the entries are represented as binary strings.

(b) Show that any row of S_4 can be transformed into any other row by a similar type of operation.

3.12 Suppose that $\pi_S : \{0, 1\}^m \to \{0, 1\}^n$ is an S-box. Prove the following facts about the function N_L.

(a) $N_L(0, 0) = 2^m$.

(b) $N_L(a, 0) = 2^m - 1$ for all integers a such that $0 \le a \le 2^m - 1$.

(c) For all integers a such that $0 \le a \le 2^m - 1$, it holds that

$$\sum_{a=0}^{2^m - 1} N_L(a, b) = 2^{2m-1} \pm 2^{m-1}.$$

(d) It holds that

$$\sum_{a=0}^{2^m - 1} \sum_{b=0}^{2^n - 1} N_L(a, b) \in \{2^{n+2m-1}, 2^{n+2m-1} + 2^{n+m-1}\}.$$

3.13 An S-box $\pi_S : \{0, 1\}^m \to \{0, 1\}^n$ is said to be *balanced* if

$$|\pi_S^{-1}(y)| = 2^{n-m}$$

for all $y \in \{0, 1\}^n$. Prove the following facts about the function N_L for a balanced S-box.

(a) $N_L(0, b) = 2^m - 1$ for all integers b such that $0 \le b \le 2^n - 1$.

(b) For all integers a such that $0 \le a \le 2^m - 1$, it holds that

$$\sum_{b=0}^{2^n - 1} N_L(a, b) = 2^{m+n-1} - 2^{m-1} + i2^n,$$

where i is an integer such that $0 \le i \le 2^{m-n}$.

3.14 Suppose that the S-box of Example 3.1 is replaced by the S-box defined by the following substitution $\pi_{S'}$:

z	0	1	2	3	4	5	6	7	8	9	A	B	C	D	E	F
$\pi_{S'}(z)$	8	4	2	1	C	6	3	D	A	5	E	7	F	B	9	0

(a) Compute the table of values N_L for this S-box.

(b) Find a linear approximation using three active S-boxes, and use the piling-up lemma to estimate the bias of the random variable

$$\mathbf{X_{16}} \oplus \mathbf{U_1^4} \oplus \mathbf{U_9^4}.$$

(c) Describe a linear attack, analogous to Algorithm 3.2, that will find eight subkey bits in the last round.

(d) Implement your attack and test it to see how many plaintexts are required in order for the algorithm to find the correct subkey bits (approximately 1000–1500 plaintexts should suffice; this attack is more efficient than Algorithm 3.3 because the bias is larger by a factor of 2, which means that the number of plaintexts can be reduced by a factor of about 4).

3.15 Suppose that the S-box of Example 3.1 is replaced by the S-box defined by the following substitution $\pi_{S''}$:

z	0	1	2	3	4	5	6	7	8	9	A	B	C	D	E	F
$\pi_{S''}(z)$	E	2	1	3	D	9	0	6	F	4	5	A	8	C	7	B

(a) Compute the table of values N_D for this S-box.

(b) Find a differential trail using four active S-boxes, namely, S_1^1, S_4^1, S_4^2 and S_4^3, that has propagation ratio $27/2048$.

(c) Describe a differential attack, analogous to Algorithm 3.3, that will find eight subkey bits in the last round.

(d) Implement your attack and test it to see how many plaintexts are required in order for the algorithm to find the correct subkey bits (approximately 100–200 plaintexts should suffice; this attack is not as efficient as Algorithm 3.3 because the propagation ratio is smaller by a factor of 2).

3.16 Suppose that we use the SPN presented in Example 3.1, but the S-box is replaced by a function π_T that is not a permutation. This means, in particular, that π_T is not surjective. Use this fact to derive a ciphertext-only attack that can be used to determine the key bits in the last round, given a sufficient number of ciphertexts which all have been encrypted using the same key.

4

Cryptographic Hash Functions

4.1 Hash Functions and Data Integrity

A cryptographic hash function can provide assurance of data integrity. A hash function is used to construct a short "fingerprint" of some data; if the data is altered, then the fingerprint will no longer be valid. Even if the data is stored in an insecure place, its integrity can be checked from time to time by recomputing the fingerprint and verifying that the fingerprint has not changed.

Let h be a hash function and let x be some data. As an illustrative example, x could be a binary string of arbitrary length. The corresponding fingerprint is defined to be $y = h(x)$. This fingerprint is often referred to as a *message digest*. A message digest would typically be a fairly short binary string; 160 bits is a common choice.

We suppose that y is stored in a secure place, but x is not. If x is changed, to x', say, then we hope that the "old" message digest, y, is not also a message digest for x'. If this is indeed the case, then the fact that x has been altered can be detected simply by computing the message digest $y' = h(x')$ and verifying that $y' \neq y$.

A particularly important application of hash functions occurs in the context of digital signature schemes, which will be studied in Chapter 7.

The motivating example discussed above assumes the existence of a single, fixed hash function. It is also useful to study families of keyed hash functions. A keyed hash function is often used as a *message authentication code*, or MAC. Suppose that Alice and Bob share a secret key, K, which determines a hash function, say h_K. For a message, say x, the corresponding authentication tag, $y = h_K(x)$, can be computed by Alice or Bob. The pair (x, y) can be transmitted over an insecure channel from Alice to Bob. When Bob receives the pair (x, y), he can verify if $y = h_K(x)$. If this condition holds, then he is confident that neither x nor y was altered by an adversary, provided that the hash family is "secure."

Notice the distinction between the assurance of data integrity provided by an unkeyed, as opposed to a keyed, hash function. In the case of an unkeyed hash function, the message digest must be securely stored so it cannot be altered. On the other hand, if Alice and Bob use a secret key K to specify the hash function they are using, then they can transmit both the data and the authentication tag over an insecure channel.

In the remainder of this chapter, we will study hash functions, as well as keyed hash families. We begin by giving definitions for a keyed hash family.

Definition 4.1:   A *hash family* is a four-tuple $(\mathcal{X}, \mathcal{Y}, \mathcal{K}, \mathcal{H})$, where the following conditions are satisfied:

1. $\mathcal{X}$ is a set of possible *messages*
2. $\mathcal{Y}$ is a finite set of possible *message digests* or *authentication tags*
3. $\mathcal{K}$, the *keyspace*, is a finite set of possible *keys*
4. For each $K \in \mathcal{K}$, there is a *hash function* $h_K \in \mathcal{H}$. Each $h_K : \mathcal{X} \to \mathcal{Y}$.

In the above definition, $\mathcal{X}$ could be a finite or infinite set; $\mathcal{Y}$ is always a finite set. If $\mathcal{X}$ is a finite set, a hash function is sometimes called a *compression function*. In this situation, we will always assume that $|\mathcal{X}| \geq |\mathcal{Y}|$, and we will often assume the stronger condition that $|\mathcal{X}| \geq 2|\mathcal{Y}|$.

A pair $(x, y) \in \mathcal{X} \times \mathcal{Y}$ is said to be *valid* under the key K if $h_K(x) = y$. Much of what we discuss in this chapter concerns methods to prevent the construction of certain types of valid pairs by an adversary.

Let $\mathcal{F}^{\mathcal{X},\mathcal{Y}}$ denote the set of all functions from $\mathcal{X}$ to $\mathcal{Y}$. Suppose that $|\mathcal{X}| = N$ and $|\mathcal{Y}| = M$. Then it is clear that $|\mathcal{F}^{\mathcal{X},\mathcal{Y}}| = M^N$. Any hash family $\mathcal{F} \subseteq \mathcal{F}^{\mathcal{X},\mathcal{Y}}$ is termed an (N, M)-*hash family*.

An *unkeyed hash function* is a function $h : \mathcal{X} \to \mathcal{Y}$, where $\mathcal{X}$ and $\mathcal{Y}$ are the same as in Definition 4.1. We could think of an unkeyed hash function simply as a hash family in which there is only one possible key, i.e., one in which $|\mathcal{K}| = 1$.

The remaining sections of this chapter are organized as follows. In Section 4.2, we introduce concepts of security for hash functions, in particular, the idea of collision resistance. We also study the exact security of "ideal" hash functions in the random oracle model in this section; and we discuss the birthday paradox, which provides an estimate of the difficulty of finding collisions for an arbitrary hash function. In Section 4.3, we introduce the important design technique of iterated hash functions. We discuss how this method is used in the design of practical hash functions, as well as in the construction of a provably secure hash function from a secure compression function. Section 4.4 provides a treatment of message authentication codes, where we again provide some general constructions and security proofs. Unconditionally secure MACs, and their construction using strongly universal hash families, are considered in Section 4.5.

4.2 Security of Hash Functions

Suppose that $h : \mathcal{X} \to \mathcal{Y}$ is an unkeyed hash function. Let $x \in \mathcal{X}$, and define $y = h(x)$. In many cryptographic applications of hash functions, it is desirable that the only way to produce a valid pair (x, y) is to first choose x, and then compute $y = h(x)$ by applying the function h to x. Other security requirements of hash functions are motivated by their applications in particular protocols, such as signature schemes (see Chapter 7). We now define three problems; if a hash function is to be considered secure, it should be the case that these three problems are difficult to solve.

Problem 4.1:   Preimage

Instance: A hash function $h : \mathcal{X} \to \mathcal{Y}$ and an element $y \in \mathcal{Y}$.

Find: $x \in \mathcal{X}$ such that $f(x) = y$.

Given a (possible) message digest y, the problem **Preimage** asks if x can be found such that $h(x) = y$. If **Preimage** can be solved for a given $y \in \mathcal{Y}$, then the pair (x, y) is a valid pair. A hash function for which **Preimage** cannot be efficiently solved is often said to be *one-way* or *preimage resistant*.

Problem 4.2:   Second Preimage

Instance: A hash function $h : \mathcal{X} \to \mathcal{Y}$ and an element $x \in \mathcal{X}$.

Find: $x' \in \mathcal{X}$ such that $x' \neq x$ and $h(x') = h(x)$.

Given a message x, the problem **Second Preimage** asks if $x' \neq x$ can be found such that $h(x') = h(x)$. Note that, if this can be done, then $(x', h(x))$ is a valid pair. A hash function for which **Second Preimage** cannot be efficiently solved is often said to be *second preimage resistant*.

Problem 4.3:   Collision

Instance: A hash function $h : \mathcal{X} \to \mathcal{Y}$.

Find: $x, x' \in \mathcal{X}$ such that $x' \neq x$ and $h(x') = h(x)$.

The problem **Collision** asks if $x' \neq x$ can be found such that $h(x') = h(x)$. A solution to this problem does not directly yield a valid pair. However, if (x, y) is a valid pair and x, x' is a solution to **Collision**, then (x', y) is also a valid pair. There are various scenarios where we want to avoid such a situation from arising. A hash function for which **Collision** cannot be efficiently solved is often said to be *collision resistant*.

Some of the questions we address in the next sections concern the difficulty of each of these three problems, as well as the relative difficulty of the three problems.

4.2.1   The Random Oracle Model

In this section, we describe a certain idealized model for a hash function, which attempts to capture the concept of an "ideal" hash function. If a hash function h is well designed, it should be the case that the only efficient way to determine the value $h(x)$ for a given x is to actually evaluate the function h at the value x. This should remain true even if many other values $h(x_1), h(x_2), \ldots$ have already been computed.

To illustrate an example where the above property does not hold, suppose that the hash function $h : \mathbb{Z}_n \times \mathbb{Z}_n \to \mathbb{Z}_n$ is a linear function, say

$$h(x, y) = ax + by \bmod n,$$

$a, b \in \mathbb{Z}_n$ and $n \geq 2$ is a positive integer. Suppose that we are given the values

$$h(x_1, y_1) = z_1$$

and

$$h(x_2, y_2) = z_2.$$

Let $r, s \in \mathbb{Z}_n$; then we have that

$$h(rx_1 + sx_2 \bmod n, ry_1 + sy_2 \bmod n) = a(rx_1 + sx_2) + b(ry_1 + sy_2) \bmod n$$
$$= r(ax_1 + by_1) + s(ax_2 + by_2) \bmod n$$
$$= rh(x_1, y_1) + sh(x_2, y_2) \bmod n.$$

Therefore, given the value of function h at two points (x_1, y_1) and (x_2, y_2), we know its value at various other points, without actually having to evaluate h at those points (and note also that we do not even need to know the values of the constants a and b in order to apply the above-described technique).

The *random oracle model*, which was introduced by Bellare and Rogaway, provides a mathematical model of an "ideal" hash function. In this model, a hash function $h : \mathcal{X} \to \mathcal{Y}$ is chosen randomly from $\mathcal{F}^{\mathcal{X}, \mathcal{Y}}$, and we are only permitted *oracle* access to the function h. This means that we are not given a formula or an algorithm to compute values of the function h. Therefore, the only way to compute a value $h(x)$ is to query the oracle. This can be thought of as looking up the value $h(x)$ in a giant book of random numbers such that, for each possible x, there is a completely random value $h(x)$.

As a consequence of the assumptions made in the random oracle model, it is obvious that the following *independence* property holds:

THEOREM 4.1   *Suppose that $h \in \mathcal{F}^{\mathcal{X}, \mathcal{Y}}$ is chosen randomly, and let $\mathcal{X}_0 \subseteq \mathcal{X}$. Suppose that the values $h(x)$ have been determined (by querying an oracle for h) if and only if $x \in \mathcal{X}_0$. Then $\mathbf{Pr}[h(x) = y] = 1/M$ for all $x \in \mathcal{X} \backslash \mathcal{X}_0$ and all $y \in \mathcal{Y}$.*

In the above theorem, the probability $\mathbf{Pr}[h(x) = y]$ is in fact a conditional probability that is computed over all functions h that take on the specified values for all $x \in \mathcal{X}_0$. Theorem 4.1 is the key property used in proofs involving complexity of problems in the random oracle model. We pursue this further in the next section.

4.2.2 Algorithms in the Random Oracle Model

In this section, we consider the the complexity of the three problems defined in Section 4.2 in the random oracle model. An algorithm in the random oracle model can be applied to any hash function, since the algorithm needs to know nothing whatsoever about the hash function (except that a method must be specified to evaluate the hash function for arbitrary values of x).

The algorithms we present and analyze are *randomized algorithms*; they can make random choices during their execution. A *Las Vegas algorithm* is a randomized algorithm which may fail to give an answer (i.e., it can terminate with the message "failure"), but if the algorithm does return an answer, then the answer must be correct.

Suppose $0 \leq \epsilon < 1$ is a real number. A randomized algorithm has *worst-case success probability* ϵ if, for every problem instance, the algorithm returns a correct answer with probability at least ϵ. A randomized algorithm has *average-case success probability* ϵ if the probability that the algorithm returns a correct answer, averaged over all problem instances of a specified size, is at least ϵ. Note that, in this latter situation, the probability that the algorithm returns a correct answer for a given problem instance can be greater than or less than ϵ.

In this section, we use the terminology (ϵ, q)-*algorithm* to denote a Las Vegas algorithm with average-case success probability ϵ, in which the number of oracle queries (i.e., evaluations of h) made by the algorithm is at most q. The success probability ϵ is the average over all possible random choices of $h \in \mathcal{F}^{\mathcal{X}, \mathcal{Y}}$, and all possible random choices of $x \in \mathcal{X}$ or $y \in \mathcal{Y}$, if x and/or y is specified as part of the problem instance.

We analyze the trivial algorithms, which evaluate $h(x)$ for q values of $x \in \mathcal{X}$, in the random oracle model. In fact, it turns out that the complexity of such an algorithm is independent of the choice of the q values of x because we are averaging over all functions $h \in \mathcal{F}^{\mathcal{X}, \mathcal{Y}}$.

We first consider an algorithm that attempts to solve **Preimage** by evaluating h at q points.

Algorithm 4.1: FINDPREIMAGE(h, y, q)

choose any $\mathcal{X}_0 \subseteq \mathcal{X}, |\mathcal{X}_0| = q$
for each $x \in \mathcal{X}_0$
$\quad$**do** $\begin{cases} \textbf{if } h(x) = y \\ \quad \textbf{then return } (x) \end{cases}$
return (failure)

THEOREM 4.2   *For any $X_0 \subseteq X$ with $|X_0| = q$, the average-case success probability of Algorithm 4.1 is $\epsilon = 1 - (1 - 1/M)^q$.*

PROOF   Let $y \in Y$ be fixed. Let $X_0 = \{x_1, \ldots, x_q\}$. For $1 \le i \le q$, let let E_i denote the event "$h(x_i) = y$." It follows from Proposition 4.1 that the E_i's are independent events, and $\mathbf{Pr}[E_i] = 1/M$ for all $1 \le i \le q$. Therefore, it holds that

$$\mathbf{Pr}[E_1 \vee E_2 \vee \cdots \vee E_q] = 1 - \left(1 - \frac{1}{M}\right)^q.$$

The success probability of Algorithm 4.1, for any fixed y, is constant. Therefore, the success probability averaged over all $y \in Y$ is identical, too. ∎

Note that the above success probability is approximately q/M provided that q is small compared to M.

We now present and analyze a very similar algorithm that attempts to solve Second Preimage.

Algorithm 4.2: FINDSECONDPREIMAGE(h, x, q)

$y \leftarrow h(x)$
choose $X_0 \subseteq X \backslash \{x\}, |X_0| = q - 1$
for each $x_0 \in X_0$
$\quad$ **do** $\begin{cases} \textbf{if } h(x_0) = y \\ \quad \textbf{then return } (x_0) \end{cases}$
return (failure)

The analysis of Algorithm 4.2 is similar to the previous algorithm. The only difference is that we require an "extra" application of h to compute $y = h(x)$ for the input value x.

THEOREM 4.3   *For any $X_0 \subseteq X \backslash \{x\}$ with $|X_0| = q - 1$, the success probability of Algorithm 4.2 is $\epsilon = 1 - (1 - 1/M)^{q-1}$.*

Next, we look at an elementary algorithm for Collision.

Algorithm 4.3: FINDCOLLISION(h, q)

choose $X_0 \subseteq X \backslash \{x\}, |X_0| = q$
for each $x \in X_0$
$\quad$ **do** $y_x \leftarrow h(x)$
if $y_x = y_{x'}$ for some $x' \ne x$
$\quad$ **then return** (x, x')
$\quad$ **else return** (failure)

In Algorithm 4.3, the test to see if $y_x = y_{x'}$ for some $x' \neq x$ could be done efficiently by sorting the y_x's, for example. This algorithm is analyzed using a probability argument analogous to the standard "birthday paradox." The *birthday paradox* says that in a group of 23 randomly chosen people, at least two will share a birthday with probability at least $1/2$. (Of course this is not a paradox, but it is probably counter-intuitive). This may not appear to be relevant to hash functions, but if we reformulate the problem, the connection will be clear. Suppose that the function h has as its domain the set of all living human beings, and for all x, $h(x)$ denotes the birthday of person x. Then the range of h consists of the 365 days in a year (366 days if we include February 29). Finding two people with the same birthday is the same thing as finding a collision for this particular hash function. In this setting, the birthday paradox is saying that Algorithm 4.3 has success probability at least $1/2$ when $q = 23$ and $M = 365$.

We now analyze Algorithm 4.3 in general, in the random oracle model. This algorithm is analogous to throwing q balls randomly into M bins and then checking to see if some bin contains at least two balls. (The q balls correspond to the q random x_i's, and the M bins correspond to the M possible elements of $\mathcal{Y}$.)

THEOREM 4.4 *For any $\mathcal{X}_0 \subseteq \mathcal{X}$ with $|\mathcal{X}_0| = q$, the success probability of Algorithm 4.3 is*

$$\epsilon = 1 - \left(\frac{M-1}{M}\right)\left(\frac{M-2}{M}\right) \cdots \left(\frac{M-q+1}{M}\right).$$

PROOF Let $\mathcal{X}_0 = \{x_1, \ldots, x_q\}$. For $1 \leq i \leq q$, let E_i denote the event

$$\text{"}h(x_i) \notin \{h(x_1), \ldots, h(x_{i-1})\}\text{."}$$

Using induction, it follows from Proposition 4.1 that $\mathbf{Pr}[E_1] = 1$ and

$$\mathbf{Pr}[E_i | E_1 \wedge E_2 \wedge \cdots \wedge E_{i-i}] = \frac{M-i+1}{M},$$

for $2 \leq i \leq q$. Therefore, we have that

$$\mathbf{Pr}[E_1 \wedge E_2 \wedge \cdots \wedge E_q] = \left(\frac{M-1}{M}\right)\left(\frac{M-2}{M}\right) \cdots \left(\frac{M-q+1}{M}\right).$$

The result follows. ∎

The above theorem shows that the probability of finding no collisions is

$$\left(1 - \frac{1}{M}\right)\left(1 - \frac{2}{M}\right) \cdots \left(1 - \frac{q-1}{M}\right) = \prod_{i=1}^{q-1}\left(1 - \frac{i}{M}\right).$$

If x is a small real number, then $1 - x \approx e^{-x}$. This estimate is derived by taking the first two terms of the series expansion

$$e^{-x} = 1 - x + \frac{x^2}{2!} - \frac{x^3}{3!} \cdots.$$

Using this estimate, the probability of finding no collisions is approximately

$$\prod_{i=1}^{q-1} \left(1 - \frac{i}{M}\right) \approx \prod_{i=1}^{q-1} e^{\frac{-i}{M}}$$

$$\approx e^{-\sum_{i=1}^{q-1} \frac{i}{M}}$$

$$= e^{\frac{-q(q-1)}{2M}}.$$

Consequently, we can estimate the probability of finding at least one collision to be

$$1 - e^{\frac{-q(q-1)}{2M}}.$$

If we denote this probability by ϵ, then we can solve for q as a function of M and ϵ:

$$e^{\frac{-q(q-1)}{2M}} \approx 1 - \epsilon$$

$$\frac{-q(q-1)}{2M} \approx \ln(1 - \epsilon)$$

$$q^2 - q \approx 2M \ln \frac{1}{1 - \epsilon}.$$

If we ignore the term $-q$, then we estimate that

$$q \approx \sqrt{2M \ln \frac{1}{1 - \epsilon}}.$$

If we take $\epsilon = .5$, then our estimate is

$$q \approx 1.17\sqrt{M}.$$

So this says that hashing just over $\sqrt{M}$ random elements of X yields a collision with a probability of 50%. Note that a different choice of ϵ leads to a different constant factor, but q will still be proportional to $\sqrt{M}$. The algorithm is a $(1/2, O(\sqrt{M}))$-algorithm.

We return to the example we mentioned earlier. Taking $M = 365$ in our estimate, we get $q \approx 22.3$. Hence, as mentioned earlier, the probability is at least $1/2$ that there will be at least one duplicated birthday among 23 randomly chosen people.

The birthday attack imposes a lower bound on the sizes of secure message digests. A 40-bit message digest would be very insecure, since a collision could be found with probability $1/2$ with just over 2^{20} (about a million) random hashes. It is usually suggested that the minimum acceptable size of a message digest is 128 bits (the birthday attack will require over 2^{64} hashes in this case). In fact, a 160-bit message digest (or larger) is usually recommended.

4.2.3 Comparison of Security Criteria

In the random oracle model, we have seen that solving **Collision** is easier than solving **Preimage** or **Second Preimage**. A related question is whether there exist reductions among the three problems which could be applied to arbitrary hash functions. It is fairly easy to see that we can reduce **Collision** to **Second Preimage** using Algorithm 4.4.

Algorithm 4.4: COLLISIONTOSECONDPREIMAGE(h)

external ORACLE2NDPREIMAGE
choose $x \in X$ uniformly at random
if (ORACLE2NDPREIMAGE$(h, x) = x'$) **and** $(x' \neq x)$ **and** $(h(x') = h(x))$
  then return (x, x')
  else return (failure)

Suppose that ORACLE2NDPREIMAGE is an (ϵ, q)-algorithm that solves **Second Preimage** for a particular, fixed hash function h. Then it is clear that COLLISIONTOSECONDPREIMAGE is an $(\epsilon, q + 2)$-algorithm that solves **Collision** for the same hash function h. This reduction does not make any assumptions about the hash function h. As a consequence of this reduction, we could say that the property of collision resistance implies the property of second preimage resistance.

We are now going to investigate the more interesting question of whether **Collision** can be reduced to **Preimage**. In other words, does collision resistance imply preimage resistance? We will prove that this is indeed the case, at least in some special situations. More specifically, we will prove that an arbitrary algorithm that solves **Preimage** with probability equal to 1 can be used to solve **Collision**.

This reduction can be accomplished with a fairly weak assumption on the relative sizes of the domain and range of the hash function. We will assume that the hash function $h : X \rightarrow Y$, where X and Y are finite sets and $|X| \geq 2|Y|$. We suppose that ORACLEPREIMAGE is a $(1, q)$ algorithm for **Preimage**. ORACLEPREIMAGE accepts as input a message digest $y \in Y$, and always finds an element ORACLEPREIMAGE$(y) \in X$ such that $h(\text{ORACLEPREIMAGE}(y)) = y$ (in particular, this implies that h is surjective). We will analyze the algorithm COLLISIONTOPREIMAGE, which is presented as Algorithm 4.5.

Algorithm 4.5: COLLISIONTOPREIMAGE(h)

external ORACLEPREIMAGE
choose $x \in X$ uniformly at random
$y \leftarrow h(x)$
if (ORACLEPREIMAGE$(h, y) = x'$) **and** $(x' \neq x)$
  then return (x, x')
  else return (failure)

We prove the following theorem.

THEOREM 4.5   *Suppose* $h : X \to Y$ *is a hash function where* $|X|$ *and* $|Y|$ *are finite and* $|X| \geq 2|Y|$. *Suppose* ORACLEPREIMAGE *is a* $(1, q)$ *algorithm for* **Preimage**, *for the fixed hash function* h. *Then* COLLISIONTOPREIMAGE *is a* $(1/2, q + 1)$-*algorithm for* **Collision**, *for the fixed hash function* h.

PROOF   Clearly COLLISIONTOPREIMAGE is a probabilistic algorithm of the Las Vegas type, since it either finds a collision or returns "failure." Thus our main task is to compute the average-case probability of success. For any $x \in X$, define $x \sim x_1$ if $h(x) = h(x_1)$. It is easy to see that $\sim$ is an equivalence relation. Define

$$[x] = \{x_1 \in X : x \sim x_1\}.$$

Each equivalence class $[x]$ consists of the inverse image of an element of Y, so the number of equivalence classes is equal to $|Y|$. Denote the set of equivalence classes by $\mathcal{C}$.

Now, suppose x is the random element of X chosen by the algorithm COLLI-SIONTOPREIMAGE. For this x, there are $|[x]|$ possible x_1's that could be returned as the output of ORACLEPREIMAGE. $|[x]| - 1$ of these x_1's are different from x and thus yield a collision. (Note that the algorithm ORACLEPREIMAGE does not know the representative of the equivalence class $[x]$ that was initially chosen by algorithm COLLISIONTOPREIMAGE.) So, given the element $x \in X$, the probability of success is $(|[x]| - 1)/|[x]|$.

The probability of success of the algorithm COLLISIONTOPREIMAGE is computed by averaging over all possible choices for x:

$$
\begin{aligned}
\mathbf{Pr}[\text{success}] &= \frac{1}{|X|} \sum_{x \in X} \frac{|[x]| - 1}{|[x]|} \\
&= \frac{1}{|X|} \sum_{C \in \mathcal{C}} \sum_{x \in C} \frac{|C| - 1}{|C|} \\
&= \frac{1}{|X|} \sum_{C \in \mathcal{C}} (|C| - 1) \\
&= \frac{1}{|X|} \left(\sum_{C \in \mathcal{C}} |C| - \sum_{C \in \mathcal{C}} 1 \right) \\
&= \frac{|X| - |Y|}{|X|} \\
&\geq \frac{|X| - |X|/2}{|X|} \\
&= \frac{1}{2}.
\end{aligned}
$$

Hence we have constructed a Las Vegas algorithm with average-case success probability at least $1/2$. ∎

4.3 Iterated Hash Functions

So far, we have considered hash functions with a finite domain (i.e., compression functions). We now study a particular technique by which a compression function, say compress, can be extended to a hash function with an infinite domain, h. A hash function h constructed by this method is called an *iterated hash function*.

In this section, we restrict our attention to hash functions whose inputs and outputs are bitstrings (i.e., strings formed of zeroes and ones). We denote the length of a bitstring x by $|x|$, and the concatenation of bitstrings x and y is denoted $x \parallel y$.

Suppose that compress $: \{0,1\}^{m+t} \to \{0,1\}^m$ is a compression function (where $t \geq 1$). The evaluation of the iterated hash function h based on compress will consist of three main steps:

preprocessing

Given an input string x, where $|x| \geq m + t + 1$, construct a string y, using a public algorithm, such that $|y| \equiv 0 \pmod{t}$. Denote

$$y = y_1 \parallel y_2 \parallel \cdots \parallel y_r,$$

where $|y_i| = t$ for $1 \leq i \leq r$.

processing

Let IV be a public initial value which is a bitstring of length m. Then compute the following:

$$z_0 \leftarrow \text{IV}$$

$$z_1 \leftarrow \text{compress}(z_0 \parallel y_1)$$

$$z_2 \leftarrow \text{compress}(z_1 \parallel y_2)$$

$$\vdots \quad \vdots \quad \vdots$$

$$z_r \leftarrow \text{compress}(z_{r-1} \parallel y_r).$$

optional output transformation

Let $g : \{0,1\}^m \to \{0,1\}^\ell$ be a public function. Define $h(x) = g(z_r)$.

The iterated hash function constructed above is $h : \bigcup_{i=m+t+1}^{\infty} \{0,1\}^i \to \{0,1\}^\ell$.

A commonly used preprocessing step is to construct the string y in the following way:

$$y = x \parallel \mathsf{pad}(x),$$

where $\mathsf{pad}(x)$ is constructed from x using a *padding function*. A padding function typically incorporates the value of $|x|$, and pads the result with additional bits (zeros, for example) so that the resulting string y has a length that is a multiple of t.

The preprocessing step must ensure that the mapping $x \mapsto y$ is an injection. (If the mapping $x \mapsto y$ is not one-to-one, then it may be possible to find $x \neq x'$ so that $y = y'$. Then $h(x) = h(x')$, and h would not be collision-resistant.) Note also that $|y| = rt \geq |x|$ because of the required injective property.

Most hash functions used in practice are in fact iterated hash functions and can be viewed as special cases of the generic construction described above. In a later section, we will describe one such hash function, the *Secure Hash Algorithm* (known as *SHA-1*) in detail. The Merkle-Damgård construction, which we discuss in the next section, is also a construction of an iterated hash function which permits a formal security proof to be given.

4.3.1   The Merkle-Damgård Construction

In this section, we present a particular method of constructing a hash function from a compression function. This construction has the property that the resulting hash function satisfies desired security properties such as collision resistance provided that the compression function does. This technique is often called the *Merkle-Damgård construction*.

Suppose $\mathsf{compress} : \{0,1\}^{m+t} \to \{0,1\}^m$ is a collision resistant compression function, where $t \geq 1$. We will use $\mathsf{compress}$ to construct a collision resistant hash function $h : \mathcal{X} \to \{0,1\}^m$, where

$$\mathcal{X} = \bigcup_{i=m+t+1}^{\infty} \{0,1\}^i.$$

We first consider the situation where $t \geq 2$.

We will treat elements of $x \in \mathcal{X}$ as bit-strings. Suppose $|x| = n \geq m + t + 1$. We can express x as the concatenation

$$x = x_1 \parallel x_2 \parallel \cdots \parallel x_k,$$

where

$$|x_1| = |x_2| = \cdots = |x_{k-1}| = t - 1$$

and

$$|x_k| = t - 1 - d,$$

where $0 \leq d \leq t - 2$. Hence, we have that

$$k = \left\lceil \frac{n}{t-1} \right\rceil.$$

We define $h(x)$ to be the output of Algorithm 4.6.

Algorithm 4.6: MERKLE-DAMGÅRD(x)

external compress
comment: compress : $\{0,1\}^{m+t} \rightarrow \{0,1\}^m$, where $t \geq 2$

$n \leftarrow |x|$
$k \leftarrow \lceil n/(t-1) \rceil$
$d \leftarrow n - k(t-1)$
for $i \leftarrow 1$ **to** $k - 1$
$\quad$ **do** $y_i \leftarrow x_i$
$y_k \leftarrow x_k \parallel 0^d$
$y_{k+1} \leftarrow$ the binary representation of d
$z_1 \leftarrow 0^{m+1} \parallel y_1$
$g_1 \leftarrow$ compress(z_1)
for $i \leftarrow 1$ **to** k
$\quad$ **do** $\begin{cases} z_{i+1} \leftarrow g_i \parallel 1 \parallel y_{i+1} \\ g_{i+1} \leftarrow \text{compress}(z_{i+1}) \end{cases}$
$h(x) \leftarrow g_{k+1}$
return $(h(x))$

Denote

$$y(x) = y_1 \parallel y_2 \parallel \cdots \parallel y_{k+1}.$$

Observe that y_k is formed from x_k by padding on the right with d zeroes, so that all the blocks y_i ($1 \leq i \leq k$) are of length $t - 1$. Also, y_{k+1} should be padded on the left with zeroes so that $|y_{k+1}| = t - 1$.

As was done in the generic construction described in Section 4.3, we hash x by first constructing $y(x)$, and then processing the blocks $y_1, y_2, \ldots, y_{k+1}$ in a particular fashion. y_{k+1} is defined in such a way that the mapping $x \mapsto y(x)$ is an injection, which we observed is necessary for the iterated hash function to be collision-resistant.

The following theorem proves that, if a collision can be found for h, then a collision can be found for compress. In other words, h is collision resistant provided that compress is collision resistant.

THEOREM 4.6 *Suppose* compress : $\{0,1\}^{m+t} \rightarrow \{0,1\}^m$ *is a collision resistant compression function, where $t \geq 2$. Then the function*

$$h : \bigcup_{i=m+t+1}^{\infty} \{0,1\}^i \rightarrow \{0,1\}^m,$$

as constructed in Algorithm 4.6, is a collision resistant hash function.

PROOF   Suppose that we can find $x \neq x'$ such that $h(x) = h(x')$. We will show how we can find a collision for compress in polynomial time.

Denote

$$y(x) = y_1 \parallel y_2 \parallel \cdots \parallel y_{k+1}$$

and

$$y(x') = y'_1 \parallel y'_2 \parallel \cdots \parallel y'_{\ell+1},$$

where x and x' are padded with d and d' 0's, respectively. Denote the g-values computed in the algorithm by $g_1, \ldots, g_{k+1}$ and $g'_1, \ldots, g'_{\ell+1}$, respectively.

We identify two cases, depending on whether $|x| \equiv |x'| \pmod{t-1}$ (or not).

case 1: $|x| \not\equiv |x'| \pmod{t-1}$.

Here $d \neq d'$ and $y_{k+1} \neq y'_{\ell+1}$. We have

$$
\begin{aligned}
\mathsf{compress}(g_k \parallel 1 \parallel y_{k+1}) &= g_{k+1} \\
&= h(x) \\
&= h(x') \\
&= g'_{\ell+1} \\
&= \mathsf{compress}(g'_\ell \parallel 1 \parallel y'_{\ell+1}),
\end{aligned}
$$

which is a collision for h because $y_{k+1} \neq y'_{\ell+1}$.

case 2: $|x| \equiv |x'| \pmod{t-1}$.

It is convenient to split this case into two subcases:

case 2a: $|x| = |x'|$.

Here we have $k = \ell$ and $y_{k+1} = y'_{k+1}$. We begin as in case 1:

$$
\begin{aligned}
\mathsf{compress}(g_k \parallel 1 \parallel y_{k+1}) &= g_{k+1} \\
&= h(x) \\
&= h(x') \\
&= g'_{k+1} \\
&= \mathsf{compress}(g'_k \parallel 1 \parallel y'_{k+1}).
\end{aligned}
$$

If $g_k \neq g'_k$, then we find a collision for compress, so assume $g_k = g'_k$. Then we have

$$
\begin{aligned}
\mathsf{compress}(g_{k-1} \parallel 1 \parallel y_k) &= g_k \\
&= g'_k \\
&= \mathsf{compress}(g'_{k-1} \parallel 1 \parallel y'_k).
\end{aligned}
$$

Either we find a collision for compress, or $g_{k-1} = g'_{k-1}$ and $y_k = y'_k$. Assuming we do not find a collision, we continue working backwards, until finally we obtain

$$\text{compress}(0^{m+1} \mid\mid y_1) = g_1$$
$$= g'_1$$
$$= \text{compress}(0^{m+1} \mid\mid y'_1).$$

If $y_1 \neq y'_1$, then we find a collision for compress, so we assume $y_1 = y'_1$. But then $y_i = y'_i$ for $1 \leq i \leq k+1$, so $y(x) = y(x')$. But this implies $x = x'$, because the mapping $x \mapsto y(x)$ is an injection. We assumed $x \neq x'$, so we have a contradiction.

case 2b: $|x| \neq |x'|$.

Without loss of generality, assume $|x'| > |x|$, so $\ell > k$. This case proceeds in a similar fashion as case 2a. Assuming we find no collisions for compress, we eventually reach the situation where

$$\text{compress}(0^{m+1} \mid\mid y_1) = g_1$$
$$= g'_{\ell-k+1}$$
$$= \text{compress}(g'_{\ell-k} \mid\mid 1 \mid\mid y'_{\ell-k+1}).$$

But the $(m+1)$st bit of

$$0^{m+1} \mid\mid y_1$$

is a 0 and the $(m+1)$st bit of

$$g'_{\ell-k} \mid\mid 1 \mid\mid y'_{\ell-k+1}$$

is a 1. So we find a collision for compress.

Since we have considered all possible cases, we have proven the desired conclusion. ∎

The construction presented in Algorithm 4.6 can be used only when $t \geq 2$. Let's now look at the situation where $t = 1$. We need to use a different construction for h. Suppose $|x| = n \geq m + 2$. We first encode x in a special way. This will be done using the function f defined as follows:

$$f(0) = 0$$
$$f(1) = 01.$$

The construction of $h(x)$ is presented as Algorithm 4.7.

Algorithm 4.7: MERKLE-DAMGÅRD2(x)

external compress
comment: compress : $\{0,1\}^{m+1} \rightarrow \{0,1\}^m$

$n \leftarrow |x|$
$y \leftarrow 11 \parallel f(x_1) \parallel f(x_2) \parallel \cdots \parallel f(x_n)$
denote $y = y_1 \parallel y_2 \parallel \cdots \parallel y_k$, where $y_i \in \{0,1\}, 1 \leq i \leq k$
$g_1 \leftarrow \text{compress}(0^m \parallel y_1)$
for $i \leftarrow 1$ **to** $k-1$
   do $g_{i+1} \leftarrow \text{compress}(g_i \parallel y_{i+1})$
return (g_k)

The encoding $x \mapsto y = y(x)$, as defined in Algorithm 4.7, satisfies two important properties:

1. If $x \neq x'$, then $y(x) \neq y(x')$ (i.e., $x \mapsto y(x)$ is an injection).
2. There do not exist two strings $x \neq x'$ and a string z such that $y(x) = z \parallel y(x')$. (In other words, no encoding is a *postfix* of another encoding. This is easily seen because each string $y(x)$ begins with 11, and there do not exist two consecutive 1's in the remainder of the string.)

THEOREM 4.7 *Suppose* compress : $\{0,1\}^{m+1} \rightarrow \{0,1\}^m$ *is a collision resistant compression function. Then the function*

$$h : \bigcup_{i=m+2}^{\infty} \{0,1\}^i \rightarrow \{0,1\}^m,$$

as constructed in Algorithm 4.7, is a collision resistant hash function.

PROOF Suppose that we can find $x \neq x'$ such that $h(x) = h(x')$. Denote

$$y(x) = y_1 y_2 \cdots y_k$$

and

$$y(x') = y_1' y_2' \cdots y_\ell'.$$

We consider two cases.

case 1: $k = \ell$.

   As in Theorem 4.6, either we find a collision for compress, or we obtain $y = y'$. But this implies $x = x'$, a contradiction.

case 2: $k \neq \ell$.

   Without loss of generality, assume $\ell > k$. This case proceeds in a similar fashion. Assuming we find no collisions for compress, we have the following sequence of equalities:

$$y_k = y'_\ell$$

$$y_{k-1} = y'_{\ell-1}$$

$$\vdots \quad \vdots$$

$$y_1 = y'_{\ell-k+1}.$$

But this contradicts the "postfix-free" property stated above.

We conclude that h is collision resistant. $\blacksquare$

We summarize the two constructions of hash functions in this section, and the number of applications of compress needed to compute h, in the following theorem.

THEOREM 4.8 *Suppose* compress $: \{0,1\}^{m+t} \to \{0,1\}^m$ *is a collision resistant compression function, where* $t \geq 1$. *Then there exists a collision resistant hash function*

$$h : \bigcup_{i=m+t+1}^{\infty} \{0,1\}^i \to \{0,1\}^m.$$

The number of times compress *is computed in the evaluation of h is at most*

$$1 + \left\lceil \frac{n}{t-1} \right\rceil \quad \text{if } t \geq 2$$
$$2n + 2 \quad \text{if } t = 1,$$

where $|x| = n$.

4.3.2 The Secure Hash Algorithm

In this section, we describe *SHA-1* (the *Secure Hash Algorithm*), which is an iterated hash function with a 160-bit message digest. *SHA-1* is built from word-oriented operations on bitstrings, where a word consists of 32 bits (or eight hexadecimal digits). The operations used in *SHA-1* are as follows:

$X \wedge Y$	bitwise "and" of X and Y
$X \vee Y$	bitwise "or" of X and Y
$X \oplus Y$	bitwise "xor" of X and Y
$\neg X$	bitwise complement of X
$X + Y$	integer addition modulo 2^{32}
$\text{ROTL}^s(X)$	circular left shift of X by s positions $(0 \leq s \leq 31)$

We first describe the padding scheme used in *SHA-1*; it is presented as Algorithm 4.8. Note that *SHA-1* requires that $|x| \leq 2^{64} - 1$. Therefore the binary representation of $|x|$, which is denoted by ℓ in Algorithm 4.8, has length at most 64 bits. If $|\ell| < 64$, then it is padded on the left with zeroes so that its length is exactly 64 bits.

Algorithm 4.8: SHA-1-PAD(x)

comment: $|x| \leq 2^{64} - 1$

$d \leftarrow (447 - |x|) \bmod 512$
$\ell \leftarrow$ the binary representation of $|x|$, where $|\ell| = 64$
$y \leftarrow x \parallel 1 \parallel 0^d \parallel \ell$

In the construction of y, we append a single 1 to x, then we concatenate enough 0's so that the length becomes congruent to 448 modulo 512, and finally we concatenate 64 bits that contain the binary representation of the (original) length of x. The resulting string y has length divisible by 512. Then we write y as a concatenation of n blocks, each having 512 bits:

$$y = M_1 \parallel M_2 \parallel \cdots \parallel M_n.$$

Define the functions $f_0, \ldots, f_{79}$ as follows:

$$f_i(B, C, D) = \begin{cases} (B \wedge C) \vee ((\neg B) \wedge D) & \text{if } 0 \leq t \leq 19 \\ B \oplus C \oplus D & \text{if } 20 \leq t \leq 39 \\ (B \wedge C) \vee (B \wedge D) \vee (C \wedge D) & \text{if } 40 \leq t \leq 59 \\ B \oplus C \oplus D & \text{if } 60 \leq t \leq 79. \end{cases}$$

Each function f_i takes three words B, C and D as input, and produces one word as output.

Define the word constants $K_0, \ldots, K_{79}$, which are used in the computation of *SHA-1*(x), as follows:

$$K_i = \begin{cases} \text{5A827999} & \text{if } 0 \leq t \leq 19 \\ \text{6ED9EBA1} & \text{if } 20 \leq t \leq 39 \\ \text{8F1BBCDC} & \text{if } 40 \leq t \leq 59 \\ \text{CA62C1D6} & \text{if } 60 \leq t \leq 79. \end{cases}$$

Now *SHA-1* is presented as Cryptosystem 4.1.

Cryptosystem 4.1: *SHA-1(x)*

external SHA-1-PAD
global $K_0, \ldots, K_{79}$
$y \leftarrow$ SHA-1-PAD(x)
denote $y = M_1 \parallel M_2 \parallel \cdots \parallel M_n$, where each M_i is a 512-bit block
$H_0 \leftarrow$ 67452301
$H_1 \leftarrow$ EFCDAB89
$H_2 \leftarrow$ 98BADCFE
$H_3 \leftarrow$ 10325476
$H_4 \leftarrow$ C3D2E1F0
for $i \leftarrow 1$ **to** n

$$
\mathbf{do}
\begin{cases}
\text{denote } M_i = W_0 \parallel W_1 \parallel \cdots \parallel W_{15}, \text{ where each } W_i \text{ is a word} \\
\textbf{for } t \leftarrow 16 \textbf{ to } 79 \\
\quad \textbf{do } W_t \leftarrow \text{ROTL}^1(W_{t-3} \oplus W_{t-8} \oplus W_{t-14} \oplus W_{t-16}) \\
A \leftarrow H_0 \\
B \leftarrow H_1 \\
C \leftarrow H_2 \\
D \leftarrow H_3 \\
E \leftarrow H_4 \\
\textbf{for } t \leftarrow 0 \textbf{ to } 79 \\
\quad \mathbf{do}
\begin{cases}
temp \leftarrow \text{ROTL}^5(A) + \mathsf{f}_t(B, C, D) + E + W_t + K_t \\
E \leftarrow D \\
D \leftarrow C \\
C \leftarrow \text{ROTL}^{30}(B) \\
B \leftarrow A \\
A \leftarrow temp
\end{cases} \\
H_0 \leftarrow H_0 + A \\
H_1 \leftarrow H_1 + B \\
H_2 \leftarrow H_2 + C \\
H_3 \leftarrow H_3 + D \\
H_4 \leftarrow H_4 + E
\end{cases}
$$

return $(H_0 \parallel H_1 \parallel H_2 \parallel H_3 \parallel H_4)$

Observe that *SHA-1* closely follows the general model of an iterated hash function. The padding scheme extends the input x by at most one extra 512-bit block. The compression function maps $160 + 512$ bits to 160 bits, where the 512 bits comprise a block of the message.

SHA-1 is the latest in a series of related iterated hash functions. The first of these hash functions, *MD4*, was proposed by Rivest in 1990. Rivest then modified *MD4* to produce *MD5* in 1992. *SHA* was proposed as a standard by NIST in 1993, and it was adopted as FIPS 180. *SHA-1* is a minor variation of *SHA*; it is FIPS 180-1.

This progression of hash functions incorporated various modifications to improve the security of the later versions of the hash functions against attacks that were found against earlier versions. For example, collisions in the compression functions of *MD4* and *MD5* were discovered. (No collisions in the "complete" hash functions are presently known.) The original *SHA* has a weakness that would allow collisions to be found in approximately 2^{61} steps. This attack is probably not feasible with current technology, but it is much more efficient than a birthday attack, which would require about 2^{80} steps. The weakness in *SHA* was rectified in *SHA-1*.

On May 30, 2001 NIST announced that a draft version of FIPS 180-2 was available for public comment and review. This proposed standard includes *SHA-1* as well as three new hash functions, which are known as *SHA-256*, *SHA-384* and *SHA-512* The suffixes "256," "384" and "512" refer to the sizes of the message digests. The three new hash functions are also iterated hash functions, but they have a more complex description than *SHA-1*.

4.4 Message Authentication Codes

We now turn our attention to message authentication codes, which are keyed hash functions satisfying certain security properties. As we will see, the security properties required by a MAC are rather different than those required by an (unkeyed) hash function.

One common way of constructing a MAC is to incorporate a secret key into an unkeyed hash function, by including it as part of the message to be hashed. This must be done carefully, however, in order to prevent certain attacks from being carried out. We illustrate the possible pitfalls with a couple of simple examples.

As a first attempt, suppose we construct a keyed hash function h_K from an unkeyed iterated hash function, say h, by defining $IV = K$ and keeping its value secret. For simplicity, suppose also that h does not have a preprocessing step or an output transformation. Such a hash function requires that every input message x have length that is a multiple of t, where compress : $\{0,1\}^{m+t} \to \{0,1\}^m$ is the compression function used to build h. Further, the key K is an m-bit key.

We show how an opponent can construct a valid MAC for a certain message, without knowing the secret key K, given any message x and its corresponding MAC, $h_K(x)$. Let x' be any bitstring of length t, and consider the message $x \parallel x'$. The MAC for this message, $h_K(x \parallel x')$, is computed to be

$$h_K(x \parallel x') = \text{compress}(h_K(x), x').$$

Since $h_K(x)$ and x' are both known, it is a simple matter for an opponent to compute $h_K(x \parallel x')$, even though K is secret. This is clearly a security problem.

Even if messages are padded, a modification of the above attack can be carried out. For example, suppose that $y = x \parallel \text{pad}(x)$ in the preprocessing step. Note that $|y| = rt$ for some integer r. Let w be any bitstring of length t, and define

$$x' = x \parallel \mathsf{pad}(x) \parallel w.$$

In the preprocessing step, we would compute

$$y' = x' \parallel \mathsf{pad}(x') = x \parallel \mathsf{pad}(x) \parallel w \parallel \mathsf{pad}(x'),$$

where $|y'| = r't$ for some integer $r' > r$.

Consider the computation of $h_K(x')$ (this is the same as computing $h(x')$ when IV $= K$). In the processing step, it is clear that $z_r = h_K(x)$. It is therefore possible for an adversary to compute the following:

$$z_{r+1} \leftarrow \mathsf{compress}(h_K(x) \parallel y_{r+1})$$
$$z_{r+2} \leftarrow \mathsf{compress}(z_{r+1} \parallel y_{r+2})$$

$$\vdots \quad \vdots \quad \vdots$$

$$z_{r'} \leftarrow \mathsf{compress}(z_{r'-1} \parallel y_{r'}),$$

and then

$$h_K(x') = z_{r'}.$$

Therefore the adversary can compute $h_K(x')$ even though he doesn't know the secret key K (and notice that the attack makes no assumptions about the length of the pad).

Keeping the above examples in mind, we formulate definitions of what it should mean for a MAC algorithm to be secure. As we saw, the objective of an opponent is to try to produce a pair (x, y) that is valid under an unknown but fixed key, K. The opponent is allowed to request (up to) q valid MACs on messages $x_1, x_2, \ldots$ of his own choosing. It is convenient to assume that there is a black box (or oracle) that answers the adversary's queries, and we will often use this terminology when it is convenient to do so.

Thus, the adversary obtains a list of valid pairs (under the unknown key K):

$$(x_1, y_1), (x_2, y_2), \ldots, (x_q, y_q)$$

by querying the oracle with messages $x_1, \ldots, x_q$. Later, when the adversary outputs the pair (x, y), it is required that $x \notin \{x_1, \ldots, x_q\}$. If, in addition, (x, y) is a valid pair, then the pair is said to be a *forgery*. If the probability that the adversary outputs a forgery is at least ϵ, then the adversary is said to be an (ϵ, q)-*forger* for the given MAC. The probability ϵ could be taken to be either an average-case probability over all the possible keys, or worst-case. To be concrete, in the following sections we will generally take ϵ to be a worst-case probability. This means that the adversary can produce a forgery with probability at least ϵ, regardless of the secret key being used.

Using this terminology, the attacks described above are $(1, 1)$-forgers.

4.4.1 Nested MACs and HMAC

A *nested MAC* builds a MAC algorithm from the composition of two (keyed) hash families. Suppose that $(\mathcal{X}, \mathcal{Y}, \mathcal{K}, \mathcal{G})$ and $(\mathcal{Y}, \mathcal{Z}, \mathcal{L}, \mathcal{H})$ are hash families. The *composition* of these hash families is the hash family $(\mathcal{X}, \mathcal{Z}, \mathcal{M}, \mathcal{G} \circ \mathcal{H})$ in which $\mathcal{M} = \mathcal{K} \times \mathcal{L}$ and

$$\mathcal{G} \circ \mathcal{H} = \{g \circ h : g \in \mathcal{G}, h \in \mathcal{H}\},$$

where $(g \circ h)_{(K,L)}(x) = h_L(g_K(x))$ for all $x \in \mathcal{X}$. In this construction, $\mathcal{Y}$ and $\mathcal{Z}$ are finite sets such that $|\mathcal{Y}| \geq |\mathcal{Z}|$; $\mathcal{X}$ could be finite or infinite. If $\mathcal{X}$ is finite, then $|\mathcal{X}| > |\mathcal{Y}|$.

We are interested in finding situations under which we can guarantee that a nested MAC is secure, assuming that the two hash families from which it is constructed satisfy appropriate security requirements. Roughly speaking, it can be shown that the nested MAC is secure provided that the following two conditions are satisfied:

1. $(\mathcal{Y}, \mathcal{Z}, \mathcal{L}, \mathcal{H})$ is secure as a MAC, given a fixed (unknown) key, and

2. $(\mathcal{X}, \mathcal{Y}, \mathcal{K}, \mathcal{G})$ is collision-resistant, given a fixed (unknown) key.

Intuitively, we are building a secure "big MAC" (namely, the nested MAC) from the composition of a secure "little MAC" (namely, $(\mathcal{Y}, \mathcal{Z}, \mathcal{L}, \mathcal{H})$) and a collision-resistant keyed hash family (namely, $(\mathcal{X}, \mathcal{Y}, \mathcal{K}, \mathcal{G})$). Let's try to make the above conditions more precise, and then present a proof of a specific security result.

The security result will in fact be comparing the relative difficulties of certain types of attacks against the three hash families. We will therefore be considering all three of the following adversaries:

- a forger for the nested MAC (a "big MAC attack"),

- a forger for the little MAC (a "little MAC attack"), and

- a collision-finder for the hash family, when the key is secret (an "unknown-key collision attack").

Here is a more careful description of each of the three adversaries: In a *big MAC attack*, a pair of keys, (K, L), is chosen and kept secret. The adversary is allowed to choose values for x and query a big MAC oracle for values of $h_L(g_K(x))$. Then the adversary attempts to output a pair (x', z) such that $z = h_L(g_K(x'))$, where x' was not one of its previous queries.

In a *little MAC attack*, a key L is chosen and kept secret. The adversary is allowed to choose values for y and query a little MAC oracle for values of $h_L(y)$. Then the adversary attempts to output a pair (y', z) such that $z = h_L(y')$, where y' was not one of its previous queries.

In an *unknown-key collision attack*, a key K is chosen and kept secret. The adversary is allowed to choose values for x and query a hash oracle for values of

$g_K(x)$. Then the adversary attempts to output a pair x', x'' such that $x' \neq x''$ and $g_K(x') = g_K(x'')$.

We will assume that there does not exist an $(\epsilon_1, q + 1)$-unknown-key collision attack for a randomly chosen function $g_K \in \mathcal{G}$. (If the key K were not secret, then this would correspond to our usual notion of collision resistance. Since we assume that K is secret, the problem facing the adversary is more difficult, and therefore we are making a weaker security assumption than collision-resistance.) We also assume that there does not exist an (ϵ_2, q)-little MAC attack for a randomly chosen function $h_L \in \mathcal{H}$, where L is secret. Finally, suppose that there exists an (ϵ, q)-big MAC attack for a randomly chosen function $(g \circ h)_{(K,L)} \in \mathcal{G} \circ \mathcal{H}$, where (K, L) is secret.

With probability at least ϵ, the big MAC attack outputs a valid pair (x, z) after making at most q queries to a big MAC oracle. Let $x_1, \ldots, x_q$ denote the queries made by the adversary, and let $z_1, \ldots, z_q$ be the corresponding responses made by the oracle. After the adversary has finished executing, we have the list of valid pairs $(x_1, z_1), \ldots, (x_q, z_q)$, as well as the possibly valid pair (x, z).

Suppose we now take the values $x_1, \ldots, x_q$, and x, and make $q + 1$ queries to a hash oracle. We obtain the list of values $y_1 = g_K(x_1), \ldots, y_q = g_K(x_q)$, and $y = g_K(x)$. Suppose it happens that $y \in \{y_1, \ldots, y_q\}$, say $y = y_i$. Then we can output the pair x, x_i as a solution to Collision. This would be a successful unknown-key collision attack. On the other hand, if $y \notin \{y_1, \ldots, y_q\}$, then we output the pair (y, z), which (possibly) is a valid pair for the little MAC. This would be a forgery constructed after (indirectly) obtaining q answers to q little MAC queries, namely $(y_1, z_1), \ldots, (y_q, z_q)$.

By the assumption we made, any unknown-key collision attack has probability at most ϵ_1 of succeeding. As well, we assumed that the big MAC attack has success probability at least ϵ. Therefore, the probability that (x, z) is a valid pair and $y \notin \{y_1, \ldots, y_q\}$ is at least $\epsilon - \epsilon_1$. The success probability of any little MAC attack is at most ϵ_2, and the success probability of the little MAC attack described above is at least $\epsilon - \epsilon_1$. Hence, it follows that $\epsilon \leq \epsilon_1 + \epsilon_2$.

We have proven the following result.

THEOREM 4.9 *Suppose* $(\mathcal{X}, \mathcal{Z}, \mathcal{M}, \mathcal{G} \circ \mathcal{H})$ *is a nested MAC. Suppose there does not exist an* $(\epsilon_1, q + 1)$-*collision attack for a randomly chosen function* $g_K \in \mathcal{G}$, *when the key* K *is secret. Further, suppose that there does not exist an* (ϵ_2, q)-*forger for a randomly chosen function* $h_L \in \mathcal{H}$, *where* L *is secret. Finally, suppose there exists an* (ϵ, q)-*forger for the nested MAC, for a randomly chosen function* $(g \circ h)_{(K,L)} \in \mathcal{G} \circ \mathcal{H}$. *Then* $\epsilon \leq \epsilon_1 + \epsilon_2$.

HMAC is a nested MAC algorithm that is a proposed FIPS standard. It constructs a MAC from an (unkeyed) hash function; we describe *HMAC* based on *SHA-1*. This version of *HMAC* uses a 512-bit key, denoted K. x is the message to be authenticated, and *ipad* and *opad* are 512-bit constants, defined in hexadecimal notation as follows:

$$ipad = 3636\cdots 36$$

$$opad = 5C5C\cdots 5C$$

Then the 160-bit MAC is defined as follows:

$$\text{HMAC}_K(x) = \text{SHA-1}((K \oplus opad) \parallel \text{SHA-1}((K \oplus ipad) \parallel x)).$$

Note that *HMAC* uses *SHA-1* with the value $K \oplus ipad$, which is prepended to x, used as the key. This application of *SHA-1* is assumed to be secure against an unknown-key collision attack. Now the key value $K \oplus ipad$ is prepended to the previously constructed message digest, and *SHA-1* is applied again. This second computation of *SHA-1* requires only one application of the compression function, and we are assuming that *SHA-1* when used in this way is secure as a MAC. If these two assumptions are valid, then Theorem 4.9 says that *HMAC* is secure as a MAC.

4.4.2   CBC-MAC

One of the most popular ways to construct a MAC is to use a block cipher in CBC mode with a fixed (public) initialization vector. In CBC mode, recall that each ciphertext block y_i is x-ored with the next plaintext block, x_{i+1}, before being encrypted with the secret key K. More formally, we start with an initialization vector, denoted by IV, and define $y_0 = \text{IV}$. Then we construct $y_1, y_2, \ldots$ using the rule

$$y_i = e_K(y_{i-1} \oplus x_i),$$

$i \geq 1$.

Suppose that $(\mathcal{P}, \mathcal{C}, \mathcal{K}, \mathcal{E}, \mathcal{D})$ is an endomorphic cryptosystem, where $\mathcal{P} = \mathcal{C} = \{0,1\}^t$. Let IV be the bitstring consisting of t zeroes, and let $K \in \mathcal{K}$ be a secret key. Finally, let $x = x_1 \parallel \cdots \parallel x_n$ be a bitstring of length tn (for some positive integer n), where each x_i is a bitstring of length t. We compute CBC-MAC(x, K) as follows..

Cryptosystem 4.2: *CBC-MAC(x, K)*

denote $x = x_1 \parallel \cdots \parallel x_n$
IV $\leftarrow 00\cdots 0$
$y_0 \leftarrow$ IV
for $i \leftarrow 1$ **to** n
  do $y_i \leftarrow e_K(y_{i-1} \oplus x_i)$
return (y_n)

The best known general attack on *CBC-MAC* is a birthday (collision) attack. We describe this attack now. Basically, we assume that the adversary can request

MACs on a large number of messages. If a duplicated MAC is found, then the adversary can construct an additional message and request its MAC. Finally, the adversary can produce a new message and its corresponding MAC (i.e., a forgery), even though he does not know the secret key. The attack works for messages of any prespecified fixed size.

Here are the details of the attack. Let $n \geq 2$ be an integer. Let $x_3, \ldots, x_n$ be fixed bitstrings of length t. Let $q \approx 1.17 \times 2^{t/2}$ be an integer, and choose any q distinct bitstrings of length t, which we denote $x_1^1, \ldots, x_1^q$. Now let $x_2^1, \ldots, x_2^q$ be randomly chosen bitstrings of length t, and define

$$x^i = x_1^i \parallel \cdots \parallel x_n^i$$

for $1 \leq i \leq q$. Note that $x^i \neq x^j$ if $i \neq j$, because $x_1^i \neq x_1^j$.

Now the adversary requests the MACs of $x^1, x^2, \ldots, x^q$. Note that $h_K(x^i) = h_K(x^j)$ if and only if $y_2^i = y_2^j$, which happens if and only if

$$y_1^i \oplus x_2^i = y_1^j \oplus x_2^j.$$

Let x_δ be any bitstring of length t. Define

$$v = x_1^i \parallel (x_2^i \oplus x_\delta) \parallel \cdots \parallel x_n^i$$

and

$$w = x_1^j \parallel (x_2^j \oplus x_\delta) \parallel \cdots \parallel x_n^j.$$

Then the adversary requests the MAC of v. It is not difficult to see that v and w have identical MACs, so the adversary is successfully able to construct the MAC of w even though he does not know the key K. This attack produces a $(1/2, O(2^{t/2}))$-forger.

It is known that *CBC-MAC* is secure if the underlying encryption satisfies appropriate security properties. That is, if certain plausible but unproved assumptions about the randomness of an encryption scheme are true, then *CBC-MAC* will be secure.

4.5 Unconditionally Secure MACs

In this section, we define universal hash families and discuss their application to the construction of unconditionally secure MACs. In our study of unconditionally secure MACs, we assume that a key is used to produce only one authentication tag. Therefore, an adversary is limited to making at most one query before he outputs a (possible) forgery. Stated another way, we will construct MACs for which we can prove that there do not exist $(0, \epsilon)$- and $(1, \epsilon)$-forgers, for appropriate values of ϵ, even when the adversary possesses infinite computing power.

For $q = 0, 1$, we define the *deception probability* Pd_q to be the maximum value of ϵ such that an (ϵ, q)-forger exists. This maximum is computed over all possible values of the secret key, K, assuming that K is chosen uniformly at random from $\mathcal{K}$. In general, we want to construct MACs for which Pd_0 and Pd_1 are small. This means that the adversary has a low probability of successfully carrying out an attack, no matter which key is used. Sometimes we will refer to the attack carried out by an $(\epsilon, 0)$-forger as *impersonation*, and the attack carried out by an $(\epsilon, 1)$-forger will be termed *substitution*.

We illustrate the above concepts by considering a small example of an unconditionally secure MAC.

Example 4.1 Suppose
$$\mathcal{X} = \mathcal{Y} = \mathbb{Z}_3$$
and
$$\mathcal{K} = \mathbb{Z}_3 \times \mathbb{Z}_3.$$
For each $K = (a, b) \in \mathcal{K}$ and each $x \in \mathcal{X}$, define
$$h_{(a,b)}(x) = ax + b \bmod 3,$$
and then define
$$\mathcal{H} = \{h_{(a,b)} : (a, b) \in \mathbb{Z}_3 \times \mathbb{Z}_3\}.$$
It will be useful to study the *authentication matrix* of the hash family $(\mathcal{X}, \mathcal{Y}, \mathcal{K}, \mathcal{H})$, which tabulates all the values $h_{(a,b)}(x)$ as follows. For each key $(a, b) \in \mathcal{K}$ and for each $x \in \mathcal{X}$, place the authentication tag $h_{(a,b)}(x)$ in row (a, b) and column x of a $|\mathcal{K}| \times |\mathcal{X}|$ matrix, say M. The array M is presented in Figure 4.1.

Let's first consider an impersonation attack. Oscar will pick a message x, and attempt to guess the "correct" authentication tag. Denote by K_0 the actual key being used (which is unknown to Oscar). Oscar will succeed in creating a forgery if he guesses the tag $y_0 = h_{K_0}(x)$. However, for any $x \in \mathcal{X}$ and $y \in \mathcal{Y}$, it is easy to verify that there are exactly three (out of nine) keys $K \in \mathcal{K}$ such that $h_K(x) = y$. (In other words, each symbol occurs three times in each column of the authentication matrix.) Thus, any pair (x, y) will be a valid pair with probability $1/3$. Hence, it follows that $Pd_0 = 1/3$.

Substitution is a bit more complicated to analyze. As a specific case, suppose Oscar queries the tag for $x = 0$, and is given the answer $y = 0$. Therefore $(x, y) = (0, 0)$ is a valid pair. This gives Oscar some information about the key: he knows that
$$K_0 \in \{(0, 0), (1, 0), (2, 0)\}.$$
Now suppose Oscar outputs the pair $(1, 1)$ as a (possible) forgery. The pair $(1, 1)$ is a forgery if and only if $K_0 = (1, 0)$. The (conditional) probability that K_0 is the key, given that $(0, 0)$ is a valid pair, is $1/3$, since the key is known to be in the set $\{(0, 0), (1, 0), (2, 0)\}$.

FIGURE 4.1
An authentication matrix

key	0	1	2
$(0,0)$	0	0	0
$(0,1)$	1	1	1
$(0,2)$	2	2	2
$(1,0)$	0	1	2
$(1,1)$	1	2	0
$(1,2)$	2	0	1
$(2,0)$	0	2	1
$(2,1)$	1	0	2
$(2,2)$	2	1	0

A similar analysis can be done for any value of x for which Oscar queries the tag y, and for any pair (x', y') (where $x' \neq x$) that Oscar outputs as his (possible) forgery. In general, knowledge of any valid pair (x, y) restricts the key to one of three possibilities. Then, for each choice of (x', y') (where $x' \neq x$), it can be verified that there is one key (out of the three possible keys) under which y' is the correct authentication tag for x'. Hence, it follows that $Pd_1 = 1/3$.   □

We now discuss how to compute the deception probabilities for an arbitrary message authentication code by examining its authentication matrix. First, we consider Pd_0. As above, let K_0 denote the key chosen by Alice and Bob. For $x \in X$ and $y \in Y$, define payoff(x, y) to be the probability that the pair (x, y) is valid. It is not difficult to see that

$$\text{payoff}(x, y) = \mathbf{Pr}[y = h_{K_0}(x)]$$
$$= \frac{|\{K \in \mathcal{K} : h_K(x) = y\}|}{|\mathcal{K}|}.$$

That is, payoff(x, y) is computed by counting the number of rows of the authentication matrix that have entry y in column x, and dividing the result by the number of possible keys.

In order to maximize his chance of success, Oscar will choose (x, y) such that payoff(x, y) is a maximum. Hence, we have the following formula:

$$Pd_0 = \max\{\text{payoff}(x, y) : x \in X, y \in Y\}. \tag{4.1}$$

Now, we turn our attention to substitution. Suppose we fix $x \in X$ and $y \in Y$ such that (x, y) is a valid pair. (x is a possible query by Oscar, and y would be the oracle's reply.) Now let $x' \in \mathcal{X}$, where $x' \neq x$. Define $\mathsf{payoff}(x', y'; x, y)$ to be the probability that (x', y') is a valid pair, given that (x, y) is a valid pair. As before, let K_0 denote the key chosen by Alice and Bob. Then we can compute the following:

$$
\begin{aligned}
\mathsf{payoff}(x', y'; x, y) &= \mathbf{Pr}[(y' = h_{K_0}(x')|y = h_{K_0}(x)] \\
&= \frac{\mathbf{Pr}[(y' = h_{K_0}(x') \wedge y = h_{K_0}(x)]}{\mathbf{Pr}[y = h_{K_0}(x)]} \\
&= \frac{|\{K \in \mathcal{K} : h_K(x') = y', h_K(x) = y\}|}{|\{K \in \mathcal{K} : h_K(x) = y\}|}.
\end{aligned}
$$

The numerator of this fraction is the number of rows of the authentication matrix that have the value y in column x, and also have the value y' in column x'; the denominator is the number of rows that have the value y in column x. Note that the denominator is non-zero because we are assuming that (x, y) is a valid pair under at least one key.

Given the valid pair (x, y), Oscar will choose (x', y') such that $\mathsf{payoff}(x, y)$ is a maximum. We want to minimize Oscar's success probability (of performing a successful substitution) over all possible valid pairs (x, y). Therefore, we define

$$
\mathcal{V} = \{(x, y) : |\{K \in \mathcal{K} : h_K(x) = y\}| \geq 1\}.
$$

$\mathcal{V}$ is just the set of all pairs that are valid pairs under at least one key.

We have the following formula:

$$
Pd_1 = \max\{\mathsf{payoff}(x', y'; x, y) : x, x' \in \mathcal{X}, y, y' \in \mathcal{Y}, (x, y) \in \mathcal{V}, x \neq x'\}.
$$

$$(4.2)$$

4.5.1 Strongly Universal Hash Families

Strongly universal hash families are used in several areas of cryptography. We begin with a definition of these important objects.

Definition 4.2:   Suppose that $(\mathcal{X}, \mathcal{Y}, \mathcal{K}, \mathcal{H})$ is an (N, M) hash family. This hash family is *strongly universal* provided that the following condition is satisfied for every $x, x' \in \mathcal{X}$ such that $x \neq x'$, and for every $y, y' \in \mathcal{Y}$:

$$
|\{K \in \mathcal{K} : h_K(x) = y, h_K(x') = y'\}| = \frac{|\mathcal{K}|}{M^2}.
$$

As an example, the reader can verify that the hash family in Example 4.1 is a strongly universal $(3, 3)$-hash family.

Strongly universal hash families immediately yield authentication codes in which Pd_0 and Pd_1 can easily be computed. We prove a theorem on the values of these deception probabilities after stating and proving a simple lemma about strongly universal hash families.

LEMMA 4.10 *Suppose that* $(\mathcal{X}, \mathcal{Y}, \mathcal{K}, \mathcal{H})$ *is a strongly universal* (N, M)-*hash family. Then*

$$|\{K \in \mathcal{K} : h_K(x) = y\}| = \frac{|\mathcal{K}|}{M},$$

for every $x \in \mathcal{X}$ *and for every* $y \in \mathcal{Y}$.

PROOF Let $x, x' \in \mathcal{X}$ and $y \in \mathcal{Y}$ be fixed, where $x \neq x'$. Then we have the following:

$$|\{K \in \mathcal{K} : h_K(x) = y\}| = \sum_{y' \in \mathcal{Y}} |\{K \in \mathcal{K} : h_K(x) = y, h_K(x') = y'\}|$$

$$= \sum_{y' \in \mathcal{Y}} \frac{|\mathcal{K}|}{M^2}$$

$$= \frac{|\mathcal{K}|}{M}.$$

∎

THEOREM 4.11 *Suppose that* $(\mathcal{X}, \mathcal{Y}, \mathcal{K}, \mathcal{H})$ *is a strongly universal* (N, M)-*hash family. Then* $(\mathcal{X}, \mathcal{Y}, \mathcal{K}, \mathcal{H})$ *is an authentication code with* $Pd_0 = Pd_1 = 1/M$.

PROOF We proved in Lemma 4.10 that

$$|\{K \in \mathcal{K} : h_K(x) = y\}| = \frac{|\mathcal{K}|}{M},$$

for every $x \in \mathcal{X}$ and for every $y \in \mathcal{Y}$. Therefore $\text{payoff}(x, y) = 1/M$ for every $x \in \mathcal{X}, y \in \mathcal{Y}$, and hence $Pd_0 = 1/M$.

Now let $x, x' \in \mathcal{X}$ such that $x \neq x'$, and let $y, y' \in \mathcal{Y}$, where $(x, y) \in \mathcal{V}$. We have that

$$\text{payoff}(x', y'; x, y) = \frac{|\{K \in \mathcal{K} : h_K(x') = y', h_K(x) = y\}|}{|\{K \in \mathcal{K} : h_K(x) = y\}|}$$

$$= \frac{|\mathcal{K}|/M^2}{|\mathcal{K}|/M}$$

$$= \frac{1}{M}.$$

Therefore $Pd_1 = 1/M$.

∎

We now give some constructions of strongly universal hash families. Our first construction generalizes Example 4.1.

THEOREM 4.12 *Let p be prime. For $a, b \in \mathbb{Z}_p$, define $f_{a,b} : \mathbb{Z}_p \to \mathbb{Z}_p$ by the rule*

$$f_{(a,b)}(x) = ax + b \bmod p.$$

Then $(\mathbb{Z}_p, \mathbb{Z}_p, \mathbb{Z}_p \times \mathbb{Z}_p, \{f_{(a,b)} : a, b \in \mathbb{Z}_p\})$ is a strongly universal (p, p)-hash family.

PROOF   Suppose that $x, x', y, y' \in \mathbb{Z}_p$, where $x \neq x'$. We will show that there is a unique key $(a, b) \in \mathbb{Z}_p \times \mathbb{Z}_p$ such that $ax + b \equiv y \pmod{p}$ and $ax' + b \equiv y'$ $\pmod{p}$. This is not difficult, as (a, b) is the solution of a system of two linear equations in two unknowns over $\mathbb{Z}_p$. Specifically,

$$a = (y' - y)(x' - x)^{-1} \bmod p, \quad \text{and}$$

$$b = y - x(y' - y)(x' - x)^{-1} \bmod p.$$

(Note that $(x' - x)^{-1} \bmod p$ exists because $x \not\equiv x' \pmod{p}$ and p is prime.)   ∎

Here is a construction for classes of strongly universal hash families in which the domain can have much larger cardinality than the range.

THEOREM 4.13   *Let ℓ be a positive integer and let p be prime. Define*

$$\mathcal{X} = \{0, 1\}^{\ell} \backslash \{(0, \dots, 0)\}.$$

For every $\vec{r} \in (\mathbb{Z}_p)^{\ell}$, define $f_{\vec{r}} : \mathcal{X} \to \mathbb{Z}_p$ by the rule

$$f_{\vec{r}}(\vec{x}) = \vec{r} \cdot \vec{x} \bmod p,$$

where $\vec{x} \in \mathcal{X}$ and

$$\vec{r} \cdot \vec{x} = \sum_{i=1}^{\ell} r_i x_i$$

is the usual inner product of vectors. Then $(\mathcal{X}, \mathbb{Z}_p, (\mathbb{Z}_p)^{\ell}, \{f_{\vec{r}} : \vec{r} \in (\mathbb{Z}_p)^{\ell}\})$ is a strongly universal $(2^{\ell} - 1, p)$-hash family.

PROOF   Let $\vec{x}, \vec{x}' \in \mathcal{X}$, $\vec{x} \neq \vec{x}'$, and let $y, y' \in \mathbb{Z}_p$. We want to show that the number of vectors $\vec{r} \in (\mathbb{Z}_p)^{\ell}$ such that $\vec{r} \cdot \vec{x} \equiv y \pmod{p}$ and $\vec{r} \cdot \vec{x}' \equiv y' \pmod{p}$ is a constant. The desired vectors $\vec{r}$ are the solution of two linear equations in ℓ unknowns over $\mathbb{Z}_p$. The two equations are linearly independent, and so the number of solutions to the linear system is $p^{\ell-2}$, which is a constant.   ∎

4.5.2   Optimality of Deception Probabilities

In this section, we prove some lower bounds on deception probabilities of unconditionally secure MACs, which show that the authentication codes derived from strongly universal hash families have minimum possible deception probabilities.

Suppose $(\mathcal{X}, \mathcal{Y}, \mathcal{K}, \mathcal{H})$ is an (N, M)-hash family. Suppose we fix a message $x \in \mathcal{X}$. Then we can compute as follows:

$$
\begin{aligned}
\sum_{y \in \mathcal{Y}} \mathsf{payoff}(x, y) &= \sum_{y \in \mathcal{Y}} \frac{|\{K \in \mathcal{K} : h_K(x) = y\}|}{|\mathcal{K}|} \\
&= \frac{|\mathcal{K}|}{|\mathcal{K}|} \\
&= 1.
\end{aligned}
$$

Hence, for every $x \in \mathcal{X}$, there exists an authentication tag y (depending on x), such that

$$
\mathsf{payoff}(x, y) \geq \frac{1}{M}.
$$

The following theorem is an easy consequence of the above computations.

THEOREM 4.14 *Suppose $(\mathcal{X}, \mathcal{Y}, \mathcal{K}, \mathcal{H})$ is an (N, M)-hash family. Then $Pd_0 \geq 1/M$. Further, $Pd_0 = 1/M$ if and only if*

$$
|\{K \in \mathcal{K} : h_K(x) = y\}| = \frac{|\mathcal{K}|}{M} \tag{4.3}
$$

for every $x \in \mathcal{X}$, $y \in \mathcal{Y}$.

Now, we turn our attention to substitution. Suppose that we fix $x, x' \in \mathcal{X}$ and $y, y' \in \mathcal{Y}$, where $x \neq x'$ and $(x, y) \in \mathcal{V}$. We have the following:

$$
\begin{aligned}
\sum_{y' \in \mathcal{Y}} \mathsf{payoff}(x', y'; x, y) &= \sum_{y' \in \mathcal{Y}} \frac{|\{K \in \mathcal{K} : h_K(x') = y', h_K(x) = y\}|}{|\{K \in \mathcal{K} : h_K(x) = y\}|} \\
&= \frac{|\{K \in \mathcal{K} : h_K(x) = y\}|}{|\{K \in \mathcal{K} : h_K(x) = y\}|} \\
&= 1.
\end{aligned}
$$

Hence, for each $(x, y) \in \mathcal{V}$ and for each x' such that $x' \neq x$, there exists an authentication tag y' such that

$$
\mathsf{payoff}(x', y'; x, y) \geq \frac{1}{M}.
$$

We have proven the following theorem.

THEOREM 4.15 *Suppose $(\mathcal{X}, \mathcal{Y}, \mathcal{K}, \mathcal{H})$ is an (N, M)-hash family. Then $Pd_1 \geq 1/M$.*

With a bit more work, we can determine necessary and sufficient conditions such that $Pd_1 = 1/M$.

THEOREM 4.16   *Suppose* $(\mathcal{X}, \mathcal{Y}, \mathcal{K}, \mathcal{H})$ *is an* (N, M)-*hash family. Then* $Pd_1 = 1/M$ *if and only if the hash family is strongly universal.*

PROOF   We proved already in Theorem 4.11 that $Pd_1 = 1/M$ if the hash family is strongly universal. We need to prove the converse now; so, we assume that $Pd_1 = 1/M$.

We will show first that $\mathcal{V} = \mathcal{X} \times \mathcal{Y}$. Let $(x', y') \in \mathcal{X} \times \mathcal{Y}$; we will show that $(x', y') \in \mathcal{V}$. Let $x \in \mathcal{X}$, $x \neq x'$. Choose any $y \in \mathcal{Y}$ such that $(x, y) \in \mathcal{V}$. From the discussion preceding Theorem 4.15, it is clear that

$$\frac{|\{K \in \mathcal{K} : h_K(x') = y', h_K(x) = y\}|}{|\{K \in \mathcal{K} : h_K(x) = y\}|} = \frac{1}{M} \tag{4.4}$$

for every $x, x' \in \mathcal{X}$, $x' \neq x$, $y, y' \in \mathcal{Y}$ such that $(x, y) \in \mathcal{V}$. Therefore

$$|\{K \in \mathcal{K} : h_K(x') = y', h_K(x) = y\}| > 0,$$

and hence

$$|\{K \in \mathcal{K} : h_K(x') = y'\}| > 0.$$

This proves that $(x', y') \in \mathcal{V}$, and hence $\mathcal{V} = \mathcal{X} \times \mathcal{Y}$.

Now, let's look again at (4.4). Let $x, x' \in \mathcal{X}$, $x \neq x'$, and let $y, y' \in \mathcal{Y}$. We now know that $(x, y) \in \mathcal{V}$ and $(x', y') \in \mathcal{V}$, so we can interchange the roles of (x, y) and (x', y') in (4.4). This yields

$$|\{K \in \mathcal{K} : h_K(x) = y\}| = |\{K \in \mathcal{K} : h_K(x') = y'\}|$$

for all such x, x', y, y'. Hence, the quantity

$$|\{K \in \mathcal{K} : h_K(x) = y\}|$$

is a constant. (In other words, the number of occurrences of any symbol y in any column x of the authentication matrix x is a constant.) Now, we can return one last time to (4.4), and it follows that the quantity

$$|\{K \in \mathcal{K} : h_K(x') = y', h_K(x) = y\}|$$

is also a constant. Therefore the hash family is strongly universal. ∎

The following corollary establishes that $Pd_0 = 1/M$ whenever $Pd_1 = 1/M$.

COROLLARY 4.17   *Suppose* $(\mathcal{X}, \mathcal{Y}, \mathcal{K}, \mathcal{H})$ *is an* (N, M)-*hash family such that* $Pd_1 = 1/M$. *Then* $Pd_0 = 1/M$.

PROOF   Under the stated hypotheses, Theorem 4.16 says that $(\mathcal{X}, \mathcal{Y}, \mathcal{K}, \mathcal{H})$ is strongly universal. Then $Pd_0 = 1/M$ from Theorem 4.11. ∎

4.6 Notes and References

For a good recent survey on hash functions, see Preneel [173]. Concepts such as preimage resistance and collision resistance have been discussed for some time; see [173] for further datails.

The random oracle model was introduced by Bellare and Rogaway in [14]; the analyses in Section 4.2.2 are based on Stinson [207].

The material from Section 4.3 is based on Damgård [56]. Similar methods were discovered by Merkle [146].

Rivest's *MD4* and *MD5* hashing algorithms are described in [179] and [180], respectively. *SHA* is the subject of FIPS publication 180 [76]; it was superceded by *SHA-1*, which is described in FIPS publication 180-1 [77]. A draft version exists of FIPS publication 180-2 [78]; this specifies the hash algorithms *SHA-256*, *SHA-384* and *SHA-512*.

An attack against two of the three rounds of *MD4* is given by den Boer and Bossalaers [61]. Later, Dobbertin [69] found collisions for *MD4*. Collisions for the compression function of *MD5* were found by den Boer and Bossalaers [62].

An algorithm to find collisions in *SHA* in roughly 2^{61} operations was found by Chabaud and Joux in 1998 [44]. By that time, *SHA* had already been upgraded to *SHA-1*, perhaps because of similar classified attacks found earlier by the NSA.

The use of cipher block chaining mode for message authentication is specified in FIPS publication 113 [75] in 1985. The security proof for this method of constructing MACs came much later: see Bellare, Kilian and Rogaway [13]. Preneel and van Oorschot [175] is a recent study of iterated message authentication codes. Provably secure methods of constructing MACs include *HMAC* (which we described in Section 4.4.1), due to Bellare, Canetti and Krawczyk [11]; and *XOR-MAC*, due to Bellare, Guerin and Rogaway [12].

Unconditionally secure authentication codes were invented in 1974 by Gilbert, MacWilliams and Sloane [92]. Much of the theory of unconditionally secureauthentication codes was developed by Simmons, who proved many fundamental results in the area; Simmons [195] is a good survey. Our treatment differs from the model introduced by Simmons in that we are considering active attacks, in which the adversary queries an oracle for an authentication tag before producing a (possible) forgery. In the model considered by Simmons, the attacks are passive attacks: the adversary observes a message with a corresponding authentication tag, but this message is not chosen by the adversary.

Universal hash families were introduced by Carter and Wegman [42, 212]. The paper [212] was the first to apply strongly universal hash families to authentication; it also introduces the concept of *almost strongly universal hash families*, which allow the key length of unconditionally secure MACs to be greatly reduced. For more on this and related topics, see the expository paper by Stinson [206]. Finally, we note that universal hash families are also used in the construction of efficient computationally secure MACs; one such MAC is *UMAC*, which is described in Black *et al.* [25].

Exercises

4.1 Suppose $h : \mathcal{X} \to \mathcal{Y}$ is an (N, M)-hash function. For any $y \in \mathcal{Y}$, let
$$h^{-1}(y) = \{x : h(x) = y\}$$
and denote $s_y = |h^{-1}(y)|$. Define
$$S = |\{\{x_1, x_2\} : h(x_1) = h(x_2)\}|.$$

Note that S counts the number of unordered pairs in $\mathcal{X}$ that collide under h.

(a) Prove that
$$\sum_{y \in \mathcal{Y}} s_y = N,$$
so the mean of the s_y's is
$$\overline{s} = \frac{N}{M}.$$

(b) Prove that
$$S = \sum_{y \in \mathcal{Y}} \binom{s_y}{2} = \frac{1}{2} \sum_{y \in \mathcal{Y}} s_y^2 - \frac{N}{2}.$$

(c) Prove that
$$\sum_{y \in \mathcal{Y}} (s_y - \overline{s})^2 = 2N + N - \frac{N^2}{M}.$$

(d) Using the result proved in part (c), prove that
$$S \geq \frac{1}{2} \left(\frac{N^2}{M} - N \right).$$

Further, show that equality is attained if and only if
$$s_y = \frac{N}{M}$$
for every $y \in \mathcal{Y}$.

4.2 As in Exercise 4.1, suppose $h : \mathcal{X} \to \mathcal{Y}$ is an (N, M)-hash function, and let
$$h^{-1}(y) = \{x : h(x) = y\}$$
for any $y \in \mathcal{Y}$. Let ϵ denote the probability that $h(x_1) = h(x_2)$, where x_1 and x_2 are random (not necessarily distinct) elements of $\mathcal{X}$. Prove that
$$\epsilon \geq \frac{1}{M},$$
with equality if and only if
$$|h^{-1}(y)| = \frac{N}{M}$$
for every $y \in \mathcal{Y}$.

4.3 Suppose that $h : \mathcal{X} \to \mathcal{Y}$ is an (N, M)-hash function, let
$$h^{-1}(y) = \{x : h(x) = y\}$$

and let $s_y = |h^{-1}(y)|$ for any $y \in \mathcal{Y}$. Suppose that we try to solve **Preimage** for the function h, using Algorithm 4.1, assuming that we have only oracle access for h. For a given $y \in \mathcal{Y}$, suppose that $\mathcal{X}_0$ is chosen to be a random subset of $\mathcal{X}$ having cardinality q.

(a) Prove that the success probability of Algorithm 4.1, given y, is

$$1 - \frac{\binom{N-s_y}{q}}{\binom{N}{q}}.$$

(b) Prove that the average success probabilty of Algorithm 4.1 (over all $y \in \mathcal{Y}$) is

$$1 - \frac{1}{M} \sum_{y \in \mathcal{Y}} \frac{\binom{N-s_y}{q}}{\binom{N}{q}}.$$

(c) In the case $q = 1$, show that the success probability in part (b) is $1/M$.

4.4 Suppose that $h : \mathcal{X} \to \mathcal{Y}$ is an (N, M)-hash function, let

$$h^{-1}(y) = \{x : h(x) = y\}$$

and let $s_y = |h^{-1}(y)|$ for any $y \in \mathcal{Y}$. Suppose that we try to solve **Second Preimage** for the function h, using Algorithm 4.2, assuming that we have only oracle access for h. For a given $x \in \mathcal{Y}$, suppose that $\mathcal{X}_0$ is chosen to be a random subset of $\mathcal{X}\backslash\{x\}$ having cardinality $q - 1$.

(a) Prove that the success probability of Algorithm 4.2, given x, is

$$1 - \frac{\binom{N-s_y}{q-1}}{\binom{N-1}{q-1}}.$$

(b) Prove that the average success probabilty of Algorithm 4.1 (over all $x \in \mathcal{X}$) is

$$1 - \frac{1}{N} \sum_{y \in \mathcal{Y}} \frac{s_y \binom{N-s_y}{q-1}}{\binom{N-1}{q-1}}.$$

(c) In the case $q = 2$, show that the success probability in part (b) is

$$\frac{\sum_{y \in \mathcal{Y}} s_y^{\,2}}{N(N-1)} - \frac{1}{N-1}.$$

4.5 If we define a hash function (or compression function) h that will hash an n-bit binary string to an m-bit binary string, we can view h as a function from $\mathbb{Z}_{2^n}$ to $\mathbb{Z}_{2^m}$. It is tempting to define h using integer operations modulo 2^m. We show in this exercise that some simple constructions of this type are insecure and should therefore be avoided.

(a) Suppose that $n = m$ and $h : \mathbb{Z}_{2^m} \to \mathbb{Z}_{2^m}$ is defined as

$$h(x) = x^2 + ax + b \bmod 2^m.$$

Prove that it is easy to solve **Second Preimage** for any $x \in \mathbb{Z}_{2^m}$ without having to solve a quadratic equation.

(b) Suppose that $n > m$ and $h : \mathbb{Z}_{2^n} \to \mathbb{Z}_{2^m}$ is defined to be a polynomial of degree d:

$$h(x) = \sum_{i=0}^{d} a_i x^i \bmod 2^m,$$

where $a_i \in \mathbb{Z}$ for $0 \le i \le d$. Prove that it is easy to solve **Second Preimage** for any $x \in \mathbb{Z}_{2^n}$ without having to solve a polynomial equation.

4.6 Suppose that $f : \{0, 1\}^m \rightarrow \{0, 1\}^m$ is a preimage resistant bijection. Define $h : \{0, 1\}^{2m} \rightarrow \{0, 1\}^m$ as follows. Given $x \in \{0, 1\}^{2m}$, write

$$x = x' \parallel x''$$

where $x', x'' \in \{0, 1\}^m$. Then define

$$h(x) = f(x' \oplus x'').$$

Prove that h is not second preimage resistant.

4.7 For $M = 365$ and $15 \leq q \leq 30$, compare the exact value of ϵ given by the formula in the statement of Theorem 4.4 with the estimate for ϵ derived in the proof of that theorem.

4.8 Suppose $h : \mathcal{X} \rightarrow \mathcal{Y}$ is a hash function where $|\mathcal{X}|$ and $|\mathcal{Y}|$ are finite and $|\mathcal{X}| \geq 2|\mathcal{Y}|$. Suppose that H is balanced (i.e.,

$$|h^{-1}(y)| = \frac{|\mathcal{X}|}{|\mathcal{Y}|}$$

for all $y \in \mathcal{Y}$). Finally, suppose ORACLEPREIMAGE is an (ϵ, q)-algorithm for **Preimage**, for the fixed hash function h. Prove that COLLISIONTOPREIMAGE is an $(\epsilon/2, q + 1)$-algorithm for **Collision**, for the fixed hash function h.

4.9 Suppose $h_1 : \{0, 1\}^{2m} \rightarrow \{0, 1\}^m$ is a collision resistant hash function.

   (a) Define $h_2 : \{0, 1\}^{4m} \rightarrow \{0, 1\}^m$ as follows:

      1. Write $x \in \{0, 1\}^{4m}$ as $x = x_1 \parallel x_2$, where $x_1, x_2 \in \{0, 1\}^{2m}$.
      2. Define $h_2(x) = h_1(h_1(x_1) \parallel h_1(x_2))$.

      Prove that h_2 is collision resistant.

   (b) For an integer $i \geq 2$, define a hash function $h_i : \{0, 1\}^{2^i m} \rightarrow \{0, 1\}^m$ recursively from h_{i-1}, as follows:

      1. Write $x \in \{0, 1\}^{2^i m}$ as $x = x_1 \parallel x_2$, where $x_1, x_2 \in \{0, 1\}^{2^{i-1} m}$.
      2. Define $h_i(x) = h_1(h_{i-1}(x_1) \parallel h_{i-1}(x_2))$.

      Prove that h_i is collision resistant.

4.10 In this exercise, we consider a simplified version of the Merkle-Damgård construction. Suppose

$$\text{compress} : \{0, 1\}^{m+t} \rightarrow \{0, 1\}^m,$$

where $t \geq 1$, and suppose that

$$x = x_1 \parallel x_2 \parallel \cdots \parallel x_k,$$

where

$$|x_1| = |x_2| = \cdots = |x_k| = t.$$

We study the following iterated hash function:

Algorithm 4.9: SIMPLIFIED MERKLE-DAMGÅRD(x, k, t)

external compress
$z_1 \leftarrow 0^m \parallel x_1$
$g_1 \leftarrow \text{compress}(z_1)$
for $i \leftarrow 1$ **to** $k - 1$
$\quad$ **do** $\begin{cases} z_{i+1} \leftarrow g_i \parallel x_{i+1} \\ g_{i+1} \leftarrow \text{compress}(z_{i+1}) \end{cases}$
$h(x) \leftarrow g_{k+1}$
return $(h(x))$

Suppose that compress is collision resistant, and suppose further that compress is zero preimage resistant, which means that it is hard to find $z \in \{0, 1\}^{m+t}$ such that compress$(z) = 0^m$. Under these assumptions, prove that h is collision resistant.

4.11 A message authentication code can be produced by using a block cipher in CFB mode instead of CBC mode. Given a sequence of plaintext blocks, $x_1 \cdots x_n$, suppose we define the initialization vector IV to be x_1. Then encrypt the sequence $x_2 \cdots x_n$ using key K in CFB mode, obtaining the ciphertext sequence $y_1 \cdots y_{n-1}$ (note that there are only $n - 1$ ciphertext blocks). Finally, define the MAC to be $e_K(y_{n-1})$. Prove that this MAC is identical to the MAC produced in Section 3.7 using CBC mode.

4.12 Suppose that $(\mathcal{P}, \mathcal{C}, \mathcal{K}, \mathcal{E}, \mathcal{D})$ is an endomorphic cryptosystem with $\mathcal{P} = \mathcal{C} = \{0, 1\}^m$. Let $n \geq$ be an integer, and define a hash family $(\mathcal{X}, \mathcal{Y}, \mathcal{K}, \mathcal{H})$, where $\mathcal{X} = (\{0, 1\}^m)^n$ and $\mathcal{Y} = \{0, 1\}^m$, as follows:

$$h_K(y_1, \ldots y_n) = e_K(y_1) \oplus \cdots \oplus e_K(y_n).$$

Prove that $(\mathcal{X}, \mathcal{Y}, \mathcal{K}, \mathcal{H})$ is not a secure message authentication code as follows.

(a) Prove the existence of a $(1, 1)$-forger for this hash family.

(b) Prove the existence of a $(1, 2)$-forger for this hash family which can forge the MAC for an arbitrary message $(y_1, \ldots, y_n)$ (this is called a *selective forgery*; the forgeries previously considered are examples of *existential forgeries*). Note that the difficult case is when $y_1 = \cdots = y_n$.

4.13 Suppose that $(\mathcal{P}, \mathcal{C}, \mathcal{K}, \mathcal{E}, \mathcal{D})$ is an endomorphic cryptosystem with $\mathcal{P} = \mathcal{C} = \{0, 1\}^m$. Let $n \geq$ be an integer, and define a hash family $(\mathcal{X}, \mathcal{Y}, \mathcal{K}, \mathcal{H})$, where $\mathcal{X} = (\{0, 1\}^m)^n$ and $\mathcal{Y} = \{0, 1\}^m$, as follows:

$$h_K(y_1, \ldots, y_n) = e_K(y_1) + 3e_K(y_2) + \cdots + (2n - 1)e_K(y_n) \bmod 2^m.$$

(a) When n is odd, prove the existence of a $(1, 2)$-forger for this hash family.

(b) When $n = 2$, prove the existence of a $(1/8, 2)$-forger for this hash family, as follows:

1. Request the MACs of (x, y) and (y, x). Suppose that $a = h_K(x, y)$ and $b = h_K(y, x)$.

2. Show that there are exactly eight ordered pairs (x', y') such that $x' = e_K(x)$, $y' = e_K(y)$ is consistent with the given MAC values a and b.

3. Choose one of these eight values for x' at random, and output the possible forgery $(x, x), x'$. Prove that this is a valid forgery with probability $1/8$.

(c) Prove the existence of a $(1, 3)$-forger for this hash family which can forge the MAC for an arbitrary message $(y_1, \ldots, y_n)$.

4.14 Suppose that $(\mathcal{X}, \mathcal{Y}, \mathcal{K}, \mathcal{H})$ is a strongly universal (N, M)-hash family.

(a) If $|\mathcal{K}| = M^2$, show that there exists a $(1, 2)$-forger for this hash family (i.e., $Pd_2 = 1$).

(b) (This generalizes the result proven in part (a).) Denote $\lambda = |\mathcal{K}|/M^2$. Prove there exists a $(1/\lambda, 2)$-forger for this hash family (i.e., $Pd_2 \geq 1/\lambda$).

4.15 Compute Pd_0 and Pd_1 for the following authentication code, represented in matrix form:

key	1	2	3	4
1	1	1	2	3
2	1	2	3	1
3	2	1	3	1
4	2	3	1	2
5	3	2	1	3
6	3	3	2	1

4.16 Let p be an odd prime. For $a, b \in \mathbb{Z}_p$, define $f_{(a,b)} : \mathbb{Z}_p \to \mathbb{Z}_p$ by the rule

$$f_{(a,b)}(x) = (x + a)^2 + b \bmod p.$$

Prove that $(\mathbb{Z}_p, \mathbb{Z}_p, \mathbb{Z}_p \times \mathbb{Z}_p, \{f_{(a,b)} : a, b \in \mathbb{Z}_p\})$ is a strongly universal (p, p)-hash family.

4.17 Let $k \geq 1$ be an integer. An (N, M) hash family, $(\mathcal{X}, \mathcal{Y}, \mathcal{K}, \mathcal{H})$, is *strongly k-universal* provided that the following condition is satisfied for all choices of k distinct elements $x_1, x_2, \ldots, x_k \in \mathcal{X}$ and for all choices of k (not necessarily distinct) elements $y_1, \ldots, y_k \in \mathcal{Y}$:

$$|\{K \in \mathcal{K} : h_K(x_i) = y_1 \text{ for } 1 \leq i \leq k\}| = \frac{|\mathcal{K}|}{M^k}.$$

(a) Prove that a strongly k-universal hash family is strongly ℓ-universal for all ℓ such that $1 \leq \ell \leq k$.

(b) Let p be prime and let $k \geq 1$ be an integer. For all k-tuples $(a_0, \ldots, a_{k-1}) \in (\mathbb{Z}_p)^k$, define $f_{(a_0,\ldots,a_{k-1})} : \mathbb{Z}_p \to \mathbb{Z}_p$ by the rule

$$f_{(a_0,\ldots,a_{k-1})}(x) = \sum_{i=0}^{k-1} a_i x^i \bmod p.$$

Prove that $(\mathbb{Z}_p, \mathbb{Z}_p, (\mathbb{Z}_p)^k, \{f_{(a_0,\ldots,a_{k-1})} : (a_0, \ldots, a_{k-1}) \in (\mathbb{Z}_p)^k\})$ is a strongly k-universal (p, p) hash family.

HINT   Use the fact that any degree d polynomial over a field has at most d roots.

5
The RSA Cryptosystem and Factoring Integers

5.1 Introduction to Public-key Cryptography

In the classical model of cryptography that we have been studying up until now, Alice and Bob secretly choose the key K. K then gives rise to an encryption rule e_K and a decryption rule d_K. In the cryptosystems we have seen so far, d_K is either the same as e_K, or easily derived from it (for example, DES decryption is identical to encryption, but the key schedule is reversed). A cryptosystem of this type is known as a *symmetric-key cryptosystem*, since exposure of either of e_K or d_K renders the system insecure.

One drawback of a symmetric-key system is that it requires the prior communication of the key K between Alice and Bob, using a secure channel, before any ciphertext is transmitted. In practice, this may be very difficult to achieve. For example, suppose Alice and Bob live far away from each other and they decide that they want to communicate electronically, using email. In a situation such as this, Alice and Bob may not have access to a reasonable secure channel.

The idea behind a *public-key cryptosystem* is that it might be possible to find a cryptosystem where it is computationally infeasible to determine d_K given e_K. If so, then the encryption rule e_K is a *public key* which could be published in a directory, for example (hence the term public-key system). The advantage of a public-key system is that Alice (or anyone else) can send an encrypted message to Bob (without the prior communication of a shared secret key) by using the public encryption rule e_K. Bob will be the only person that can decrypt the ciphertext, using the decryption rule d_K, which is called the *private key*.

Consider the following analogy: Alice places an object in a metal box, and then locks it with a combination lock left there by Bob. Bob is the only person who can open the box since only he knows the combination.

The idea of a public-key cryptosystem was put forward by Diffie and Hellman in 1976. Then, in 1977, Rivest, Shamir, and Adleman invented the well-known *RSA Cryptosystem* which we study in this chapter. Several public-key systems have since been proposed, whose security rests on different computational problems. Of these, the most important are the *RSA Cryptosystem* (and variations of it), in which the security is based on the difficulty of factoring large integers; and the *ElGamal Cryptosystem* (and variations such as *Elliptic Curve Cryptosystems*) in which the security is based on the discrete logarithm problem. We discuss the *RSA Cryptosystem* and its variants in this chapter, while *ElGamal Cryptosystems* are studied in Chapter 6.

Prior to Diffie and Hellman, the idea of public-key cryptography had already been proposed by James Ellis in January 1970, in a paper entitled "The possibility of non-secret encryption." (The phrase "non-secret encryption" can be read as "public-key cryptography.") James Ellis was a member of the Communication-Electronics Security Group (CESG), which is a special section of the British Government Communications Headquarters (GCHQ). This paper was not published in the open literature, and was one of five papers released by the GCHQ officially in December 1997. Also included in these five papers was a 1973 paper written by Clifford Cocks, entitled "A note on non-secret encryption," in which a public-key cryptosystem is described that is essentially the same as the *RSA Cryptosystem*.

One very important observation is that a public-key cryptosystem can never provide unconditional security. This is because an opponent, on observing a ciphertext y, can encrypt each possible plaintext in turn using the public encryption rule e_K until he finds the unique x such that $y = e_K(x)$. This x is the decryption of y. Consequently, we study the computational security of public-key systems.

It is helpful conceptually to think of a public-key system in terms of an abstraction called a trapdoor one-way function. We informally define this notion now.

Bob's public encryption function, e_K, should be easy to compute. We have just noted that computing the inverse function (i.e., decrypting) should be hard (for anyone other than Bob). Recall from Section 4.2 that a function that is easy to compute but hard to invert is often called a one-way function. In the context of encryption, we desire that e_K be an injective one-way function so that decryption can be performed. Unfortunately, although there are many injective functions that are believed to be one-way, there currently do not exist such functions that can be proved to be one-way.

Here is an example of a function which is believed to be one-way. Suppose n is the product of two large primes p and q, and let b be a positive integer. Then define $f : \mathbb{Z}_n \to \mathbb{Z}_n$ to be

$$f(x) = x^b \bmod n.$$

(If $\gcd(b, \phi(n)) = 1$, then this is in fact an RSA encryption function; we will have much more to say about it later.)

If we are to construct a public-key cryptosystem, then it is not sufficient to find an injective one-way function. We do not want e_K to be one-way from Bob's point of view, because he needs to be able to decrypt messages that he receives in an efficient way. Thus, it is necessary that Bob possesses a *trapdoor*, which consists of secret information that permits easy inversion of e_K. That is, Bob can decrypt efficiently because he has some extra secret knowledge, namely, K, which provides him with the decryption function d_K. So, we say that a function is a *trapdoor one-way function* if it is a one-way function, but it becomes easy to invert with the knowledge of a certain trapdoor.

Let's consider the function $f(x) = x^b \bmod n$ considered above. We will see in Section 5.3 that the inverse function f^{-1} has a similar form: $f(x) = x^a \bmod n$ for an appropriate value of a. The trapdoor is an efficient method for computing the correct exponent a (given b), which makes use of the factorization of n.

It is often convenient to specify a family of trapdoor one-way functions, say $\mathcal{F}$. Then a function $f \in \mathcal{F}$ is chosen at random and used as the public encryption function; the inverse function, f^{-1}, is the private decryption function. This is analogous to choosing a random key from a specified keyspace, as we did with symmetric-key cryptosystems.

The rest of this chapter is organized as follows. Section 5.2 introduces several important number-theoretic results. In Section 5.3, we begin our study of the *RSA Cryptosystem*. Section 5.4 presents some important methods of primality testing. Section 5.5 is a short section on the existence of square roots modulo n. Then we present several algorithms for factoring in Section 5.6. Section 5.7 considers other attacks against the *RSA Cryptosystem*, and the *Rabin Cryptosystem* is described in Section 5.8. Finally, semantic security of RSA-like cryptosystems is the topic of Section 5.9.

5.2   More Number Theory

Before describing how the *RSA Cryptosystem* works, we need to discuss some more facts concerning modular arithmetic and number theory. Two fundamental tools that we require are the EUCLIDEAN ALGORITHM and the Chinese remainder theorem.

5.2.1   The Euclidean Algorithm

We already observed in Chapter 1 that $\mathbb{Z}_n$ is a ring for any positive integer n. We also proved there that $b \in \mathbb{Z}_n$ has a multiplicative inverse if and only if $\gcd(b, n) = 1$, and that the number of positive integers less than n and relatively prime to n is $\phi(n)$.

The set of residues modulo n that are relatively prime to n is denoted $\mathbb{Z}_n^*$. It is not hard to see that $\mathbb{Z}_n^*$ forms an abelian group under multiplication. We already have stated that multiplication modulo n is associative and commutative, and that 1 is the multiplicative identity. Any element in $\mathbb{Z}_n^*$ will have a multiplicative inverse (which is also in $\mathbb{Z}_n^*$). Finally, $\mathbb{Z}_n^*$ is closed under multiplication since xy is relatively prime to n whenever x and y are relatively prime to n (prove this!).

At this point, we know that any $b \in \mathbb{Z}_n^*$ has a multiplicative inverse, b^{-1}, but we do not yet have an efficient algorithm to compute b^{-1}. Such an algorithm exists; it is called the EXTENDED EUCLIDEAN ALGORITHM. However, we first describe the EUCLIDEAN ALGORITHM, in its basic form, which can be used to compute the greatest common divisor of two positive integers, say a and b. The EUCLIDEAN ALGORITHM sets r_0 to be a and r_1 to be b, and performs the following sequence of divisions:

$$
\begin{aligned}
r_0 &= q_1 r_1 + r_2, & 0 < r_2 < r_1 \\
r_1 &= q_2 r_2 + r_3, & 0 < r_3 < r_2 \\
&\vdots \\
r_{m-2} &= q_{m-1} r_{m-1} + r_m, & 0 < r_m < r_{m-1} \\
r_{m-1} &= q_m r_m.
\end{aligned}
$$

A pseudocode description of the EUCLIDEAN ALGORITHM is presented as Algorithm 5.1.

Algorithm 5.1: EUCLIDEAN ALGORITHM(a, b)

$r_0 \leftarrow a$
$r_1 \leftarrow b$
$m \leftarrow 1$
while $r_m \neq 0$
$\textbf{do} \begin{cases} q_m \leftarrow \lfloor \frac{r_{m-1}}{r_m} \rfloor \\ r_{m+1} \leftarrow r_{m-1} - q_m r_m \\ m \leftarrow m + 1 \end{cases}$
$m \leftarrow m - 1$
return $(q_1, \ldots, q_m; r_m)$
comment: $r_m = \gcd(a, b)$

REMARK   We will make use of the list $(q_1, \ldots, q_m)$ that is computed during the execution of Algorithm 5.1 in a later section of this chapter.   ∎

In Algorithm 5.1, it is not hard to show that

$$
\gcd(r_0, r_1) = \gcd(r_1, r_2) = \cdots = \gcd(r_{m-1}, r_m) = r_m.
$$

Hence, it follows that $\gcd(r_0, r_1) = r_m$.

Since the EUCLIDEAN ALGORITHM computes greatest common divisors, it can be used to determine if a positive integer $b < n$ has a multiplicative inverse modulo n, by calling EUCLIDEAN ALGORITHM(n, b) and checking to see if $r_m = 1$. However, it does not compute the value of $b^{-1} \bmod n$ (if it exists).

Now, suppose we define two sequences of numbers,

$$t_0, t_1, \ldots, t_m \quad \text{and} \quad s_0, s_1, \ldots, s_m,$$

according to the following recurrences (where the q_j's are defined as in Algorithm 5.1):

$$t_j = \begin{cases} 0 & \text{if } j = 0 \\ 1 & \text{if } j = 1 \\ t_{j-2} - q_{j-1}t_{j-1} & \text{if } j \geq 2 \end{cases}$$

and

$$s_j = \begin{cases} 1 & \text{if } j = 0 \\ 0 & \text{if } j = 1 \\ s_{j-2} - q_{j-1}s_{j-1} & \text{if } j \geq 2. \end{cases}$$

Then we have the following useful result.

THEOREM 5.1 *For $0 \leq j \leq m$, we have that $r_j = s_j r_0 + t_j r_1$, where the r_j's are defined as in Algorithm 5.1, and the s_j's and t_j's are defined in the above recurrence.*

PROOF   The proof is by induction on j. The assertion is trivially true for $j = 0$ and $j = 1$. Assume the assertion is true for $j = i - 1$ and $i - 2$, where $i \geq 2$; we will prove the assertion is true for $j = i$. By induction, we have that

$$r_{i-2} = s_{i-2}r_0 + t_{i-2}r_1$$

and

$$r_{i-1} = s_{i-1}r_0 + t_{i-1}r_1.$$

Now, we compute:

$$
\begin{aligned}
r_i &= r_{i-2} - q_{i-1}r_{i-1} \\
&= s_{i-2}r_0 + t_{i-2}r_1 - q_{i-1}(s_{i-1}r_0 + t_{i-1}r_1) \\
&= (s_{i-2} - q_{i-1}s_{i-1})r_0 + (t_{i-2} - q_{i-1}t_{i-1})r_1 \\
&= s_i r_0 + t_i r_1.
\end{aligned}
$$

Hence, the result is true, for all integers $j \geq 0$, by induction.   ∎

In Algorithm 5.2, we present the EXTENDED EUCLIDEAN ALGORITHM, which takes two integers a and b as input and computes integers r, s and t such that $r = \gcd(a, b)$ and $sa + tb = r$. In this version of the algorithm, we do not keep track of all the q_j's, r_j's, s_j's and t_j's; it suffices to record only the "last" two terms in each of these sequences at any point in the algorithm.

Algorithm 5.2: EXTENDED EUCLIDEAN ALGORITHM(a, b)

$a_0 \leftarrow a$
$b_0 \leftarrow b$
$t_0 \leftarrow 0$
$t \leftarrow 1$
$s_0 \leftarrow 1$
$s \leftarrow 0$
$q \leftarrow \lfloor \frac{a_0}{b_0} \rfloor$
$r \leftarrow a_0 - qb_0$
while $r > 0$

$$\textbf{do} \begin{cases} temp \leftarrow t_0 - qt \\ t_0 \leftarrow t \\ t \leftarrow temp \\ temp \leftarrow s_0 - qs \\ s_0 \leftarrow s \\ s \leftarrow temp \\ a_0 \leftarrow b_0 \\ b_0 \leftarrow r \\ q \leftarrow \lfloor \frac{a_0}{b_0} \rfloor \\ r \leftarrow a_0 - qb_0 \end{cases}$$

return (r, s, t)
comment: $r = \gcd(a, b)$ and $sa + tb = r$

The next corollary is an immediate consequence of Theorem 5.1.

COROLLARY 5.2 *Suppose* $\gcd(r_0, r_1) = 1$. *Then* $r_1^{-1} \bmod r_0 = t_m \bmod r_0$.

PROOF   From Theorem 5.1, we have that

$$1 = \gcd(r_0, r_1) = s_m r_0 + t_m r_1.$$

Reducing this equation modulo r_0, we obtain

$$t_m r_1 \equiv 1 \pmod{r_0}.$$

The result follows.   ∎

We present a small example to illustrate, in which we show the values of all the s_j's, t_j's, q_j's and r_j's.

Example 5.1 Suppose we wish to calculate $28^{-1} \bmod 75$. Then we compute the following:

i	r_i	q_i	s_i	t_i
0	75		1	0
1	28	2	0	1
2	19	1	1	-2
3	9	2	-1	3
4	1	9	3	-8

Therefore, we have found that

$$3 \times 75 - 8 \times 28 = 1.$$

Applying Corollary 5.2, we see that

$$28^{-1} \bmod 75 = -8 \bmod 75 = 67.$$

$\square$

The EXTENDED EUCLIDEAN ALGORITHM immediately yields the value b^{-1} modulo a (if it exists). In fact, the multiplicative inverse $b^{-1} \bmod a = t \bmod a$; this follows immediately from Corollary 5.2. However, a more efficient way to compute multiplicative inverses is to remove the s's from Algorithm 5.2, and to reduce the t's modulo a during each iteration of the main loop. We obtain the following algorithm:

Algorithm 5.3: MULTIPLICATIVE INVERSE(a, b)

$a_0 \leftarrow a$
$b_0 \leftarrow b$
$t_0 \leftarrow 0$
$t \leftarrow 1$
$q \leftarrow \lfloor \frac{a_0}{b_0} \rfloor$
$r \leftarrow a_0 - q b_0$
while $r > 0$

$\textbf{do} \begin{cases} temp \leftarrow (t_0 - qt) \bmod a \\ t_0 \leftarrow t \\ t \leftarrow temp \\ a_0 \leftarrow b_0 \\ b_0 \leftarrow r \\ q \leftarrow \lfloor \frac{a_0}{b_0} \rfloor \\ r \leftarrow a_0 - q b_0 \end{cases}$

if $r \neq 1$
   then b has no inverse modulo n
   else return (t)

5.2.2   The Chinese Remainder Theorem

The Chinese remainder theorem is really a method of solving certain systems of congruences. Suppose $m_1, \ldots, m_r$ are pairwise relatively prime positive integers (that is, $\gcd(m_i, m_j) = 1$ if $i \neq j$). Suppose $a_1, \ldots, a_r$ are integers, and consider the following system of congruences:

$$x \equiv a_1 \pmod{m_1}$$

$$x \equiv a_2 \pmod{m_2}$$

$$\vdots$$

$$x \equiv a_r \pmod{m_r}.$$

The Chinese remainder theorem asserts that this system has a unique solution modulo $M = m_1 \times m_2 \times \cdots \times m_r$. We will prove this result in this section, and also describe an efficient algorithm for solving systems of congruences of this type.

It is convenient to study the function $\chi : \mathbb{Z}_M \to \mathbb{Z}_{m_1} \times \cdots \times \mathbb{Z}_{m_r}$, which we define as follows:

$$\chi(x) = (x \bmod m_1, \ldots, x \bmod m_r).$$

Example 5.2   Suppose $r = 2$, $m_1 = 5$ and $m_2 = 3$, so $M = 15$. Then the function χ has the following values:

$$
\begin{array}{llllll}
\chi(0) & = & (0,0) & \chi(1) & = & (1,1) \\
\chi(3) & = & (3,0) & \chi(4) & = & (4,1) \\
\chi(6) & = & (1,0) & \chi(7) & = & (2,1) \\
\chi(9) & = & (4,0) & \chi(10) & = & (0,1) \\
\chi(12) & = & (2,0) & \chi(13) & = & (3,1)
\end{array}
\qquad
\begin{array}{lll}
\chi(2) & = & (2,2) \\
\chi(5) & = & (0,2) \\
\chi(8) & = & (3,2) \\
\chi(11) & = & (1,2) \\
\chi(14) & = & (4,2).
\end{array}
$$

$\square$

Proving the Chinese remainder theorem amounts to proving that the function χ is a bijection. In Example 5.2 this is easily seen to be the case. In fact, we will be able to give an explicit general formula for the inverse function χ^{-1}.

For $1 \leq i \leq r$, define

$$M_i = \frac{M}{m_i}.$$

Then it is not difficult to see that

$$\gcd(M_i, m_i) = 1$$

for $1 \le i \le r$. Next, for $1 \le i \le r$, define

$$y_i = M_i^{-1} \bmod m_i.$$

(This inverse exists because $\gcd(M_i, m_i) = 1$, and it can be found using Algorithm 5.3.) Note that

$$M_i y_i \equiv 1 \pmod{m_i}$$

for $1 \le i \le r$.

Now, define a function $\rho : \mathbb{Z}_{m_1} \times \cdots \times \mathbb{Z}_{m_r} \to \mathbb{Z}_M$ as follows:

$$\rho(a_1, \ldots, a_r) = \sum_{i=1}^{r} a_i M_i y_i \bmod M.$$

We will show that the function $\rho = \chi^{-1}$, i.e., it provides an explicit formula for solving the original system of congruences.

Denote $X = \rho(a_1, \ldots, a_r)$, and let $1 \le j \le r$. Consider a term $a_i M_i y_i$ in the above summation, reduced modulo m_j: If $i = j$, then

$$a_i M_i y_i \equiv a_i \pmod{m_i}$$

because

$$M_i y_i \equiv 1 \pmod{m_i}.$$

On the other hand, if $i \ne j$, then

$$a_i M_i y_i \equiv 0 \pmod{m_j}$$

because $m_j \mid M_i$ in this case. Thus, we have that

$$X \equiv \sum_{i=1}^{r} a_i M_i y_i \pmod{m_j}$$

$$\equiv a_j \pmod{m_j}.$$

Since this is true for all j, $1 \le j \le r$, X is a solution to the system of congruences.

At this point, we need to show that the solution X is unique modulo M. But this can be done by simple counting. The function χ is a function from a domain of cardinality M to a range of cardinality M. We have just proved that χ is a surjective (i.e., onto) function. Hence, χ must also be injective (i.e., one-to-one), since the domain and range have the same cardinality. It follows that χ is a bijection and $\chi^{-1} = \rho$. Note also that χ^{-1} is a linear function of its arguments $a_1, \ldots, a_r$.

Here is a bigger example to illustrate.

Example 5.3   Suppose $r = 3$, $m_1 = 7$, $m_2 = 11$ and $m_3 = 13$. Then $M = 1001$. We compute $M_1 = 143$, $M_2 = 91$ and $M_3 = 77$, and then $y_1 = 5$, $y_2 = 4$ and $y_3 = 12$. Then the function $\chi^{-1} : \mathbb{Z}_7 \times \mathbb{Z}_{11} \times \mathbb{Z}_{13} \to \mathbb{Z}_{1001}$ is the following:

$$\chi^{-1}(a_1, a_2, a_3) = (715a_1 + 364a_2 + 924a_3) \bmod 1001.$$

For example, if $x \equiv 5 \pmod 7$, $x \equiv 3 \pmod{11}$ and $x \equiv 10 \pmod{13}$, then this formula tells us that

$$x = (715 \times 5 + 364 \times 3 + 924 \times 10) \bmod 1001$$

$$= 13907 \bmod 1001$$

$$= 894.$$

This can be verified by reducing 894 modulo 7, 11 and 13.   ▯

For future reference, we record the results of this section as a theorem.

THEOREM 5.3   *(Chinese remainder theorem)*   *Suppose $m_1, \ldots, m_r$ are pairwise relatively prime positive integers, and suppose $a_1, \ldots, a_r$ are integers. Then the system of r congruences $x \equiv a_i \pmod{m_i}$ $(1 \le i \le r)$ has a unique solution modulo $M = m_1 \times \cdots \times m_r$, which is given by*

$$x = \sum_{i=1}^{r} a_i M_i y_i \bmod M,$$

where $M_i = M/m_i$ and $y_i = M_i^{-1} \bmod m_i$, for $1 \le i \le r$.

5.2.3   Other Useful Facts

We next mention another result from elementary group theory, called Lagrange's theorem, that will be relevant in our treatment of the *RSA Cryptosystem*. For a (finite) multiplicative group G, define the *order* of an element $g \in G$ to be the smallest positive integer m such that $g^m = 1$. The following result is fairly simple, but we will not prove it here.

THEOREM 5.4   *(Lagrange)*   *Suppose G is a multiplicative group of order n, and $g \in G$. Then the order of g divides n.*

For our purposes, the following corollaries are essential.

COROLLARY 5.5   *If $b \in \mathbb{Z}_n^*$, then $b^{\phi(n)} \equiv 1 \pmod n$.*

PROOF $\mathbb{Z}_n^*$ is a multiplicative group of order $\phi(n)$.   ∎

COROLLARY 5.6 (Fermat) *Suppose p is prime and $b \in \mathbb{Z}_p$. Then $b^p \equiv b$ (mod p).*

PROOF If p is prime, then $\phi(p) = p - 1$. So, for $b \not\equiv 0$ (mod p), the result follows from Corollary 5.5. For $b \equiv 0$ (mod p), the result is also true since $0^p \equiv 0$ (mod p). ∎

At this point, we know that if p is prime, then $\mathbb{Z}_p^*$ is a group of order $p - 1$, and any element in $\mathbb{Z}_p^*$ has order dividing $p - 1$. In fact, if p is prime, then the group $\mathbb{Z}_p^*$ is a *cyclic group*: there exists an element $\alpha \in \mathbb{Z}_p^*$ having order equal to $p - 1$. We will not prove this very important fact, but we do record it for future reference:

THEOREM 5.7 *If p is prime, then $\mathbb{Z}_p^*$ is a cyclic group.*

An element α having order $p-1$ modulo p is called a *primitive element* modulo p. Observe that α is a primitive element modulo p if and only if

$$\{\alpha^i : 0 \leq i \leq p - 2\} = \mathbb{Z}_p^*.$$

Now, suppose p is prime and α is a primitive element modulo p. Any element $\beta \in \mathbb{Z}_p^*$ can be written as $\beta = \alpha^i$, where $0 \leq i \leq p - 2$, in a unique way. It is not difficult to prove that the order of $\beta = \alpha^i$ is

$$\frac{p - 1}{\gcd(p - 1, i)}.$$

Thus β is itself a primitive element if and only if $\gcd(p - 1, i) = 1$. It follows that the number of primitive elements modulo p is $\phi(p - 1)$.

We do a small example to illustrate.

Example 5.4 Suppose $p = 13$. The results proven above establish that there are exactly four primitive elements modulo 13. First, by computing successive powers of 2, we can verify that 2 is a primitive element modulo 13:

$$2^0 \bmod 13 = 1$$

$$2^1 \bmod 13 = 2$$

$$2^2 \bmod 13 = 4$$

$$2^3 \bmod 13 = 8$$

$$2^4 \bmod 13 = 3$$

$$2^5 \bmod 13 = 6$$

$$2^6 \bmod 13 = 12$$

$$2^7 \bmod 13 = 11$$

$$2^8 \bmod 13 = 9$$

$$2^9 \bmod 13 = 5$$

$$2^{10} \bmod 13 = 10$$

$$2^{11} \bmod 13 = 7.$$

The element 2^i is primitive if and only if $\gcd(i, 12) = 1$, i.e., if and only if $i = 1, 5, 7$ or 11. Hence, the primitive elements modulo 13 are $2, 6, 7$ and 11. $\Box$

In the above example, we computed all the powers of 2 in order to verify that it was a primitive element modulo 13. If p is a large prime, however, it would take a long time to compute $p - 1$ powers of an element $\alpha \in \mathbb{Z}_p^*$. Fortunately, if the factorization of $p - 1$ is known, then we can verify whether $\alpha \in \mathbb{Z}_p^*$ is a primitive element much more quickly, by making use of the following result.

THEOREM 5.8   *Suppose that p is prime and $\alpha \in \mathbb{Z}_p^*$. Then α is a primitive element modulo p if and only if $\alpha^{(p-1)/q} \not\equiv 1 \pmod{p}$ for all primes q such that $q \mid (p-1)$.*

PROOF   If α is a primitive element modulo p, then $\alpha^i \not\equiv 1 \pmod{p}$ for all i such that $1 \le i \le p - 2$, so the result follows.

Conversely, suppose that $\alpha \in \mathbb{Z}_p^*$ is not a primitive element modulo p. Let d be the order of α. Then $d \mid (p-1)$ by Lagrange's theorem, and $d < p-1$ because α is not primitive. Then $(p - 1)/d$ is an integer exceeding 1. Let q be a prime divisor of $(p-1)/d$. Then d is a divisor of the integer $(p-1)/q$. Since $\alpha^d \equiv 1 \pmod{p}$ and $d \mid (p-1)/q$, it follows that $\alpha^{(p-1)/q} \equiv 1 \pmod{p}$. ∎

The factorization of 12 is $12 = 2^2 \times 3$. Therefore, in the previous example, we could verify that 2 is a primitive element modulo 13 by verifying that $2^6 \not\equiv 1 \pmod{13}$ and $2^4 \not\equiv 1 \pmod{13}$.

5.3 The RSA Cryptosystem

We can now describe the *RSA Cryptosystem*. This cryptosystem uses computations in $\mathbb{Z}_n$, where n is the product of two distinct odd primes p and q. For such an integer n, note that $\phi(n) = (p-1)(q-1)$. The formal description of the cryptosystem is as follows.

Cryptosystem 5.1: *RSA Cryptosystem*

Let $n = pq$, where p and q are primes. Let $\mathcal{P} = \mathcal{C} = \mathbb{Z}_n$, and define

$$\mathcal{K} = \{(n, p, q, a, b) : ab \equiv 1 \ (\text{mod} \ \phi(n))\}.$$

For $K = (n, p, q, a, b)$, define

$$e_K(x) = x^b \ \text{mod} \ n$$

and

$$d_K(y) = y^a \ \text{mod} \ n$$

$(x, y \in \mathbb{Z}_n)$. The values n and b comprise the public key, and the values p, q and a form the private key.

Let's verify that encryption and decryption are inverse operations. Since

$$ab \equiv 1 \ (\text{mod} \ \phi(n)),$$

we have that

$$ab = t\phi(n) + 1$$

for some integer $t \geq 1$. Suppose that $x \in \mathbb{Z}_n^*$; then we have

$$(x^b)^a \equiv x^{t\phi(n)+1} \ (\text{mod} \ n)$$

$$\equiv (x^{\phi(n)})^t x \ (\text{mod} \ n)$$

$$\equiv 1^t x \ (\text{mod} \ n)$$

$$\equiv x \ (\text{mod} \ n),$$

as desired. We leave it as an Exercise to show that $(x^b)^a \equiv x \ (\text{mod} \ n)$ if $x \in \mathbb{Z}_n \backslash \mathbb{Z}_n^*$.

Here is a small (insecure) example of the *RSA Cryptosystem*.

Example 5.5 Suppose Bob chooses $p = 101$ and $q = 113$. Then $n = 11413$ and $\phi(n) = 100 \times 112 = 11200$. Since $11200 = 2^6 5^2 7$, an integer b can be used as an encryption exponent if and only if b is not divisible by 2, 5 or 7. (In practice, however, Bob will not factor $\phi(n)$. He will verify that $\gcd(\phi(n), b) = 1$ using Algorithm 5.3, and compute b^{-1} at the same time.) Suppose Bob chooses $b = 3533$. Then

$$b^{-1} \bmod 11200 = 6597.$$

Hence, Bob's secret decryption exponent is $a = 6597$.

Bob publishes $n = 11413$ and $b = 3533$ in a directory. Now, suppose Alice wants to encrypt the plaintext 9726 to send to Bob. She will compute

$$9726^{3533} \bmod 11413 = 5761$$

and send the ciphertext 5761 over the channel. When Bob receives the ciphertext 5761, he uses his secret decryption exponent to compute

$$5761^{6597} \bmod 11413 = 9726.$$

(At this point, the encryption and decryption operations might appear to be very complicated, but we will discuss efficient algorithms for these operations in the next section.)   ▯

The security of the *RSA Cryptosystem* is based on the belief that the encryption function $e_K(x) = x^b \bmod n$ is a one-way function, so it will be computationally infeasible for an opponent to decrypt a ciphertext. The trapdoor that allows Bob to decrypt a ciphertext is the knowledge of the factorization $n = pq$. Since Bob knows this factorization, he can compute $\phi(n) = (p-1)(q-1)$, and then compute the decryption exponent a using the EXTENDED EUCLIDEAN ALGORITHM. We will say more about the security of the *RSA Cryptosystem* later on.

5.3.1 Implementing RSA

There are many aspects of the *RSA Cryptosystem* to discuss, including the details of setting up the cryptosystem, the efficiency of encrypting and decrypting, and security issues. In order to set up the system, Bob uses the RSA PARAMETER GENERATION algorithm, presented informally as Algorithm 5.4. How Bob carries out the steps of this algorithm will be discussed later in this chapter.

Algorithm 5.4: RSA PARAMETER GENERATION

1. Generate two large primes, p and q
2. $n \leftarrow pq$ and $\phi(n) \leftarrow (p-1)(q-1)$
3. Choose a random b $(1 < b < \phi(n))$ such that $\gcd(b, \phi(n)) = 1$
4. $a \leftarrow b^{-1} \bmod \phi(n)$
5. The public key is (n, b) and the private key is (p, q, a).

One obvious attack on the *RSA Cryptosystem* is for a cryptanalyst to attempt to factor n. If this can be done, it is a simple manner to compute $\phi(n) = (p - 1)(q - 1)$ and then compute the decryption exponent a from b exactly as Bob did. (It has been conjectured that breaking the *RSA Cryptosystem* is polynomially equivalent[1] to factoring n, but this remains unproved.)

If the *RSA Cryptosystem* is to be secure, it is certainly necessary that $n = pq$ must be large enough that factoring it will be computationally infeasible. Current factoring algorithms are able to factor numbers having up to 512 bits in their binary representation (for more information on factoring, see Section 5.6). It is generally recommended that, to be on the safe side, one should choose each of p and q to be 512-bit primes; then n will be a 1024-bit modulus. Factoring a number of this size is well beyond the capability of the best current factoring algorithms.

Leaving aside for the moment the question of how to find 512-bit primes, let us look now at the arithmetic operations of encryption and decryption. An encryption (or decryption) involves performing one exponentiation modulo n. Since n is very large, we must use multiprecision arithmetic to perform computations in $\mathbb{Z}_n$, and the time required will depend on the number of bits in the binary representation of n.

Suppose that x and y are positive integers having k and ℓ bits respectively in their binary representations; i.e., $k = \lfloor \log_2 x \rfloor + 1$ and $\ell = \lfloor \log_2 y \rfloor + 1$. Assume that $k \geq \ell$. Using standard "grade-school" arithmetic techniques, it is not difficult to obtain big-oh upper bounds on the amount of time to perform various operations on x and y. We summarize these results now (and we do not claim that these are the best possible bounds).

- $x + y$ can be computed in time $O(k)$
- $x - y$ can be computed in time $O(k)$
- xy can be computed in time $O(k\ell)$
- $\lfloor x/y \rfloor$ can be computed in time $O(\ell(k - \ell))$. $O(k\ell)$ is a weaker bound.
- $\gcd(x, y)$ can be computed in time $O(k^3)$.

In reference to the last item, GCD's can be computed using Algorithm 5.1. It can be shown that the number of iterations required in the EUCLIDEAN AL-GORITHM is $O(k)$ (see the Exercises). Each iteration performs a long division requiring time $O(k^2)$; so, the complexity of a GCD computation is seen to be $O(k^3)$. (Actually, a more careful analysis can be used to show that the complexity is, in fact, $O(k^2)$.)

Now we turn to modular arithmetic, i.e., operations in $\mathbb{Z}_n$. Suppose that n is a k-bit integer, and $0 \leq m_1, m_2 \leq n - 1$. Also, let c be a positive integer. We have the following:

- Computing $(m_1 + m_2) \bmod n$ can be done in time $O(k)$.

---

[1] Two problems are said to be *polynomially equivalent* if the existence of a polynomial-time algorithm for either problem implies the existence of a polynomial-time algorithm for the other problem.

- Computing $(m_1 - m_2) \bmod n$ can be done in time $O(k)$.
- Computing $(m_1 m_2) \bmod n$ can be done in time $O(k^2)$.
- Computing $(m_1)^{-1} \bmod n$ can be done in time $O(k^3)$.
- Computing $(m_1)^c \bmod n$ can be done in time $O((\log c) \times k^2)$.

Most of the above results are not hard to prove. The first three operations (modular addition, subtraction and multiplication) can be accomplished by doing the corresponding integer operation and then performing a single reduction modulo n. Modular inversion (i.e., computing multiplicative inverses) is done using Algorithm 5.3. The complexity is analyzed in a similar fashion as a GCD computation.

We now consider *modular exponentiation*, i.e., computation of a function of the form $x^c \bmod n$. Both the encryption and the decryption operations in the *RSA Cryptosystem* are modular exponentiations. Computation of $x^c \bmod n$ can be done using $c - 1$ modular multiplications; however, this is very inefficient if c is large. Note that c might be as big as $\phi(n) - 1$, which is almost as big as n and exponentially large compared to k.

The well-known SQUARE-AND-MULTIPLY ALGORITHM reduces the number of modular multiplications required to compute $x^c \bmod n$ to at most 2ℓ, where ℓ is the number of bits in the binary representation of c. It follows that $x^c \bmod n$ can be computed in time $O(\ell k^2)$. If we assume that $c < n$ (as it is in the definition of the *RSA Cryptosystem*), then we see that RSA encryption and decryption can both be done in time $O((\log n)^3)$, which is a polynomial function of the number of bits in one plaintext (or ciphertext) character.

The SQUARE-AND-MULTIPLY ALGORITHM assumes that the exponent c is represented in binary notation, say

$$c = \sum_{i=0}^{\ell-1} c_i 2^i,$$

where $c_i = 0$ or 1, $0 \le i \le \ell - 1$. The algorithm to compute $z = x^c \bmod n$ is presented as Algorithm 5.5.

Algorithm 5.5: SQUARE-AND-MULTIPLY(x, c, n)

$z \leftarrow 1$
for $i \leftarrow \ell - 1$ **downto** 0
$\quad$ **do** $\begin{cases} z \leftarrow z^2 \bmod n \\ \textbf{if } c_i = 1 \\ \quad \textbf{then } z \leftarrow (z \times x) \bmod n \end{cases}$
return (z)

The proof of correctness of this algorithm is left as an Exercise. It is easy to count the number of modular multiplications in the algorithm. There are always ℓ squarings performed. The number of modular multiplications of the type $z \leftarrow$

$(z \times x) \bmod n$ is equal to the number of 1's in the binary representation of c. This is an integer between 0 and ℓ. Thus, the total number of modular multiplications is at least ℓ and at most 2ℓ, as stated above.

We will illustrate the use of the SQUARE-AND-MULTIPLY ALGORITHM by returning to Example 5.5.

Example 5.5 *(Cont.)* Recall that $n = 11413$, and the public encryption exponent is $b = 3533$. Alice encrypts the plaintext 9726 by computing $9726^{3533} \bmod 11413$, using the SQUARE-AND-MULTIPLY ALGORITHM, as follows:

i	b_i	z
11	1	$1^2 \times 9726 = 9726$
10	1	$9726^2 \times 9726 = 2659$
9	0	$2659^2 = 5634$
8	1	$5634^2 \times 9726 = 9167$
7	1	$9167^2 \times 9726 = 4958$
6	1	$4958^2 \times 9726 = 7783$
5	0	$7783^2 = 6298$
4	0	$6298^2 = 4629$
3	1	$4629^2 \times 9726 = 10185$
2	1	$10185^2 \times 9726 = 105$
1	0	$105^2 = 11025$
0	1	$11025^2 \times 9726 = 5761$

Hence, as stated earlier, the ciphertext is 5761. ☐

To this point we have discussed the RSA encryption and decryption operations. Regarding RSA PARAMETER GENERATION, methods to construct the primes p and q (Step 1) will be discussed in the next section. Step 2 is straightforward and can be done in time $O((\log n)^2)$. Steps 3 and 4 utilize Algorithm 5.3, which has complexity $O((\log n)^2)$.

5.4 Primality Testing

In setting up the *RSA Cryptosystem*, it is necessary to generate large "random primes." In practice, the way this is done is to generate large random numbers, and then test them for primality using a randomized polynomial-time Monte Carlo algorithm such as the SOLOVAY-STRASSEN ALGORITHM or the MILLER-RABIN ALGORITHM, both of which we will present in this section. These algorithms are fast (i.e., an integer n can be tested in time that is polynomial in $\log_2 n$, the number of bits in the binary representation of n), but there is a possibility that the algorithm may claim that n is prime when it is not. However, by running the algorithm enough times, the error probability can be reduced below any desired threshold. (We will discuss this in more detail a bit later.)

The other pertinent question is how many random integers (of a specified size) will need to be tested until we find one that is prime. Suppose we define $\pi(N)$ to be the number of primes that are less than or equal to N. A famous result in number theory, called the *Prime number theorem*, states that $\pi(N)$ is approximately $N/\ln N$. Hence, if an integer p is chosen at random between 1 and N, then the probability that it is prime is about $1/\ln N$. For a 1024 bit modulus $n = pq$, p and q will be chosen to be 512 bit primes. A random 512 bit integer will be prime with probability approximately $1/\ln 2^{512} \approx 1/355$. That is, on average, given 355 random 512 bit integers p, one one of them will be prime (of course, if we restrict our attention to odd integers, the probability doubles, to about $2/355$). So we can in fact generate sufficiently large random numbers that are "probably prime," and hence parameter generation for the *RSA Cryptosystem* is indeed practical. We proceed to describe how this is done.

A *decision problem* is a problem in which a question is to be answered "yes" or "no." Recall that a randomized algorithm is any algorithm that uses random numbers (in contrast, an algorithm that does not use random numbers is called a *deterministic algorithm*). The following definitions pertain to randomized algorithms for decision problems.

Definition 5.1:   A *yes-biased Monte Carlo algorithm* is a randomized algorithm for a decision problem in which a "yes" answer is (always) correct, but a "no" answer may be incorrect. A *no-biased* Monte Carlo algorithm is defined in the obvious way. We say that a yes-biased Monte Carlo algorithm has *error probability* equal to ϵ if, for any instance in which the answer is "yes," the algorithm will give the (incorrect) answer "no" with probability at most ϵ. (This probability is computed over all possible random choices made by the algorithm when it is run with a given input.)

REMARK   A Las Vegas algorithm may not give an answer, but any answer it gives is correct. In contrast, a Monte Carlo algorithm always gives an answer, but the answer may be incorrect.   ∎

The decision problem called Composites is presented as Problem 5.1.

Problem 5.1:   Composites

Instance: A positive integer $n \geq 2$.

Question: Is n composite?

Note that an algorithm for a decision problem only has to answer "yes" or "no." In particular, in the case of the problem **Composites**, we do not require the algorithm to find a factorization in the case that n is composite.

We will first describe the SOLOVAY-STRASSEN ALGORITHM, which is a yes-biased Monte Carlo algorithm for **Composites** with error probability $1/2$. Hence, if the algorithm answers "yes," then n is composite; conversely, if n is composite, then the algorithm answers "yes" with probability at least $1/2$.

Although the MILLER-RABIN ALGORITHM (which we will discuss later) is faster than the SOLOVAY-STRASSEN ALGORITHM, we first look at the SOLOVAY-STRASSEN ALGORITHM because it is easier to understand conceptually and because it involves some number-theoretic concepts that will be useful in later chapters of the book. We begin by developing some further background from number theory before describing the algorithm.

Definition 5.2: Suppose p is an odd prime and a is an integer. a is defined to be a *quadratic residue* modulo p if $a \not\equiv 0 \pmod{p}$ and the congruence $y^2 \equiv a \pmod{p}$ has a solution $y \in \mathbb{Z}_p$. a is defined to be a *quadratic non-residue* modulo p if $a \not\equiv 0 \pmod{p}$ and a is not a quadratic residue modulo p.

Example 5.6 In $\mathbb{Z}_{11}$, we have that $1^2 = 1$, $2^2 = 4$, $3^2 = 9$, $4^2 = 5$, $5^2 = 3$, $6^2 = 3$, $7^2 = 5$, $8^2 = 9$, $9^2 = 4$, and $(10)^2 = 1$. Therefore the quadratic residues modulo 11 are $1, 3, 4, 5$ and 9, and the quadratic non-residues modulo 11 are $2, 6, 7, 8$ and 10. $\qquad\qquad\square$

Suppose that p is an odd prime and a is a quadratic residue modulo p. Then there exists $y \in \mathbb{Z}_p^*$ such that $y^2 \equiv a \pmod{p}$. Clearly, $(-y)^2 \equiv a \pmod{p}$, and $y \not\equiv -y \pmod{p}$ because p is odd. Now consider the quadratic congruence $x^2 - a \equiv 0 \pmod{p}$. This congruence can be factored as $(x - y)(x + y) \equiv 0 \pmod{p}$, which is the same thing as saying that $p \mid (x - y)(x + y)$. Now, because p is prime, it follows that $p \mid (x - y)$ or $p \mid (x + y)$. In other words, $x \equiv \pm y \pmod{p}$, and we conclude that there are exactly two solutions (modulo p) to the congruence $x^2 - a \equiv 0 \pmod{p}$. Moreover, these two solutions are negatives of each other modulo p.

We now study the problem of determining whether an integer a is quadratic residue modulo p. The decision problem **Quadratic Residues** (Problem 5.2) is defined in the obvious way. Notice that this problem just asks for a "yes" or "no" answer: it does not require us to compute square roots in the case when a is a quadratic residue modulo p.

Problem 5.2: Quadratic Residues

Instance: An odd prime p, and an integer a.

Question: Is a a quadratic residue modulo p?

We prove a result, known as Euler's criterion, that will give rise to a polynomial-time deterministic algorithm for **Quadratic Residues**.

THEOREM 5.9 *(Euler's Criterion)   Let p be an odd prime. Then a is a quadratic residue modulo p if and only if*

$$a^{(p-1)/2} \equiv 1 \pmod{p}.$$

PROOF   First, suppose $a \equiv y^2 \pmod{p}$. Recall from Corollary 5.6 that if p is prime, then $a^{p-1} \equiv 1 \pmod{p}$ for any $a \not\equiv 0 \pmod{p}$. Thus we have

$$a^{(p-1)/2} \equiv (y^2)^{(p-1)/2} \pmod{p}$$

$$\equiv y^{p-1} \pmod{p}$$

$$\equiv 1 \pmod{p}.$$

Conversely, suppose $a^{(p-1)/2} \equiv 1 \pmod{p}$. Let b be a primitive element modulo p. Then $a \equiv b^i \pmod{p}$ for some positive integer i. Then we have

$$a^{(p-1)/2} \equiv (b^i)^{(p-1)/2} \pmod{p}$$

$$\equiv b^{i(p-1)/2} \pmod{p}.$$

Since b has order $p - 1$, it must be the case that $p - 1$ divides $i(p - 1)/2$. Hence, i is even, and then the square roots of a are $\pm b^{i/2} \bmod p$.   ∎

Theorem 5.9 yields a polynomial-time algorithm for **Quadratic Residues**, by using the SQUARE-AND-MULTIPLY ALGORITHM for exponentiation modulo p. The complexity of the algorithm will be $O((\log p)^3)$.

We now need to give some further definitions from number theory.

Definition 5.3:   Suppose p is an odd prime. For any integer a, define the *Legendre symbol* $\left(\frac{a}{p}\right)$ as follows:

$$\left(\frac{a}{p}\right) = \begin{cases} 0 & \text{if } a \equiv 0 \pmod{p} \\ 1 & \text{if } a \text{ is a quadratic residue modulo } p \\ -1 & \text{if } a \text{ is a quadratic non-residue modulo } p. \end{cases}$$

We have already seen that $a^{(p-1)/2} \equiv 1 \pmod{p}$ if and only if a is a quadratic residue modulo p. If a is a multiple of p, then it is clear that $a^{(p-1)/2} \equiv 0 \pmod{p}$. Finally, if a is a quadratic non-residue modulo p, then $a^{(p-1)/2} \equiv -1 \pmod{p}$ because

$$(a^{(p-1)/2})^2 \equiv a^{p-1} \equiv 1 \pmod{p}$$

and $a^{(p-1)/2} \not\equiv 1 \pmod{p}$. Hence, we have the following result, which provides an efficient algorithm to evaluate Legendre symbols:

THEOREM 5.10   *Suppose p is an odd prime. Then*

$$\left(\frac{a}{p}\right) \equiv a^{(p-1)/2} \pmod{p}.$$

Next, we define a generalization of the Legendre symbol.

Definition 5.4:   Suppose n is an odd positive integer, and the prime power factorization of n is

$$n = \prod_{i=1}^{k} p_i^{e_i}.$$

Let a be an integer. The *Jacobi symbol* $\left(\frac{a}{n}\right)$ is defined to be

$$\left(\frac{a}{n}\right) = \prod_{i=1}^{k} \left(\frac{a}{p_i}\right)^{e_i}.$$

Example 5.7   Consider the Jacobi symbol $\left(\frac{6278}{9975}\right)$. The prime power factorization of 9975 is $9975 = 3 \times 5^2 \times 7 \times 19$. Thus we have

$$\left(\frac{6278}{9975}\right) = \left(\frac{6278}{3}\right)\left(\frac{6278}{5}\right)^2\left(\frac{6278}{7}\right)\left(\frac{6278}{19}\right)$$

$$= \left(\frac{2}{3}\right)\left(\frac{3}{5}\right)^2\left(\frac{6}{7}\right)\left(\frac{8}{19}\right)$$

$$= (-1)(-1)^2(-1)(-1)$$

$$= -1.$$

$\square$

Suppose $n > 1$ is odd. If n is prime, then $\left(\frac{a}{n}\right) \equiv a^{(n-1)/2} \pmod{n}$ for any a. On the other hand, if n is composite, it may or may not be the case that $\left(\frac{a}{n}\right) \equiv a^{(n-1)/2} \pmod{n}$. If this congruence holds, then n is called an *Euler pseudo-prime* to the base a. For example, 91 is an Euler pseudo-prime to the base 10, because

$$\left(\frac{10}{91}\right) = -1 \equiv 10^{45} \pmod{91}.$$

It can be shown that, for any odd composite n, n is an Euler pseudoprime to the base a for at most half of the integers $a \in \mathbb{Z}_n^*$ (see the Exercises). It is also easy to see that $\left(\frac{a}{n}\right) = 0$ if and only if $\gcd(a, n) > 1$ (therefore, if $1 \leq a \leq n-1$ and $\left(\frac{a}{n}\right) = 0$, it must be the case that n is composite). These two facts show that the SOLOVAY-STRASSEN ALGORITHM, which we present as Algorithm 5.6, is a yes-biased Monte Carlo algorithm with error probability at most $1/2$.

Algorithm 5.6: SOLOVAY-STRASSEN(n)

choose a random integer a such that $1 \leq a \leq n - 1$
$x \leftarrow \left(\frac{a}{n} \right)$
if $x = 0$
   then return ("n is composite")
$y \leftarrow a^{(n-1)/2} \pmod{n}$
if $x \equiv y \pmod{n}$
   then return ("n is prime")
   else return ("n is composite")

At this point it is not clear that Algorithm 5.6 is a polynomial-time algorithm. We already know how to evaluate $a^{(n-1)/2} \bmod n$ in time $O((\log n)^3)$, but how do we compute Jacobi symbols efficiently? It might appear to be necessary to first factor n, since the Jacobi symbol $\left(\frac{a}{n} \right)$ is defined in terms of the factorization of n. But, if we could factor n, we would already know if it is prime; so this approach ends up in a vicious circle.

Fortunately, we can evaluate a Jacobi symbol without factoring n by using some results from number theory, the most important of which is a generalization of the law of quadratic reciprocity (property 4 below). We now enumerate these properties without proof:

1. If n is a positive odd integer and $m_1 \equiv m_2 \pmod{n}$, then
$$\left(\frac{m_1}{n} \right) = \left(\frac{m_2}{n} \right).$$

2. If n is a positive odd integer, then
$$\left(\frac{2}{n} \right) = \begin{cases} 1 & \text{if } n \equiv \pm 1 \pmod{8} \\ -1 & \text{if } n \equiv \pm 3 \pmod{8}. \end{cases}$$

3. If n is a positive odd integer, then
$$\left(\frac{m_1 m_2}{n} \right) = \left(\frac{m_1}{n} \right) \left(\frac{m_2}{n} \right).$$
In particular, if $m = 2^k t$ and t is odd, then
$$\left(\frac{m}{n} \right) = \left(\frac{2}{n} \right)^k \left(\frac{t}{n} \right).$$

4. Suppose m and n are positive odd integers. Then
$$\left(\frac{m}{n} \right) = \begin{cases} -\left(\frac{n}{m} \right) & \text{if } m \equiv n \equiv 3 \pmod{4} \\ \left(\frac{n}{m} \right) & \text{otherwise.} \end{cases}$$

Example 5.8 As an illustration of the application of these properties, we evaluate the Jacobi symbol $\left(\frac{7411}{9283}\right)$ as follows:

$$
\begin{aligned}
\left(\frac{7411}{9283}\right) &= -\left(\frac{9283}{7411}\right) && \text{by property 4} \\[2mm]
&= -\left(\frac{1872}{7411}\right) && \text{by property 1} \\[2mm]
&= -\left(\frac{2}{7411}\right)^4 \left(\frac{117}{7411}\right) && \text{by property 3} \\[2mm]
&= -\left(\frac{117}{7411}\right) && \text{by property 2} \\[2mm]
&= -\left(\frac{7411}{117}\right) && \text{by property 4} \\[2mm]
&= -\left(\frac{40}{117}\right) && \text{by property 1} \\[2mm]
&= -\left(\frac{2}{117}\right)^3 \left(\frac{5}{117}\right) && \text{by property 3} \\[2mm]
&= \left(\frac{5}{117}\right) && \text{by property 2} \\[2mm]
&= \left(\frac{117}{5}\right) && \text{by property 4} \\[2mm]
&= \left(\frac{2}{5}\right) && \text{by property 1} \\[2mm]
&= -1 && \text{by property 2}
\end{aligned}
$$

Notice that we successively apply properties 4, 1, 3, and 2 in this computation.
◻

In general, by applying these four properties in the same manner as was done in the example above, it is possible to compute a Jacobi symbol $\left(\frac{m}{n}\right)$ in polynomial time. The only arithmetic operations that are required are modular reductions and factoring out powers of two. Note that if an integer is represented in binary notation, then factoring out powers of two amounts to determining the number of trailing zeroes. So, the complexity of the algorithm is determined by the number of modular reductions that must be done. It is not difficult to show that at most $O(\log n)$ modular reductions are performed, each of which can be done in time $O((\log n)^2)$. This shows that the complexity is $O((\log n)^3)$, which is polynomial in $\log n$. (In fact, the complexity can be shown to be $O((\log n)^2)$ by more precise analysis.)

Suppose that we have generated a random number n and tested it for primality using the SOLOVAY-STRASSEN ALGORITHM. If we have run the algorithm m times, what is our confidence that n is prime? It is tempting to conclude that the probability that such an integer n is prime is $1 - 2^{-m}$. This conclusion is often stated in both textbooks and technical articles, but it cannot be inferred from the given data.

We need to be careful about our use of probabilities. Suppose we define the following random variables: **a** denotes the event

"a random odd integer n of a specified size is composite,"

and **b** denotes the event

"the algorithm answers 'n is prime' m times in succession."

It is certainly the case that the probability $\mathbf{Pr}[\mathbf{b}|\mathbf{a}] \leq 2^{-m}$. However, the probability that we are really interested is $\mathbf{Pr}[\mathbf{a}|\mathbf{b}]$, which is usually not the same as $\mathbf{Pr}[\mathbf{b}|\mathbf{a}]$.

We can compute $\mathbf{Pr}[\mathbf{a}|\mathbf{b}]$ using Bayes' theorem (Theorem 2.1). In order to do this, we need to know $\mathbf{Pr}[\mathbf{a}]$. Suppose $N \leq n \leq 2N$. Applying the Prime number theorem, the number of (odd) primes between N and $2N$ is approximately

$$\frac{2N}{\ln 2N} - \frac{N}{\ln N} \approx \frac{N}{\ln N}$$

$$\approx \frac{n}{\ln n}.$$

Since there are $N/2 \approx n/2$ odd integers between N and $2N$, we will use the estimate

$$\mathbf{Pr}[\mathbf{a}] \approx 1 - \frac{2}{\ln n}.$$

Then we can compute as follows:

$$\mathbf{Pr}[\mathbf{a}|\mathbf{b}] = \frac{\mathbf{Pr}[\mathbf{b}|\mathbf{a}]\mathbf{Pr}[\mathbf{a}]}{\mathbf{Pr}[\mathbf{b}]}$$

$$= \frac{\mathbf{Pr}[\mathbf{b}|\mathbf{a}]\mathbf{Pr}[\mathbf{a}]}{\mathbf{Pr}[\mathbf{b}|\mathbf{a}]\mathbf{Pr}[\mathbf{a}] + \mathbf{Pr}[\mathbf{b}|\mathbf{\bar{a}}]\mathbf{Pr}[\mathbf{\bar{a}}]}$$

$$\approx \frac{\mathbf{Pr}[\mathbf{b}|\mathbf{a}]\left(1 - \frac{2}{\ln n}\right)}{\mathbf{Pr}[\mathbf{b}|\mathbf{a}]\left(1 - \frac{2}{\ln n}\right) + \frac{2}{\ln n}}$$

$$= \frac{\mathbf{Pr}[\mathbf{b}|\mathbf{a}](\ln n - 2)}{\mathbf{Pr}[\mathbf{b}|\mathbf{a}](\ln n - 2) + 2}$$

$$\leq \frac{2^{-m}(\ln n - 2)}{2^{-m}(\ln n - 2) + 2}$$

$$= \frac{\ln n - 2}{\ln n - 2 + 2^{m+1}}.$$

FIGURE 5.1
Error probabilities for the SOLOVAY-STRASSEN ALGORITHM

m	2^{-m}	bound on error probability
1	.500	.989
2	.250	.978
5	$.312 \times 10^{-1}$	.847
10	$.977 \times 10^{-3}$	.147
20	$.954 \times 10^{-6}$	$.168 \times 10^{-3}$
30	$.931 \times 10^{-9}$	$.164 \times 10^{-6}$
50	$.888 \times 10^{-15}$	$.157 \times 10^{-12}$
100	$.789 \times 10^{-30}$	$.139 \times 10^{-27}$

Note that in this computation, $\overline{\mathbf{a}}$ denotes the event

"a random odd integer n is prime."

It is interesting to compare the two quantities $(\ln n - 2)/(\ln n - 2 + 2^{m+1})$ and 2^{-m} as a function of m. Suppose that $n \approx 2^{512} \approx e^{355}$, since these are the sizes of primes p and q used to construct an RSA modulus. Then the first function is roughly $353/(353 + 2^{m+1})$. We tabulate the two functions for some values of m in Figure 5.1.

Although $353/(353 + 2^{m+1})$ approaches zero exponentially quickly, it does not do so as quickly as 2^{-m}. In practice, however, one would take m to be something like 50 or 100, which will reduce the probability of error to a very small quantity.

We conclude with another Monte Carlo algorithm for **Composites** which is called the MILLER-RABIN ALGORITHM (also known as the "strong pseudo-prime test"). This algorithm is presented as Algorithm 5.7.

Algorithm 5.7: MILLER-RABIN(n)

write $n - 1 = 2^k m$, where m is odd
choose a random integer $a, 1 \leq a \leq n - 1$
$b \leftarrow a^m \bmod n$
if $b \equiv 1 \pmod{n}$
   then return ("n is prime")
for $i \leftarrow 0$ **to** $k - 1$
   do $\begin{cases} \textbf{if } b \equiv -1 \pmod{n} \\ \quad \textbf{then return } (\text{"}n \text{ is prime"}) \\ \quad \textbf{else } b \leftarrow b^2 \bmod n \end{cases}$
return ("n is composite")

Algorithm 5.7 is clearly a polynomial-time algorithm: an elementary analysis shows that its complexity is $O((\log n)^3)$, as is the SOLOVAY-STRASSEN ALGO-RITHM. In fact, the MILLER-RABIN ALGORITHM performs better in practice than the SOLOVAY-STRASSEN ALGORITHM.

We show now that this algorithm cannot answer "n is composite" if n is prime, i.e., the algorithm is yes-biased.

THEOREM 5.11   *The* MILLER-RABIN ALGORITHM *for* **Composites** *is a yes-biased Monte Carlo algorithm.*

PROOF   We will prove this by assuming that Algorithm 5.7 answers "n is composite" for some prime integer n, and obtain a contradiction. Since the algorithm answers "n is composite," it must be the case that $a^m \not\equiv 1 \pmod{n}$. Now consider the sequence of values b tested in the algorithm. Since b is squared in each iteration of the **for** loop, we are testing the values $a^m, a^{2m}, \ldots, a^{2^{k-1}m}$. Since the algorithm answers "n is composite," we conclude that

$$a^{2^i m} \not\equiv -1 \pmod{n}$$

for $0 \leq i \leq k - 1$.

Now, using the assumption that n is prime, Fermat's theorem (Corollary 5.6) tells us that

$$a^{2^k m} \equiv 1 \pmod{n}$$

since $n - 1 = 2^k m$. Then $a^{2^{k-1}m}$ is a square root of 1 modulo n. Because n is prime, there are only two square roots of 1 modulo n, namely, $\pm 1 \bmod n$. We have that

$$a^{2^{k-1}m} \not\equiv -1 \pmod{n},$$

so it follows that

$$a^{2^{k-1}m} \equiv 1 \pmod{n}.$$

Then $a^{2^{k-2}m}$ must be a square root of 1. By the same argument,

$$a^{2^{k-2}m} \equiv 1 \pmod{n}.$$

Repeating this argument, we eventually obtain

$$a^m \equiv 1 \pmod{n},$$

which is a contradiction, since the algorithm would have answered "n is prime" in this case.   ∎

It remains to consider the error probability of the MILLER-RABIN ALGO-RITHM. Although we will not prove it here, the error probability can be shown to be at most $1/4$.

5.5 Square Roots Modulo n

In this section, we briefly discuss several useful results related to the existence of square roots modulo n. Throughout this section, we will suppose that n is odd and $\gcd(n, a) = 1$. The first question we will consider is the number of solutions $y \in \mathbb{Z}_n$ to the congruence $y^2 \equiv a \pmod{n}$. We already know from Section 5.4 that this congruence has either zero or two solutions if n is prime, depending on whether $\left(\frac{a}{n}\right) = -1$ or $\left(\frac{a}{n}\right) = 1$.

Our next theorem extends this characterization to (odd) prime powers. A proof is outlined in the Exercises.

THEOREM 5.12 *Suppose that p is an odd prime, e is a positive integer, and $\gcd(a, p) = 1$. Then the congruence $y^2 \equiv a \pmod{p^e}$ has no solutions if $\left(\frac{a}{p}\right) = -1$, and two solutions (modulo p^e) if $\left(\frac{a}{p}\right) = 1$.*

Notice that Theorem 5.12 tells us that the existence of square roots of a modulo p^e can be determined by evaluating the Legendre symbol $\left(\frac{a}{p}\right)$.

It is not difficult to extend Theorem 5.12 to the case of an arbitrary odd integer n. The following result is basically an application of the Chinese remainder theorem.

THEOREM 5.13 *Suppose that $n > 1$ is an odd integer having factorization*

$$n = \prod_{i=1}^{\ell} p_i^{e_i},$$

where the p_i's are distinct primes and the e_i's are positive integers. Suppose further that $\gcd(a, n) = 1$. Then the congruence $y^2 \equiv a \pmod{n}$ has 2^ℓ solutions modulo n if $\left(\frac{a}{p_i}\right) = 1$ for all $i \in \{1, \ldots, \ell\}$, and no solutions, otherwise.

PROOF It is clear that $y^2 \equiv a \pmod{n}$ if and only if $y^2 \equiv a \pmod{p_i^{e_i}}$ for all $i \in \{1, \ldots, \ell\}$. If $\left(\frac{a}{p_i}\right) = -1$ for some i, then the congruence $y^2 \equiv a \pmod{p_i^{e_i}}$ has no solutions, and hence $y^2 \equiv a \pmod{n}$ has no solutions.

Now suppose that $\left(\frac{a}{p_i}\right) = 1$ for all $i \in \{1, \ldots, \ell\}$. It follows from Theorem 5.12 that each congruence $y^2 \equiv a \pmod{p_i^{e_i}}$ has two solutions modulo $p_i^{e_i}$, say $y \equiv b_{i,1}$ or $b_{i,2} \pmod{p_i^{e_i}}$. For $1 \leq i \leq \ell$, let $b_i \in \{b_{i,1}, b_{i,2}\}$. Then the system of congruences $y \equiv b_i \pmod{p_i^{e_i}}$ $(1 \leq i \leq \ell)$ has a unique solution modulo n, which can be found using the Chinese remainder theorem. There are 2^ℓ ways to choose the ℓ-tuple $(b_1, \ldots, b_\ell)$, and therefore there are 2^ℓ solutions modulo n to the congruence $y^2 \equiv a \pmod{n}$. ∎

Suppose that $x^2 \equiv y^2 \equiv a \pmod{n}$, where $\gcd(a, n) = 1$. Let $z = zy^{-1}$ mod n. It follows that $z^2 \equiv 1 \pmod{n}$. Conversely, if $z^2 \equiv 1 \pmod{n}$, then $(xz)^2 \equiv x^2 \pmod{n}$ for any x. It is therefore possible to obtain all 2^{ℓ} square roots of an element $a \in \mathbb{Z}_n^*$ by taking all 2^{ℓ} products of one given square root of a with the 2^{ℓ} square roots of 1. We will make use of this observation later in this chapter.

5.6 Factoring Algorithms

The most obvious way to attack the *RSA Cryptosystem* is to attempt to factor the public modulus. There is a huge amount of literature on factoring algorithms, and a thorough treatment would require more pages than we have in this book. We will just try to give a brief overview here, including an informal discussion of the best current factoring algorithms and their use in practice. The three algorithms that are most effective on very large numbers are the *quadratic sieve*, the *elliptic curve factoring algorithm* and the *number field sieve*. Other well-known algorithms that were precursors include Pollard's rho-method and $p-1$ algorithm, Williams' $p+1$ algorithm, the continued fraction algorithm, and, of course, trial division.

Throughout this section, we suppose that the integer n that we wish to factor is odd. If n is composite, then it is easy to see that n has a prime factor $p \leq \lfloor \sqrt{n} \rfloor$. Therefore, the simple method of *trial division*, which consists of dividing n by every odd integer up to $\lfloor \sqrt{n} \rfloor$, suffices to determine if n is prime or composite. If $n < 10^{12}$, say, this is a perfectly reasonable factorization method, but for larger n we generally need to use more sophisticated techniques.

When we say that we want to factor n, we could ask for a complete factorization into primes, or we might be content with finding any non-trivial factor. In most of the algorithms we study, we are just searching for an arbitrary non-trivial factor. In general, we obtain factorizations of the form $n = n_1 n_2$, where $1 < n_1 < n$ and $1 < n_2 < n$. If we desire a complete factorization of n into primes, we could test n_1 and n_2 for primality using a randomized primality test, and then factor one or both of them further if they are not prime.

5.6.1 The Pollard $p - 1$ Algorithm

As an example of a simple algorithm that can sometimes be applied to larger integers, we describe Pollard's $p - 1$ *algorithm*, which dates from 1974. This algorithm, presented as Algorithm 5.8, has two inputs: the (odd) integer n to be factored, and a prespecified "bound," B.

Algorithm 5.8: POLLARD $p - 1$ FACTORING ALGORITHM(n, B)

$a \leftarrow 2$
for $j \leftarrow 2$ **to** B
  do $a \leftarrow a^j \bmod n$
$d \leftarrow \gcd(a - 1, n)$
if $1 < d < n$
  then return (d)
  else return ("failure")

Here is what is taking place in the $p - 1$ algorithm: Suppose p is a prime divisor of n, and suppose that $q \leq B$ for every prime power $q \mid (p - 1)$. Then it must be the case that

$$(p - 1) \mid B!$$

At the end of the **for** loop, we have that

$$a \equiv 2^{B!} \pmod{n}.$$

Since $p \mid n$, it must be the case that

$$a \equiv 2^{B!} \pmod{p}.$$

Now,

$$2^{p-1} \equiv 1 \pmod{p}$$

by Fermat's theorem. Since $(p - 1) \mid B!$, it follows that

$$a \equiv 1 \pmod{p},$$

and hence $p \mid (a - 1)$. Since we also have that $p \mid n$, we see that $p \mid d$, where $d = \gcd(a - 1, n)$. The integer d will be a non-trivial divisor of n (unless $a = 1$). Having found a non-trivial factor d, we would then proceed to attempt to factor d and n/d if they expected to be composite.

Here is an example to illustrate.

Example 5.9 Suppose $n = 15770708441$. If we apply Algorithm 5.8 with $B = 180$, then we find that $a = 11620221425$ and d is computed to be 135979. In fact, the complete factorization of n into primes is

$$15770708441 = 135979 \times 115979.$$

In this example, the factorization succeeds because 135978 has only "small" prime factors:

$$135978 = 2 \times 3 \times 131 \times 173.$$

Hence, by taking $B \geq 173$, it will be the case that $135978 \mid B!$, as desired.   ☐

In the $p - 1$ algorithm, there are $B - 1$ modular exponentiations, each requiring at most $2 \log_2 B$ modular multiplications using the SQUARE-AND-MULTIPLY ALGORITHM. The gcd can be computed in time $O((\log n)^3)$ using the EXTENDED EUCLIDEAN ALGORITHM. Hence, the complexity of the algorithm is $O(B \log B (\log n)^2 + (\log n)^3)$. If the integer B is $O((\log n)^i)$ for some fixed integer i, then the algorithm is indeed a polynomial-time algorithm (as a function of $\log n$); however, for such a choice of B the probability of success will be very small. On the other hand, if we increase the size of B drastically, say to $\sqrt{n}$, then the algorithm is guaranteed to be successful, but it will be no faster than trial division.

Thus, the drawback of this method is that it requires n to have a prime factor p such that $p - 1$ has only "small" prime factors. It would be very easy to construct an RSA modulus $n = pq$ which would resist factorization by this method. One would start by finding a large prime p_1 such that $p = 2p_1 + 1$ is also prime, and a large prime q_1 such that $q = 2q_1 + 1$ is also prime (using one of the Monte Carlo primality testing algorithms discussed in Section 5.4). Then the RSA modulus $n = pq$ will be resistant to factorization using the $p - 1$ method.

The more powerful elliptic curve algorithm, developed by Lenstra in the mid-1980's, is in fact a generalization of the $p - 1$ method. The success of the elliptic curve method depends on the more likely situation that an integer "close to" p has only "small" prime factors. Whereas the $p - 1$ method depends on a relation that holds in the group $\mathbb{Z}_p$, the elliptic curve method involves groups defined on elliptic curves modulo p.

5.6.2   The Pollard Rho Algorithm

Let p be the smallest prime divisor of n. Suppose there exist two integers $x, x' \in \mathbb{Z}_n$, such that $x \neq x'$ and $x \equiv x' \pmod{p}$. Then $p \leq \gcd(x - x', n) < n$, so we obtain a non-trivial factor of n by computing a greatest common divisor. (Note that the value of p does not need to be known ahead of time in order for this method to work.)

Suppose we try to factor n by first choosing a random subset $X \subseteq \mathbb{Z}_n$, and then computing $\gcd(x - x', n)$ for all distinct values $x, x' \in X$. This method will be successful if and only if the mapping $x \mapsto x \bmod p$ yields at least one collision for $x \in X$. This situation can be analyzed using the birthday paradox described in Section 4.2.2: if $|X| \approx 1.17\sqrt{p}$, then there is a 50% probability that there is at least one collision, and hence a non-trivial factor of n will be found. However, in order to find a collision of the form $x \bmod p = x' \bmod p$, we need to compute $\gcd(x - x', n)$. (We cannot explicitly compute the values $x \bmod p$ for $x \in X$, and sort the resulting list, as suggested in Section 4.2.2, because the value of p is not known.) This means that we would expect to compute more than $\binom{|X|}{2} > p/2$ gcd's before finding a factor of n.

The POLLARD RHO ALGORITHM incorporates a variation of this technique that requires fewer gcd computations and less memory. Suppose that the function

f is a polynomial with integer coefficients, e.g., $f(x) = x^2 + a$, where a is a small constant ($a = 1$ is a commonly used value). Let's assume that the mapping $x \mapsto f(x) \bmod p$ behaves like a random mapping. (It is of course not "random," which means that what we are presenting is a heuristic analysis rather than a rigorous proof.) Let $x_1 \in \mathbb{Z}_n$, and consider the sequence $x_1, x_2, \ldots$, where

$$x_j = f(x_{j-1}) \bmod n,$$

for all $j \geq 2$. Let m be an integer, and define $X = \{x_1, \ldots, x_m\}$. To simplify matters, suppose that X consists of m distinct residues modulo n. Hopefully it will be the case that X is a random subset of m elements of $\mathbb{Z}_n$.

We are looking for two distinct values $x_i, x_j \in X$ such that $\gcd(x_j - x_i, n) > 1$. Each time we compute a new term x_j in the sequence, we could compute $\gcd(x_j - x_i, n)$ for all $i < j$. However, it turns out that we can reduce the number of gcd computations greatly. We describe how this can be done.

Suppose that $x_i \equiv x_j \pmod{p}$. Using the fact that f is a polynomial with integer coefficients, we have that $f(x_i) \equiv f(x_j) \pmod{p}$. Recall that $x_{i+1} = f(x_i) \bmod n$ and $x_{j+1} = f(x_j) \bmod n$. Then

$$x_{i+1} \bmod p = (f(x_i) \bmod n) \bmod p = f(x_i) \bmod p,$$

because $p \mid n$. Similarly,

$$x_{j+1} \bmod p = f(x_j) \bmod p.$$

Therefore, $x_{i+1} \equiv x_{j+1} \pmod{p}$. Repeating this argument, we obtain the following important result:

If $x_i \equiv x_j \pmod{p}$, then $x_{i+\delta} \equiv x_{j+\delta} \pmod{p}$ for all integers $\delta \geq 0$.

Denoting $\ell = j - i$, it follows that $x_{i'} \equiv x_{j'} \pmod{p}$ if $j' > i' \geq i$ and $j' - i' \equiv 0 \pmod{\ell}$.

Suppose that we construct a graph G on vertex set $\mathbb{Z}_p$, where for all $i \geq 1$, we have a directed edge from $x_i \bmod p$ to $x_{i+1} \bmod p$. There must exist a first pair x_i, x_j with $i < j$ such that $x_i \equiv x_j \pmod{p}$. By the observation made above, it is easily seen that the graph G consists of a "tail"

$$x_1 \bmod p \rightarrow x_2 \bmod p \rightarrow \cdots \rightarrow x_i \bmod p,$$

and an infinitely repeated cycle of length ℓ, having vertices

$$x_i \bmod p \rightarrow x_{i+1} \bmod p \rightarrow \cdots \rightarrow x_j \bmod p = x_i \bmod p.$$

Thus G looks like the Greek letter ρ, which is the reason for the name "rho algorithm."

We illustrate the above with an example.

Example 5.10 Suppose that $n = 7171 = 71 \times 101$, $f(x) = x^2 + 1$ and $x_1 = 1$.
The sequence of x_i's begins as follows:

$$\begin{array}{ccccccc}
1 & 2 & 5 & 26 & 677 & 6557 & 4105 \\
6347 & 4903 & 2218 & 219 & 4936 & 4210 & 4560 \\
4872 & 375 & 4377 & 4389 & 2016 & 5471 & 88
\end{array}$$

The above values, when reduced modulo 71, are as follows:

$$\begin{array}{ccccccc}
1 & 2 & 5 & 26 & 38 & 25 & 58 \\
28 & 4 & 17 & 6 & 37 & 21 & 16 \\
44 & 20 & 46 & 58 & 28 & 4 & 17
\end{array}$$

The first collision in the above list is

$$x_7 \bmod 71 = x_{18} \bmod 71 = 58.$$

Therefore the graph G consists of a tail of length seven and a cycle of length 11.
$\square$

We have already mentioned that our goal is to discover two terms $x_i \equiv x_j$
(mod p) with $i < j$, by computing a greatest common divisor. It is not necessary
that we discover the first occurrence of a collision of this type. In order to simplify
and improve the algorithm, we restrict our search for collisions by taking $j = 2i$.
The resulting algorithm is presented as Algorithm 5.9.

Algorithm 5.9: POLLARD RHO FACTORING ALGORITHM(n, x_1)

external f
$x \leftarrow x_1$
$x' \leftarrow f(x) \bmod n$
$p \leftarrow \gcd(x - x', n)$
while $p = 1$
$\quad$**do** $\begin{cases} \textbf{comment: in the } i\text{th iteration, } x = x_i \text{ and } x' = x_{2i} \\[4pt] x \leftarrow f(x) \bmod n \\ x' \leftarrow f(x') \bmod n \\ x' \leftarrow f(x') \bmod n \\ p \leftarrow \gcd(x - x', n) \end{cases}$
if $p = n$
$\quad$**then return** ("failure")
$\quad$**else return** (p)

This algorithm is not hard to analyze. If $x_i \equiv x_j$ (mod p), then it is also the
case that $x_{i'} \equiv x_{2i'}$ (mod p) for all i' such that $i' \equiv 0$ (mod ℓ) and $i' \geq i$.
Among the ℓ consecutive integers $i, \ldots, j - 1$, there must be one that is divisible

by ℓ. Therefore the smallest value i' that satisfies the two conditions above is at most $j - 1$. Hence, the number of iterations required to find a factor p is at most j, which is expected to be at most $\sqrt{p}$.

In Example 5.10, the first collision modulo 71 occurs for $i = 7$, $j = 18$. The smallest integer $i' \geq 7$ that is divisible by 11 is $i' = 11$. Therefore Algorithm 5.9 will discover the factor 71 of n when it computes $\gcd(x_{11} - x_{22}, n) = 71$.

In general, since $p < \sqrt{n}$, the expected complexity of the algorithm is $O(n^{1/4})$ (ignoring logarithmic factors). However, we again emphasize that this is a heuristic analysis, and not a mathematical proof. On the other hand, the actual performance of the algorithm in practice is similar to this estimate.

It is possible that Algorithm 5.9 could fail to find a nontrivial factor of n. This happens if and only if the first values x and x' which satisfy $x \equiv x' \pmod{p}$ actually satisfy $x \equiv x' \pmod{n}$ (this is equivalent to $x = x'$, because x and x' are reduced modulo n). We would estimate heuristically that the probability of this situation occurring is roughly p/n, which is quite small when n is large, because $p < \sqrt{n}$. If the algorithm does fail in this way, it is a simple matter to run it again with a different initial value or a different choice for the function f.

The reader might wish to run Algorithm 5.9 on a larger value of n. When $n = 15770708441$ (the same value of n considered in Example 5.9), $x_1 = 1$ and $f(x) = x^2 + 1$, it can be verified that $x_{422} = 2261992698$, $x_{211} = 7149213937$, and

$$\gcd(x_{422} - x_{211}, n) = 135979.$$

5.6.3 Dixon's Random Squares Algorithm

Many factoring algorithms are based on the following very simple idea. Suppose we can find $x \not\equiv \pm y \pmod{n}$ such that $x^2 \equiv y^2 \pmod{n}$. Then

$$n \mid (x - y)(x + y)$$

but neither of $x - y$ or $x + y$ is divisible by n. It therefore follows that $\gcd(x + y, n)$ is a non-trivial factor of n (and similarly, $\gcd(x - y, n)$ is also a non-trivial factor of n).

As an example, it is easy to verify that $10^2 \equiv 32^2 \pmod{77}$. By computing $\gcd(10 + 32, 77) = 7$, we discover the factor 7 of 77.

The random squares algorithm uses a *factor base*, which is a set $\mathcal{B}$ of the b smallest primes, for an appropriate value b. We first obtain several integers z such that all the prime factors of $z^2 \bmod n$ occur in the factor base $\mathcal{B}$. (How this is done will be discussed a bit later.) The idea is to then take the product of a subset of these z's in such a way that every prime in the factor base is used an even number of times. This then gives us a congruence of the desired type $x^2 \equiv y^2 \pmod{n}$, which (we hope) will lead to a factorization of n.

We illustrate with a carefully contrived example.

Example 5.11 Suppose $n = 15770708441$ (this was the same n that we used in Example 5.9). Let $b = 6$; then $\mathcal{B} = \{2, 3, 5, 7, 11, 13\}$. Consider the three congruences:

$$8340934156^2 \equiv 3 \times 7 \pmod{n}$$

$$12044942944^2 \equiv 2 \times 7 \times 13 \pmod{n}$$

$$2773700011^2 \equiv 2 \times 3 \times 13 \pmod{n}.$$

If we take the product of these three congruences, then we have

$$(8340934156 \times 12044942944 \times 2773700011)^2 \equiv (2 \times 3 \times 7 \times 13)^2 \pmod{n}.$$

Reducing the expressions inside the parentheses modulo n, we have

$$9503435785^2 \equiv 546^2 \pmod{n}.$$

Then, using the EUCLIDEAN ALGORITHM, we compute

$$\gcd(9503435785 - 546, 15770708441) = 115759,$$

finding the factor 115759 of n.   ☐

Suppose $\mathcal{B} = \{p_1, \ldots, p_b\}$ is the factor base. Let c be slightly larger than b (say $c = b + 4$), and suppose we have obtained c congruences:

$$z_j^2 \equiv p_1^{\alpha_{1j}} \times p_2^{\alpha_{2j}} \cdots \times p_b^{\alpha_{bj}} \pmod{n},$$

$1 \leq j \leq c$. For each j, consider the vector

$$a_j = (\alpha_{1j} \bmod 2, \ldots, \alpha_{bj} \bmod 2) \in (\mathbb{Z}_2)^b.$$

If we can find a subset of the a_j's that sum modulo 2 to the vector $(0, \ldots, 0)$, then the product of the corresponding z_j's will use each factor in $\mathcal{B}$ an even number of times.

We illustrate by returning to Example 5.11, where there exists a dependence even though $c < b$ in this case.

Example 5.11 **(Cont.)** The three vectors a_1, a_2, a_3 are as follows:

$$a_1 = (0, 1, 0, 1, 0, 0)$$

$$a_2 = (1, 0, 0, 1, 0, 1)$$

$$a_3 = (1, 1, 0, 0, 0, 1).$$

It is easy to see that

$$a_1 + a_2 + a_3 = (0, 0, 0, 0, 0, 0) \bmod 2.$$

This gives rise to the congruence we saw earlier that successfully factored n.   ☐

Observe that finding a subset of the c vectors $a_1, \ldots, a_c$ that sums modulo 2 to the all-zero vector is nothing more than finding a linear dependence (over $\mathbb{Z}_2$) of these vectors. Provided $c > b$, such a linear dependence must exist, and it can be found easily using the standard method of Gaussian elimination. The reason why we take $c > b + 1$ is that there is no guarantee that any given congruence $x^2 \equiv y^2$ (mod n) will yield the factorization of n. However, we argue heuristically that $x \equiv \pm y$ (mod n) at most 50% of the time, as follows. Suppose that $x^2 \equiv y^2 \equiv a$ (mod n), where $\gcd(a, n) = 1$. Theorem 5.13 tells us that a has 2^ℓ square roots modulo n, where ℓ is the number of prime divisors of n. If $\ell \geq 2$, then a has at least four square roots. Hence, if we assume that x and y are "random," we can then conclude that $x \equiv \pm y$ (mod n) with probability $2/2^\ell \leq 1/2$.

Now, if $c > b + 1$, we can obtain several such congruences of the form $x^2 \equiv y^2$ (mod n) (arising from different linear dependencies among the a_j's). Hopefully, at least one of the resulting congruences will yield a congruence of the form $x^2 \equiv y^2 \bmod n$ where $x \not\equiv \pm y$ (mod n), and a non-trivial factor of n will be obtained.

We now discuss how to obtain integers z such that the values $z^2 \bmod n$ factor completely over a given factor base $\mathcal{B}$. There are several methods of doing this. One way is simply to choose the z's at random; this approach yields the so-called *random squares algorithm*. However, it is particularly useful to try integers of the form $j + \lceil \sqrt{kn} \rceil$, $j = 0, 1, 2, \ldots$, $k = 1, 2, \ldots$. These integers tend to be small when squared and reduced modulo n, and hence they have a higher than average probability of factoring over $\mathcal{B}$. Another useful trick is to try integers of the form $z = \lfloor \sqrt{kn} \rfloor$. When squared and reduced modulo n, these integers are a bit less than n. This means that $-z^2 \bmod n$ is small and can perhaps be factored over $\mathcal{B}$. Therefore, if we include -1 in $\mathcal{B}$, we can factor $z^2 \bmod n$ over $\mathcal{B}$.

We illustrate these techniques with a small example.

Example 5.12   Suppose that $n = 1829$ and $\mathcal{B} = \{-1, 2, 3, 5, 7, 11, 13\}$. We compute $\sqrt{n} = 42.77$, $\sqrt{2n} = 60.48$, $\sqrt{3n} = 74.07$ and $\sqrt{4n} = 85.53$. Suppose we take $z = 42, 43, 64, 65, 74, 75, 85, 86$. We obtain several factorizations of $z^2 \bmod n$ over $\mathcal{B}$. In the following table, all congruences are modulo n:

$$
\begin{aligned}
z_1{}^2 &\equiv 42^2 &&\equiv -65 &&\equiv (-1) \times 5 \times 13 \\
z_2{}^2 &\equiv 43^2 &&\equiv 20 &&\equiv 2^2 \times 5 \\
z_3{}^2 &\equiv 61^2 &&\equiv 63 &&\equiv 3^2 \times 7 \\
z_4{}^2 &\equiv 74^2 &&\equiv -11 &&\equiv (-1) \times 11 \\
z_5{}^2 &\equiv 85^2 &&\equiv -91 &&\equiv (-1) \times 7 \times 13 \\
z_6{}^2 &\equiv 86^2 &&\equiv 80 &&\equiv 2^4 \times 5.
\end{aligned}
$$

We therefore have six factorizations, which yield six vectors in $(\mathbb{Z}_2)^7$. This is not enough to guarantee a dependence relation, but it turns out to be sufficient in this particular case. The six vectors are as follows:

$$a_1 = (1, 0, 0, 1, 0, 0, 1)$$
$$a_2 = (0, 0, 0, 1, 0, 0, 0)$$
$$a_3 = (0, 0, 0, 0, 1, 0, 0)$$
$$a_4 = (1, 0, 0, 0, 0, 1, 0)$$
$$a_5 = (1, 0, 0, 0, 1, 0, 1)$$
$$a_6 = (0, 0, 0, 1, 0, 0, 0).$$

Clearly $a_2 + a_6 = (0, 0, 0, 0, 0, 0, 0)$; however, the reader can check that this dependence relation does not yield a factorization of n. A dependence relation that does work is

$$a_1 + a_2 + a_3 + a_5 = (0, 0, 0, 0, 0, 0, 0).$$

The congruence that we obtain is

$$(42 \times 43 \times 61 \times 85)^2 \equiv (2 \times 3 \times 5 \times 7 \times 13)^2 \pmod{1829}.$$

This simplifies to give

$$1459^2 \equiv 901^2 \pmod{1829}.$$

It is then straightforward to compute

$$\gcd(1459 + 901, 1829) = 59,$$

and thus we have obtained a nontrivial factor of n.   ⬚

   An important general question is how large the factor base should be (as a function of the integer n that we are attempting to factor) and what the complexity of the algorithm is. In general, there is a trade-off: if $b = |\mathcal{B}|$ is large, then it is more likely that an integer $z^2 \bmod n$ factors over $\mathcal{B}$. But the larger b is, the more congruences we need to accumulate before we are able to find a dependence relation. A good choice for b can be determined with the help of some results from number theory. We discuss some of the main ideas now. This will be a heuristic analysis in which we will be assuming that the integers z are chosen randomly.

   Suppose that n and m are positive integers. We say that n is *m-smooth* provided that every prime factor of n is less than or equal to m. $\Psi(n, m)$ is defined to be the number of positive integers less than or equal to n that are m-smooth. An important result in number theory says that, if $n \gg m$, then

$$\frac{\Psi(n, m)}{n} \approx \frac{1}{u^u},$$

where $u = \log n / \log m$. Observe that $\Psi(n, m)/n$ represents the probability that a random integer in the set $\{1, \ldots, n\}$ is m-smooth.

Suppose that $n \approx 2^r$ and $m \approx 2^s$. Then

$$u = \frac{\log n}{\log m} \approx \frac{r}{s}.$$

Division of an r-bit integer by an s-bit integer can be done in time $O(rs)$. From this, it is possible to show that we can determine if an integer in the set $\{1, \ldots, n\}$ is m-smooth in time $O(rsm)$ if we assume that $r < m$ (see the Exercises).

Our factor base $\mathcal{B}$ consists of all the primes less than or equal to m. Therefore, applying the Prime number theorem, we have that

$$|\mathcal{B}| = b = \pi(m) \approx \frac{m}{\ln m}.$$

We need to find slightly more than b m-smooth squares modulo n in order for the algorithm to succeed. We expect to test bu^u integers in order to find b of them that are m-smooth. Therefore, the expected time to find the necessary m-smooth squares is $O(bu^u \times rsm)$. We have that b is $O(m/s)$, so the running time of the first part of the algorithm is $O(rm^2 u^u)$.

In the second part of the algorithm, we need to reduce the associated matrix modulo 2, construct our congruence of the form $x^2 \equiv y^2 \pmod{n}$, and apply the EUCLIDEAN ALGORITHM. It can be checked without too much difficulty that these steps can be done in time that is polynomial in r and m, say $O(r^i m^j)$, where i and j are positive integers. (On average, this second part of the algorithm must be done at most twice, because the probability that a congruence does not provide a factor of n is at most $1/2$. This contributes a constant factor of at most 2, which is absorbed into the big-oh.)

At this point, we know that the total running time of the algorithm can be written in the form $O(rm^2 u^u + r^i m^j)$. Recall that $n \approx 2^r$ is given, and we are trying to choose $m \approx 2^s$ to optimize the running time. It turns out that a good choice for m is to take $s \approx \sqrt{r \log_2 r}$. Then

$$u \approx \frac{r}{s} \approx \sqrt{\frac{r}{\log_2 r}}.$$

Now we can compute

$$\log_2 u^u = u \log_2 u$$

$$\approx \sqrt{\frac{r}{\log_2 r}} \log_2 \left(\sqrt{\frac{r}{\log_2 r}} \right)$$

$$< \sqrt{\frac{r}{\log_2 r}} \log_2 \sqrt{r}$$

$$= \sqrt{\frac{r}{\log_2 r}} \times \frac{\log_2 r}{2}$$

$$= \frac{\sqrt{r \log_2 r}}{2}.$$

It follows that

$$u^u \leq 2^{0.5 \sqrt{r \log_2 r}}.$$

We also have that

$$m \approx 2^{\sqrt{r \log_2 r}}$$

and

$$r = 2^{\log_2 r}.$$

Hence, the total running time can be expressed in the form

$$O \left(2^{\log_2 r + 2\sqrt{r \log_2 r} + 0.5\sqrt{r \log_2 r}} + 2^{i \log_2 r + j \sqrt{r \log_2 r}} \right),$$

which is easily seen to be

$$O \left(2^{c\sqrt{r \log_2 r}} \right)$$

for some constant c. Using the fact that $r \approx \log_2 n$, we obtain a running time of

$$O \left(2^{c\sqrt{\log_2 n \log_2 \log_2 n}} \right).$$

Often the running time is expressed in terms of logarithms and exponentials to the base e. A more precise analysis, using an optimal choice of m, leads to the following commonly stated expected running time:

$$O \left(e^{(1+o(1))\sqrt{\ln n \ln \ln n}} \right).$$

5.6.4   Factoring Algorithms in Practice

One specific, well-known algorithm that has been widely used in practice is the quadratic sieve due to Pomerance. The name "quadratic sieve" comes from a sieving procedure (which we will not describe here) that is used to determine the values $z^2 \bmod n$ that factor over $\mathcal{B}$. The number field sieve is a more recent factoring algorithm from the late 1980's. It also factors n by constructing a congruence $x^2 \equiv y^2 \pmod{n}$, but it does so by means of computations in rings of algebraic integers. In recent years, the number field sieve has become the algorithm of choice for factoring large integers,

The asymptotic running times of the quadratic sieve, elliptic curve and number field sieve are as follows:

quadratic sieve	$O\left(e^{(1+o(1))\sqrt{\ln n \ln \ln n}}\right)$
elliptic curve	$O\left(e^{(1+o(1))\sqrt{2 \ln p \ln \ln p}}\right)$
number field sieve	$O\left(e^{(1.92+o(1))(\ln n)^{1/3}(\ln \ln n)^{2/3}}\right)$

The notation $o(1)$ denotes a function of n that approaches 0 as $n \to \infty$, and p denotes the smallest prime factor of n.

In the worst case, $p \approx \sqrt{n}$ and the asymptotic running times of the quadratic sieve and elliptic curve algorithms are essentially the same. But in such a situation, quadratic sieve is faster than elliptic curve. The elliptic curve method is more useful if the prime factors of n are of differing size. One very large number that was factored using the elliptic curve method was the Fermat number $2^{2^{11}} - 1$, which was factored in 1988 by Brent.

For factoring RSA moduli (where $n = pq$, p, q are prime, and p and q are roughly the same size), the quadratic sieve was the most-used algorithm up until the mid-1990's. The number field sieve is the most recently developed of the three algorithms. Its advantage over the other algorithms is that its asymptotic running time is faster than either quadratic sieve or the elliptic curve. The number field sieve has proven to be faster for numbers having more than about 125–130 digits. An early use of the number field sieve was in 1990, when Lenstra, Lenstra, Manasse, and Pollard factored $2^{2^9} - 1$ into three primes having 7, 49 and 99 digits.

Some notable milestones in factoring have included the following factorizations. In 1983, the quadratic sieve successfully factored a 69-digit number that was a (composite) factor of $2^{251} - 1$ (a computation which was done by Davis, Holdridge, and Simmons). Progress continued throughout the 1980's, and by 1989, numbers having up to 106 digits were factored by this method by Lenstra and Manasse, by distributing the computations to hundreds of widely separated workstations (they called this approach "factoring by electronic mail").

In April 1994, a 129-digit number known as RSA-129 was factored by Atkins, Graff, Lenstra, and Leyland using the quadratic sieve. (The numbers RSA-100, RSA-110, . . . , RSA-500 are a list of RSA moduli publicized on the Internet as "challenge" numbers for factoring algorithms. Each number RSA-d is a d-digit number that is the product of two primes of approximately the same length.) The factorization of RSA-129 required 5000 MIPS-years of computing time donated by over 600 researchers around the world.

RSA-130 was factored in 1996; RSA-140 was factored in February 1999; and the factorization of RSA-155 was completed on August 22, 1999. These three factorizations were accomplished using the number field sieve.

The last of these factorizations, RSA-155, involves a 512-bit modulus (note that 155 digits is roughly the same as 512 bits). It required about 8400 MIPS-years of computation involving about 300 PCs and workstations in six countries. This factorization provided convincing evidence that 512-bit RSA moduli, which were widely used in commercial implementations of RSA, should not be considered to be secure.

A factorization of a different 155-digit RSA modulus was also accomplished in order to solve the tenth and last stage of the "cipher challenge" in Simon Singh's "The Code Book." This factorization was performed by a group of five Swedish researchers using the number field sieve, and was completed in October 2000.

Extrapolating trends in factoring into the future, it has been suggested that 768-bit moduli will be factored by 2010, and 1024-bit moduli will be factored by the year 2018.

5.7 Other Attacks on RSA

In this section, we address the following question: are there possible attacks on the *RSA Cryptosystem* other than factoring n? For example, it is at least conceivable that there could exist a method of decrypting RSA ciphertexts that does not involve finding the factorization of the modulus n.

5.7.1 Computing $\phi(n)$

We first observe that computing $\phi(n)$ is no easier than factoring n. For, if n and $\phi(n)$ are known, and n is the product of two primes p, q, then n can be easily factored, by solving the two equations

$$n = pq$$

$$\phi(n) = (p-1)(q-1)$$

for the two "unknowns" p and q. This is easily accomplished, as follows. If we substitute $q = n/p$ into the second equation, we obtain a quadratic equation in the unknown value p:

$$p^2 - (n - \phi(n) + 1)p + n = 0. \tag{5.1}$$

The two roots of equation (5.1) will be p and q, the factors of n. Hence, if a cryptanalyst can learn the value of $\phi(n)$, then he can factor n and break the system. In other words, computing $\phi(n)$ is no easier than factoring n.

Here is an example to illustrate.

Example 5.13 Suppose $n = 84773093$, and the adversary has learned that $\phi(n) = 84754668$. This information gives rise to the following quadratic equation:

$$p^2 - 18426p + 84773093 = 0.$$

This can be solved by the quadratic formula, yielding the two roots 9539 and 8887. These are the two factors of n. □

5.7.2 The Decryption Exponent

We will now prove the very interesting result that, if the decryption exponent a is known, then n can be factored in polynomial time by means of a randomized algorithm. Therefore we can say that computing a is (essentially) no easier than factoring n. (However, this does not rule out the possibility of breaking the *RSA Cryptosystem* without computing a.) Notice that this result is of much more than theoretical interest. It tells us that if a is revealed (accidently or otherwise), then it is not sufficient for Bob to choose a new encryption exponent; he must also choose a new modulus n.

The algorithm we are going to describe is a randomized algorithm of the Las Vegas type (see Section 4.2.2 for the definition). Here, we consider Las Vegas algorithms having worst-case success probability at least $1 - \epsilon$. Therefore, for any problem instance, the algorithm may fail to give an answer with probability at most ϵ.

If we have such a Las Vegas algorithm, then we simply run the algorithm over and over again until it finds an answer. The probability that the algorithm will return "no answer" m times in succession is ϵ^m. The average (i.e., expected) number of times the algorithm must be run in order to obtain an answer is $1/(1-\epsilon)$ (see the Exercises).

We will describe a Las Vegas algorithm that will factor n with probability at least $1/2$ when given the values a, b and n as input. Hence, if the algorithm is run m times, then n will be factored with probability at least $1 - 1/2^m$.

The algorithm is based on certain facts concerning square roots of 1 modulo n, where $n = pq$ is the product of two distinct odd primes. $x^2 \equiv 1 \pmod{p}$ and Theorem 5.13 asserts that there are four square roots of 1 modulo n. Two of these square roots are $\pm 1 \bmod n$; these are called the *trivial* square roots of 1 modulo n. The other two square roots are called *non-trivial*; they are also negatives of each other modulo n.

Here is a small example to illustrate.

Example 5.14 Suppose $n = 403 = 13 \times 31$. The four square roots of 1 modulo 403 are $1, 92, 311$ and 402. The square root 92 is obtained by solving the system

$$x \equiv 1 \pmod{13},$$

$$x \equiv -1 \pmod{31}.$$

using the Chinese remainder theorem. The other non-trivial square root is $403 - 92 = 311$. It is the solution to the system

$$x \equiv -1 \pmod{13},$$
$$x \equiv 1 \pmod{31}.$$

▯

Suppose x is a non-trivial square root of 1 modulo n. Then

$$x^2 \equiv 1^2 \pmod{n}$$

but

$$x \not\equiv \pm 1 \pmod{n}.$$

Then, as in the random squares factoring algorithm, we can find the factors of n by computing $\gcd(x + 1, n)$ and $\gcd(x - 1, n)$. In Example 5.14 above,

$$\gcd(93, 403) = 31$$

and

$$\gcd(312, 403) = 13.$$

Algorithm 5.10 attempts to factor n by finding a non-trivial square root of 1 modulo n. Before analyzing the algorithm, we first do an example to illustrate its application.

Example 5.15 Suppose $n = 89855713$, $b = 34986517$ and $a = 82330933$, and the random value $w = 5$. We have

$$ab - 1 = 2^3 \times 360059073378795.$$

Then

$$w^r \bmod n = 85877701.$$

It happens that

$$85877701^2 \equiv 1 \pmod{n}.$$

Therefore the algorithm will return the value

$$x = \gcd(85877702, n) = 9103.$$

This is one factor of n; the other is $n/9103 = 9871$.

▯

Algorithm 5.10: RSA-FACTOR(n, a, b)

comment: we are assuming that $ab \equiv 1 \pmod{\phi(n)}$

write $ab - 1 = 2^s r$, r odd
choose w at random such that $1 \le w \le n - 1$
$x \leftarrow \gcd(w, n)$
if $1 < x < n$
  then return (x)
comment: x is a factor of n

$v \leftarrow w^r \bmod n$
if $v \equiv 1 \pmod{n}$
  then return ("failure")
while $v \not\equiv 1 \pmod{n}$
  do $\begin{cases} v_0 \leftarrow v \\ v \leftarrow v^2 \bmod n \end{cases}$
if $v_0 \equiv -1 \pmod{n}$
  then return ("failure")
  else $\begin{cases} x \leftarrow \gcd(v_0 + 1, n) \\ \textbf{return } (x) \end{cases}$
comment: x is a factor of n

Let's now proceed to the analysis of Algorithm 5.10. First, observe that if we are lucky enough to choose w to be a multiple of p or q, then we can factor n immediately. If w is relatively prime to n, then we compute $w^r, w^{2r}, w^{4r}, \ldots$, by successive squaring, until

$$w^{2^t r} \equiv 1 \pmod{n}$$

for some t. Since

$$ab - 1 = 2^s r \equiv 0 \pmod{\phi(n)},$$

we know that $w^{2^s r} \equiv 1 \pmod{n}$. Hence, the **while** loop terminates after at most s iterations. At the end of the **while** loop, we have found a value v_0 such that $(v_0)^2 \equiv 1 \pmod{n}$ but $v_0 \not\equiv 1 \pmod{n}$. If $v_0 \equiv -1 \pmod{n}$, then the algorithm fails; otherwise, v_0 is a non-trivial square root of 1 modulo n and we are able to factor n.

The main task facing us now is to prove that the algorithm succeeds with probability at least $1/2$. There are two ways in which the algorithm can fail to factor n:

1. $w^r \equiv 1 \pmod{n}$, or
2. $w^{2^t r} \equiv -1 \pmod{n}$ for some t, $0 \le t \le s - 1$.

We have $s + 1$ congruences to consider. If a random value w is a solution to at least one of these $s + 1$ congruences, then it is a "bad" choice, and the algorithm fails. So we proceed by counting the number of solutions to each of these congruences.

First, consider the congruence $w^r \equiv 1 \pmod{n}$. The way to analyze a congruence such as this is to consider solutions modulo p and modulo q separately, and then combine them using the Chinese remainder theorem. Observe that $x \equiv 1 \pmod{n}$ if and only if $x \equiv 1 \pmod{p}$ and $x \equiv 1 \pmod{q}$.

So, we first consider $w^r \equiv 1 \pmod{p}$. Since p is prime, $\mathbb{Z}_p^*$ is a cyclic group by Theorem 5.7. Let g be a primitive element modulo p. We can write $w = g^u$ for a unique integer u, $0 \le u \le p - 2$. Then we have

$$w^r \equiv 1 \pmod{p},$$

$$g^{ur} \equiv 1 \pmod{p}, \text{ and hence}$$

$$(p - 1) \mid ur.$$

Let us write

$$p - 1 = 2^i p_1$$

where p_1 is odd, and

$$q - 1 = 2^j q_1$$

where q_1 is odd. Since

$$\phi(n) = (p - 1)(q - 1) \mid (ab - 1) = 2^s r,$$

we have that

$$2^{i+j} p_1 q_1 \mid 2^s r.$$

Hence

$$i + j \le s$$

and

$$p_1 q_1 \mid r.$$

Now, the condition $(p - 1) \mid ur$ becomes $2^i p_1 \mid ur$. Since $p_1 \mid r$ and r is odd, it is necessary and sufficient that $2^i \mid u$. Hence, $u = k2^i$, $0 \le k \le p_1 - 1$, and the number of solutions to the congruence $w^r \equiv 1 \pmod{p}$ is p_1.

By an identical argument, the congruence $w^r \equiv 1 \pmod{q}$ has exactly q_1 solutions. We can combine any solution modulo p with any solution modulo q to obtain a unique solution modulo n, using the Chinese remainder theorem. Consequently, the number of solutions to the congruence $w^r \equiv 1 \pmod{n}$ is $p_1 q_1$.

The next step is to consider a congruence $w^{2^t r} \equiv -1 \pmod{n}$ for a fixed value t (where $0 \le t \le s - 1$). Again, we first look at the congruence modulo p and then modulo q (note that $w^{2^t r} \equiv -1 \pmod{n}$ if and only if $w^{2^t r} \equiv -1 \pmod{p}$ and $w^{2^t r} \equiv -1 \pmod{q}$). First, consider $w^{2^t r} \equiv -1 \pmod{p}$. Writing $w = g^u$, as above, we get

$$g^{u2^t r} \equiv -1 \pmod{p}.$$

Since $g^{(p-1)/2} \equiv -1 \pmod{p}$, we have that

$$u2^t r \equiv \frac{p-1}{2} \pmod{p-1}$$

$$(p-1) \mid \left(u2^t r - \frac{p-1}{2}\right)$$

$$2(p-1) \mid (u2^{t+1} r - (p-1)).$$

Since $p - 1 = 2^i p_1$, we get

$$2^{i+1} p_1 \mid (u2^{t+1} r - 2^i p_1).$$

Taking out a common factor of p_1, this becomes

$$2^{i+1} \mid \left(\frac{u2^{t+1} r}{p_1} - 2^i\right).$$

Now, if $t \geq i$, then there can be no solutions since $2^{i+1} \mid 2^{t+1}$ but $2^{i+1} \nmid 2^i$. On the other hand, if $t \leq i-1$, then u is a solution if and only if u is an odd multiple of 2^{i-t-1} (note that r/p_1 is an odd integer). So, the number of solutions in this case is

$$\frac{p-1}{2^{i-t-1}} \times \frac{1}{2} = 2^t p_1.$$

By similar reasoning, the congruence $w^{2^t r} \equiv -1 \pmod{q}$ has no solutions if $t \geq j$, and $2^t q_1$ solutions if $t \leq j-1$. Using the Chinese remainder theorem, we see that the number of solutions of $w^{2^t r} \equiv -1 \pmod{n}$ is

$$\begin{array}{ll} 0 & \text{if } t \geq \min\{i, j\} \\ 2^{2t} p_1 q_1 & \text{if } t \leq \min\{i, j\} - 1. \end{array}$$

Now, t can range from 0 to $s - 1$. Without loss of generality, suppose $i \leq j$; then the number of solutions is 0 if $t \geq i$. The total number of "bad" choices for w is at most

$$p_1 q_1 + p_1 q_1 (1 + 2^2 + 2^4 + \cdots + 2^{2i-2}) = p_1 q_1 \left(1 + \frac{2^{2i} - 1}{3}\right)$$

$$= p_1 q_1 \left(\frac{2}{3} + \frac{2^{2i}}{3}\right).$$

Recall that $p - 1 = 2^i p_1$ and $q - 1 = 2^j q_1$. Now, $j \geq i \geq 1$, so $p_1 q_1 < n/4$. We also have that

$$2^{2i} p_1 q_1 \leq 2^{i+j} p_1 q_1 = (p-1)(q-1) < n.$$

Hence, we obtain

$$p_1 q_1 \left(\frac{2}{3} + \frac{2^{2i}}{3}\right) < \frac{n}{6} + \frac{n}{3}$$

$$= \frac{n}{2}.$$

Since at most $(n-1)/2$ choices for w are "bad," it follows that at least $(n-1)/2$ choices are "good" and hence the probability of success of the algorithm is at least $1/2$.

5.7.3 Wiener's Low Decryption Exponent Attack

As always, suppose that $n = pq$ where p and q are prime; then $\phi(n) = (p-1)(q-1)$. In this section, we present an attack, due to M. Wiener, that succeeds in computing the secret decryption exponent, a, whenever the following hypotheses are satisfied:

$$3a < n^{1/4} \quad \text{and} \quad q < p < 2q. \tag{5.2}$$

If n has ℓ bits in its binary representation, then the attack will work when a has fewer than $\ell/4 - 1$ bits in its binary representation and p and q are not too far apart.

Note that Bob might be tempted to choose his decryption exponent to be small in order to speed up decryption. If he uses Algorithm 5.5 to compute $y^a \bmod n$, then the running time of decryption will be reduced by roughly 75% if he chooses a value of a that satisfies (5.2). The results we prove in this section show that this method of reducing decryption time should be avoided.

Since $ab \equiv 1 \pmod{\phi(n)}$, it follows that there is an integer t such that

$$ab - t\phi(n) = 1.$$

Therefore we have that

$$\left| \frac{b}{\phi(n)} - \frac{t}{a} \right| = \frac{1}{a\phi(n)}.$$

Since $n = pq > q^2$, we have that $q < \sqrt{n}$. Hence,

$$0 < n - \phi(n) = p + q - 1 < 2q + q - 1 < 3q < 3\sqrt{n}.$$

Now, we see that

$$\left| \frac{b}{n} - \frac{t}{a} \right| = \left| \frac{ba - tn}{an} \right|$$

$$= \left| \frac{1 + t(\phi(n) - n)}{an} \right|$$

$$< \frac{3t\sqrt{n}}{an}$$

$$= \frac{3t}{a\sqrt{n}}.$$

Since $t < a$, we have that $3t < 3a < n^{1/4}$, and hence

$$\left| \frac{b}{n} - \frac{t}{a} \right| < \frac{1}{an^{1/4}}.$$

Finally, since $3a < n^{1/4}$, we have that

$$\left| \frac{b}{n} - \frac{t}{a} \right| < \frac{1}{3a^2}.$$

Therefore the fraction t/a is a very close approximation to the fraction b/n. From the theory of continued fractions, it is known that any approximation of b/n that is this close must be one of the convergents of the continued fraction expansion of b/n (see Theorem 5.14). This expansion can be obtained from the EUCLIDEAN ALGORITHM, as we describe now.

A (finite) *continued fraction* is an m-tuple of non-negative integers, say

$$[q_1, \ldots, q_m],$$

which is shorthand for the following expression:

$$q_1 + \cfrac{1}{q_2 + \cfrac{1}{q_3 + \cdots + \frac{1}{q_m}}}.$$

Suppose a and b are positive integers such that $\gcd(a, b) = 1$, and suppose that the output of Algorithm 5.1 is the m-tuple $(q_1, \ldots, q_m)$. Then it is not hard to see that $a/b = [q_1, \ldots, q_m]$. We say that $[q_1, \ldots, q_m]$ is the *continued fraction expansion* of a/b in this case. Now, for $1 \le j \le m$, define $C_j = [q_1, \ldots, q_j]$. C_j is said to be the jth *convergent* of $[q_1, \ldots, q_m]$. Each C_j can be written as a rational number c_j/d_j, where the c_j's and d_j's satisfy the following recurrences:

$$c_j = \begin{cases} 1 & \text{if } j = 0 \\ q_1 & \text{if } j = 1 \\ q_j c_{j-1} + c_{j-2} & \text{if } j \ge 2 \end{cases}$$

and

$$d_j = \begin{cases} 0 & \text{if } j = 0 \\ 1 & \text{if } j = 1 \\ q_j d_{j-1} + d_{j-2} & \text{if } j \ge 2. \end{cases}$$

Example 5.16 We compute the continued fraction expansion of $34/99$. The EUCLIDEAN ALGORITHM proceeds as follows:

$$\begin{aligned} 34 &= 0 \times 99 + 34 \\ 99 &= 2 \times 34 + 31 \\ 34 &= 1 \times 31 + 3 \\ 31 &= 10 \times 3 + 1 \\ 1 &= 3 \times 1. \end{aligned}$$

Hence, the continued fraction expansion of $34/99$ is $[0, 2, 1, 10, 3]$, i.e.,

$$\frac{34}{99} = 0 + \cfrac{1}{2 + \cfrac{1}{1 + \cfrac{1}{10 + \frac{1}{3}}}}.$$

The convergents of this continued fraction are as follows:

$$[0] = 0$$
$$[0, 2] = 1/2$$
$$[0, 2, 1] = 1/3$$
$$[0, 2, 1, 10] = 11/32, \quad \text{and}$$
$$[0, 2, 1, 10, 3] = 34/99.$$

The reader can verify that these convergents can be computed using the recurrence relations given above.   ▯

The convergents of a continued fraction expansion of a rational number satisfy many interesting properties. For our purposes, the most important property is the following.

THEOREM 5.14 *Suppose that* $\gcd(a, b) = \gcd(c, d) = 1$ *and*

$$\left| \frac{a}{b} - \frac{c}{d} \right| < \frac{1}{2d^2}.$$

Then c/d *is one of the convergents of the continued fraction expansion of* a/b.

Now we can apply this result to the *RSA Cryptosystem*. We already observed that, if condition (5.2) holds, then the unknown fraction t/a is a close approximation to b/n. Theorem 5.14 tells us that t/a must be one of the convergents of the continued fraction expansion of b/n. Since the value of b/n is public information, it is a simple matter to compute its convergents. All we need is a method to test each convergent to see if it is the "right" one.

But this is also not difficult to do. If t/a is a convergent of n/b, then we can compute the value of $\phi(n)$ to be $\phi(n) = (ab - 1)/t$. Once n and $\phi(n)$ are known, we can factor n by solving the quadratic equation (5.1) for p. We do not know ahead of time which convergent of n/b will yield the factorization of n, so we try each one in turn until the factorization of n is found. If we do not succeed in factoring n by this method, then it must be the case that the hypotheses (5.2) are not satisfied.

A pseudocode description of WIENER'S ALGORITHM is presented as Algorithm 5.11.

Algorithm 5.11: WIENER'S ALGORITHM(n, b)

$(q_1, \ldots, q_m; r_m) \leftarrow$ EUCLIDEAN ALGORITHM(n, b)
$c_0 \leftarrow 1$
$c_1 \leftarrow q_1$
$d_0 \leftarrow 0$
$d_1 \leftarrow 1$
for $j \leftarrow 1$ **to** m

$\text{do} \begin{cases} n' \leftarrow (d_j b - 1)/c_j \\ \textbf{comment: } n' = \phi(n) \text{ if } c_j/d_j \text{ is the correct convergent} \\ \\ \textbf{if } n' \text{ is an integer} \\ \quad \textbf{then} \begin{cases} \text{let } p \text{ and } q \text{ be the roots of the equation} \\ \quad x^2 - (n - n' + 1)x + n = 0 \\ \textbf{if } p \text{ and } q \text{ are positive integers less than } n \\ \quad \textbf{then return } (p, q) \end{cases} \\ j \leftarrow j + 1 \\ c_j \leftarrow q_j c_{j-1} + c_{j-2} \\ d_j \leftarrow q_j d_{j-1} + d_{j-2} \end{cases}$

return ("failure")

We present an example to illustrate.

Example 5.17 Suppose that $n = 160523347$ and $b = 60728973$. The continued fraction expansion of b/n is

$$[0, 2, 1, 1, 1, 4, 12, 102, 1, 1, 2, 3, 2, 2, 36].$$

The first few convergents are

$$0, \frac{1}{2}, \frac{1}{3}, \frac{2}{5}, \frac{3}{8}, \frac{14}{37}.$$

The reader can verify that the first five convergents do not produce a factorization of n. However, the convergent $14/37$ yields

$$n' = \frac{37 \times 60728973 - 1}{14} = 160498000.$$

Now, if we solve the equation

$$x^2 - 25348x + 160523347 = 0,$$

then we find the roots $x = 12347, 13001$. Therefore we have discovered the factorization

$$60523347 = 12347 \times 13001.$$

Notice that, for the modulus $n = 160523347$, WIENER'S ALGORITHM will work for

$$a < \frac{n^{1/4}}{3} \approx 37.52.$$

$$\square$$

5.8 The Rabin Cryptosystem

In this section, we describe the *Rabin Cryptosystem*, which is computationally secure against a chosen-plaintext attack provided that the modulus $n = pq$ cannot be factored. Therefore, the *Rabin Cryptosystem* provides an example of a provably secure cryptosystem: assuming that the problem of factoring is computationally infeasible, the *Rabin Cryptosystem* is secure. We present the *Rabin Cryptosystem* now.

Cryptosystem 5.2: *Rabin Cryptosystem*

Let $n = pq$, where p and q are primes and $p, q \equiv 3 \pmod{4}$. Let $\mathcal{P} = \mathcal{C} = \mathbb{Z}_n^*$, and define

$$\mathcal{K} = \{(n, p, q)\}.$$

For $K = (n, p, q)$, define

$$e_K(x) = x^2 \bmod n$$

and

$$d_K(y) = \sqrt{y} \bmod n.$$

The value n is the public key, while p and q are the private key.

REMARK   The requirement that $p, q \equiv 3 \pmod{4}$ can be omitted. As well, the cryptosystem still "works" if we take $\mathcal{P} = \mathcal{C} = \mathbb{Z}_n$ instead of $\mathbb{Z}_n^*$. However, the more restrictive description we use simplifies some aspects of computation and analysis of the cryptosystem.   ∎

One drawback of the *Rabin Cryptosystem* is that the encryption function e_K is not an injection, so decryption cannot be done in an unambiguous fashion. We prove this as follows. Suppose that y is a valid ciphertext; this means that $y = x^2 \bmod n$ for some $x \in \mathbb{Z}_n^*$. Theorem 5.13 proves that there are four square roots of y modulo n, which are the four possible plaintexts that encrypt to y. In general, there will be no way for Bob to distinguish which of these four possible plaintexts is the "right" plaintext, unless the plaintext contains sufficient redundancy to eliminate three of these four possible values.

Let us look at the decryption problem from Bob's point of view. He is given a ciphertext y and wants to determine x such that

$$x^2 \equiv y \pmod{n}.$$

This is a quadratic equation in $\mathbb{Z}_n$ in the unknown x, and decryption requires extracting square roots modulo n. This is equivalent to solving the two congruences

$$z^2 \equiv y \pmod{p}$$

and

$$z^2 \equiv y \pmod{q}.$$

We can use Euler's criterion to determine if y is a quadratic residue modulo p (and modulo q). In fact, y will be a quadratic residue modulo p and modulo q if encryption was performed correctly. Unfortunately, Euler's criterion does not help us find the square roots of y; it yields only an answer "yes" or "no."

When $p \equiv 3 \pmod{4}$, there is a simple formula to compute square roots of quadratic residues modulo p. Suppose y is a quadratic residue modulo p, where $p \equiv 3 \pmod{4}$. Then we have that

$$(\pm y^{(p+1)/4})^2 \equiv y^{(p+1)/2} \pmod{p}$$
$$\equiv y^{(p-1)/2}C \pmod{p}$$
$$\equiv y \pmod{p}.$$

Here we have again made use of Euler's criterion, which says that if y is a quadratic residue modulo p, then $y^{(p-1)/2} \equiv 1 \pmod{p}$. Hence, the two square roots of y modulo p are $\pm y^{(p+1)/4} \bmod p$. In a similar fashion, the two square roots of y modulo q are $\pm y^{(q+1)/4} \bmod q$. It is then straightforward to obtain the four square roots of y modulo n using the Chinese remainder theorem.

REMARK For $p \equiv 1 \pmod{4}$, there is no known polynomial-time deterministic algorithm to compute square roots of quadratic residues modulo p. (There is a polynomial-time Las Vegas algorithm, however.) This is why we stipulated that $p, q \equiv 3 \pmod{4}$ in the definition of the *Rabin Cryptosystem*. ∎

Example 5.18 Let's illustrate the encryption and decryption procedures for the *Rabin Cryptosystem* with a toy example. Suppose $n = 77 = 7 \times 11$. Then the encryption function is

$$e_K(x) = x^2 \bmod 77$$

and the decryption function is

$$d_K(y) = \sqrt{y} \bmod 77.$$

Suppose Bob wants to decrypt the ciphertext $y = 23$. It is first necessary to find the square roots of 23 modulo 7 and modulo 11. Since 7 and 11 are both congruent to 3 modulo 4, we use our formula:

$$23^{(7+1)/4} \equiv 2^2 \equiv 4 \pmod 7$$

and

$$23^{(11+1)/4} \equiv 1^3 \equiv 1 \pmod{11}.$$

Using the Chinese remainder theorem, we compute the four square roots of 23 modulo 77 to be $\pm 10, \pm 32 \bmod 77$. Therefore, the four possible plaintexts are $x = 10, 32, 45$ and 67. It can be verified that each of these plaintexts yields the value 23 when squared and reduced modulo 77. This proves that 23 is indeed a valid ciphertext. □

5.8.1 Security of the Rabin Cryptosystem

We now discuss the (provable) security of the *Rabin Cryptosystem*. The security proof uses a Turing reduction, which is defined as follows.

Definition 5.5:   Suppose that G and H are problems. A *Turing reduction* from G to H is an algorithm SOLVEG with the following properties:

1. SOLVEG assumes the existence of an arbitrary algorithm SOLVEH that solves the problem H

2. SOLVEG can call the algorithm SOLVEH and make use of any values it outputs, but SOLVEG cannot make any assumption about the actual computations performed by SOLVEH (in other words, SOLVEH is treated as a "black box" and is termed an oracle)

3. SOLVEG is a polynomial-time algorithm

4. SOLVEG correctly solves the problem G

If there is a Turing reduction from G to H, we denote this by writing $G \propto_T H$.

A Turing reduction $G \propto_T H$ does not necessarily yield a polynomial-time algorithm to solve G. It actually proves the truth of the following implication:

If there exists a polynomial-time algorithm to solve H, then there exists a polynomial-time algorithm to solve G.

This is because any algorithm SOLVEH that solves H can be "plugged into" the algorithm SOLVEG, thereby producing an algorithm that solves G. Clearly this resulting algorithm will be a polynomial-time algorithm if and only if SOLVEH is a polynomial-time algorithm.

We will provide an explicit example of a Turing reduction: We will prove that a decryption oracle RABIN DECRYPT can be incorporated into a Las Vegas algorithm that factors the modulus n with probability at least $1/2$. In other words, we show that **Factoring** $\propto_T$ **Rabin decryption**, where the Turing reduction is itself a randomized algorithm. In the following algorithm, we assume that n is the product of two distinct primes p and q; and RABIN DECRYPT is an oracle that performs Rabin decryption, returning one of the four possible plaintexts corresponding to a given ciphertext.

Algorithm 5.12: RABIN ORACLE FACTORING(n)

external RABIN DECRYPT
choose a random integer $r \in \mathbb{Z}_n^*$
$y \leftarrow r^2 \bmod n$
$x \leftarrow$ RABIN DECRYPT(y)
if $x \equiv \pm r \pmod{n}$
  then return ("failure")
  else $\begin{cases} p \leftarrow \gcd(x + r, n) \\ q \leftarrow n/p \\ \textbf{return } ("n = p \times q") \end{cases}$

There are several points of explanation needed. First, observe that y is a valid ciphertext and RABIN DECRYPT(y) will return one of four possible plaintexts as the value of x. In fact, it holds that $x \equiv \pm r \pmod{n}$ or $x \equiv \pm \omega r \pmod{n}$, where ω is one of the non-trivial square roots of 1 modulo n. In the second case, we have $x^2 \equiv r^2 \pmod{n}$, $x \not\equiv \pm r \pmod{n}$. Hence, computation of $\gcd(z + r, n)$ must yield either p or q, and the factorization of n is accomplished.

Let's compute the probability of success of this algorithm, over all choices for the random value $r \in \mathbb{Z}_n^*$. For a residue $r \in \mathbb{Z}_n^*$, define

$$[r] = \{\pm r \bmod n, \pm \omega r \bmod n\}.$$

Clearly any two residues in $[r]$ yield the same y-value in Algorithm 5.12, and the value of x that is output by the oracle RABIN DECRYPT is also in $[r]$. We have already observed that Algorithm 5.12 succeeds if and only if $x \equiv \pm \omega r \pmod{n}$. The oracle does not know which of four possible r-values was used to construct y, and r was chosen at random before the oracle RABIN DECRYPT is called. Hence, the probability that $x \equiv \pm \omega r \pmod{n}$ is $1/2$. We conclude that the probability of success of Algorithm 5.12 is $1/2$.

We have shown that the *Rabin Cryptosystem* is provably secure against a chosen plaintext attack. However, the system is completely insecure against a chosen ciphertext attack. In fact, Algorithm 5.12 can be used to break the *Rabin Cryptosystem* in a chosen ciphertext attack! In the chosen ciphertext attack, the (hypothetical) oracle RABIN DECRYPT is replaced by an actual decryption algorithm. (Informally, the security proof says that a decryption oracle can be used to factor n; and a chosen ciphertext attack assumes that a decryption oracle exists. Together, these break the cryptosystem!)

5.9 Semantic Security of RSA

To this point in the text, we have assumed that an adversary trying to break a cryptosystem is actually trying to determine the secret key (in the case of a symmetric-key cryptosystem) or the private key (in the case of a public-key cryptosystem). If Oscar can do this, then the system is completely broken. However, it is possible that the goal of an adversary is somewhat less ambitious. Even if Oscar cannot find the secret or private key, he still may be able to gain more information than we would like. If we want to be assured that the cryptosystem is "secure," we should take into account these more modest goals that an adversary might have.

Here is a short list of potential adversarial goals:

total break

The adversary is able to determine Bob's private key (in the case of a public-key cryptosystem) or the secret key (in the case of a symmetric-key cryptosystem). Therefore he can decrypt any ciphertext that has been encrypted using the given key.

partial break

With some non-negligible probability, the adversary is able to decrypt a previously unseen ciphertext (without knowing the key). Or, the adversary can determine some specific information about the plaintext, given the ciphertext.

distinguishability of ciphertexts

With some probability exceeding $1/2$, the adversary is able to distinguish between encryptions of two given plaintexts, or between an encryption of a given plaintext and a random string.

In the next sections, we will consider some possible attacks against RSA-like cryptosystems that achieve some of these types of goals. We also describe how to construct a public-key cryptosystem in which the adversary cannot (in polynomial time) distinguish ciphertexts, provided that certain computational assumptions hold. Such cryptosystems are said to achieve *semantic security*. Achieving semantic security is quite difficult, because we are providing protection against a very weak, and therefore easy to achieve, adversarial goal.

5.9.1 Partial Information Concerning Plaintext Bits

A weakness of some cryptosystems is the fact that partial information about the plaintext might be "leaked" by the ciphertext. This represents a type of partial break of the system, and it happens, in fact, in the *RSA Cryptosystem*. Suppose we are given a ciphertext, $y = x^b \bmod n$, where x is the plaintext. Since $\gcd(b, \phi(n)) = 1$, it must be the case that b is odd. Therefore the Jacobi symbol

$$\left(\frac{y}{n}\right) = \left(\frac{x}{n}\right)^b = \left(\frac{x}{n}\right).$$

Hence, given the ciphertext y, anyone can efficiently compute $\left(\frac{x}{n}\right)$ without decrypting the ciphertext. In other words, an RSA encryption "leaks" some information concerning the plaintext x, namely, the value of the Jacobi symbol $\left(\frac{x}{n}\right)$.

In this section, we consider some other specific types of partial information that could be leaked by a cryptosystem:

1. given $y = e_K(x)$, compute *parity*(y), where *parity*(y) denotes the low-order bit of x (i.e., *parity*$(y) = 0$ if x is even and *parity*$(y) = 1$ if x is odd).

2. given $y = e_K(x)$, compute *half*(y), where *half*$(y) = 0$ if $0 \leq x < n/2$ and *half*$(y) = 1$ if $n/2 < x \leq n - 1$.

We will prove that the *RSA Cryptosystem* does not leak these types of information provided that RSA encryption is secure. More precisely, we show that the problem of RSA decryption can be Turing reduced to the problem of computing *half*(y). This means that the existence of a polynomial-time algorithm that computes *half*(y) implies the existence of a polynomial-time algorithm for RSA decryption. In other words, computing certain partial information about the plaintext, namely *half*(y), is no easier than decrypting the ciphertext to obtain the whole plaintext.

We will now show how to compute $x = d_K(y)$, given a hypothetical algorithm (oracle) HALF which computes *half*(y). The algorithm is presented as Algorithm 5.13.

Algorithm 5.13: ORACLE RSA DECRYPTION(n, b, y)

external HALF
$k \leftarrow \lfloor \log_2 n \rfloor$
for $i \leftarrow 0$ **to** k
  do $\begin{cases} h_i \leftarrow \text{HALF}(n, b, y) \\ y \leftarrow (y \times 2^b) \bmod n \end{cases}$
$lo \leftarrow 0$
$hi \leftarrow n$
for $i \leftarrow 0$ **to** k
  do $\begin{cases} mid \leftarrow (hi + lo)/2 \\ \textbf{if } h_i = 1 \\ \quad \textbf{then } lo \leftarrow mid \\ \quad \textbf{else } hi \leftarrow mid \end{cases}$
return $(\lfloor hi \rfloor)$

We explain what is happening in the above algorithm. First, we note that the RSA encryption function satisfies the following multiplicative property in $\mathbb{Z}_n$:

$$e_K(x_1)e_K(x_2) = e_K(x_1 x_2).$$

Now, using the fact that

$$y = e_K(x) = x^b \bmod n,$$

it is easily seen in the ith iteration of the first **for** loop that

$$h_i = half(y \times (e_K(2))^i) = half(e_K(x \times 2^i)),$$

for $0 \leq i \leq \lfloor \log_2 n \rfloor$. We observe that

$$half(e_K(x)) = 0 \Leftrightarrow x \in \left[0, \frac{n}{2}\right)$$

$$half(e_K(2x)) = 0 \Leftrightarrow x \in \left[0, \frac{n}{4}\right) \cup \left[\frac{n}{2}, \frac{3n}{4}\right)$$

$$half(e_K(4x)) = 0 \Leftrightarrow x \in \left[0, \frac{n}{8}\right) \cup \left[\frac{n}{4}, \frac{3n}{8}\right) \cup \left[\frac{n}{2}, \frac{5n}{8}\right) \cup \left[\frac{3n}{4}, \frac{7n}{8}\right),$$

and so on. Hence, we can find x by a binary search technique, which is done in the second **for** loop. Here is a small example to illustrate.

Example 5.19   Suppose $n = 1457$, $b = 779$, and we have a ciphertext $y = 722$. Then suppose, using our oracle HALF, that we obtain the following values for h_i:

i	0	1	2	3	4	5	6	7	8	9	10
h_i	1	0	1	0	1	1	1	1	1	0	0

Then the binary search proceeds as shown in Figure 5.2. Hence, the plaintext is $x = \lfloor 999.55 \rfloor = 999$.   □

FIGURE 5.2
Binary search for RSA decryption

i	lo	mid	hi
0	0.00	728.50	1457.00
1	728.50	1092.75	1457.00
2	728.50	910.62	1092.75
3	910.62	1001.69	1092.75
4	910.62	956.16	1001.69
5	956.16	978.92	1001.69
6	978.92	990.30	1001.69
7	990.30	996.00	1001.69
8	996.00	998.84	1001.69
9	998.84	1000.26	1001.69
10	998.84	999.55	1000.26
	998.84	999.55	999.55

The complexity of Algorithm 5.13 is easily seen to be

$$O((\log n)^3) + O(\log n) \times \text{the complexity of HALF}.$$

Therefore we will obtain a polynomial-time algorithm for RSA decryption if HALF is a polynomial-time algorithm.

It is a simple matter to observe that computing $parity(y)$ is polynomially equivalent to computing $half(y)$. This follows from the following two easily proved identities involving RSA encryption (see the exercises):

$$half(y) = parity((y \times e_K(2)) \bmod n) \tag{5.3}$$

$$parity(y) = half((y \times e_K(2^{-1})) \bmod n), \tag{5.4}$$

and from the above-mentioned multiplicative rule, $e_K(x_1)e_K(x_2) = e_K(x_1 x_2)$. Hence, from the results proved above, it follows that the existence of a polynomial-time algorithm to compute *parity* implies the existence of a polynomial-time algorithm for RSA decryption.

We have provided evidence that computing *parity* or *half* is difficult, provided that RSA decryption is difficult. However, the proofs we have presented do not rule out the possibility that it might be possible to find an efficient algorithm that computes *parity* or *half* with 75% accuracy, say. There are also many other types of plaintext information that could possibly be considered, and we certainly cannot deal with all possible types of information using separate proofs. Therefore the results of this section only provide evidence of security against certain types of attacks.

5.9.2   Optimal Asymmetric Encryption Padding

What we really want is to find a method of designing a cryptosystem that allows us to prove (assuming some plausible computational assumptions) that no information of any kind regarding the plaintext is revealed in polynomial time by examining the ciphertext. It can be shown that this is equivalent to showing that an adversary cannot distinguish ciphertexts. Therefore we consider the problem of Ciphertext Distinguishability, which is defined as follows:

Problem 5.3:   Ciphertext Distinguishability

Instance: An encryption function $f : X \to X$; two plaintexts $x_1, x_2 \in X$; and a ciphertext $y = f(x_i)$, where $i \in \{1, 2\}$.

Question: Is $i = 1$?

Problem 5.3 is of course trivial if the encryption function f is deterministic, since it suffices to compute $f(x_1)$ and $f(x_2)$ and see which one yields the ciphertext y. Hence, if Ciphertext Distinguishability is going to computationally infeasible, then it will be necessary for the encryption process to be randomized. We now present some concrete methods to realize this objective. We describe Cryptosystem 5.3, which is based on an arbitrary *trapdoor one-way permutation*, which is a (bijective) trapdoor one-way function from a set X to itself. If $f : X \to X$ is a trapdoor one-way permutation, then the inverse permutation is denoted, as usual, by f^{-1}. f is the encryption function, and f^{-1} is the decryption function of the public-key cryptosystem.

Cryptosystem 5.3:   *Semantically Secure Public-key Cryptosystem*

Let m, k be positive integers; let $\mathcal{F}$ be a family of trapdoor one-way permutations such that $f : \{0,1\}^k \to \{0,1\}^k$ for all $f \in \mathcal{F}$; and let $G : \{0,1\}^k \to \{0,1\}^m$ be a random oracle. Let $\mathcal{P} = \{0,1\}^m$ and $\mathcal{C} = \{0,1\}^k \times \{0,1\}^m$, and define
$$\mathcal{K} = \{(f, f^{-1}, G) : f \in \mathcal{F}\}.$$
For $K = (f, f^{-1}, G)$, let $r \in \{0,1\}^k$ be chosen randomly, and define
$$e_K(x) = (y_1, y_2) = (f(r), G(r) \oplus x),$$
where $x, y_1 \in \{0,1\}^k$, $y_2 \in \{0,1\}^m$. Further, define
$$d_K(y_1, y_2) = G(f^{-1}(y_1)) \oplus y_2$$
($y_1 \in \{0,1\}^k$, $y_2 \in \{0,1\}^m$). The functions f and G are the public key; the function f^{-1} is the private key.

In the case of the *RSA Cryptosystem*, we would take $n = pq$, $X = \mathbb{Z}_n$, $f(x) = x^b \bmod n$ and $f^{-1}(x) = x^a \bmod n$, where $ab \equiv 1 \pmod{\phi(n)}$. Cryptosystem 5.3 also employs a certain random function, G. Actually, G will be modeled by a random oracle, which was defined in Section 4.2.1.

We observe that Cryptosystem 5.3 is quite efficient: it requires little additional computation as compared to the underlying public-key cryptosystem based on f. In practice, the function G could be built from a secure hash function, such as *SHA-1*, in a very efficient manner. The main drawback of Cryptosystem 5.3 is the data expansion: m bits of plaintext are encrypted to form $k + m$ bits of ciphertext. If f is based on the RSA encryption function, for example, then it will be necessary to take $k \geq 1024$ in order for the system to be secure.

It is easy to see that there must be some data expansion in any semantically secure cryptosystem due to the fact that encryption is randomized. However, there are more efficient schemes that are still provably secure. The most important of these, *Optimal Asymmetric Encryption Padding*, will be discussed later in this section (it is presented as Cryptosystem 5.4). We begin with Cryptosystem 5.3, however, because it is conceptually simpler and easier to analyze.

An intuitive argument that Cryptosystem 5.3 is semantically secure in the random oracle model goes as follows: In order to determine any information about the plaintext x, we need to have some information about $G(r)$. Assuming that G is a random oracle, the only way to ascertain any information about the value of $G(r)$ is to first compute $r = f^{-1}(y_1)$. (It is not sufficient to compute some partial information about r; it is necessary to have complete information about r in order to obtain any information about $G(r)$.) However, if f is one-way, then r cannot be computed in a reasonable amount of time by an adversary who does not know the trapdoor, f^{-1}.

The preceding argument might be fairly convincing, but it is not a proof. If we are going to massage this argument into a proof, we need to describe a reduction, from the problem of inverting the function f to the problem of Ciphertext Distinguishability. When f is randomized, as in Cryptosystem 5.3, it may not be feasible to solve Problem 5.3 if there are sufficiently many possible encryptions of a given plaintext.

We are going to describe a reduction that is more general than the Turing reductions considered previously. We will assume the existence of an algorithm DISTINGUISH that solves the problem of Ciphertext Distinguishability for two plaintexts x_1 and x_2, and then we will modify this algorithm in such a way that we obtain an algorithm to invert f. The algorithm DISTINGUISH need not be a "perfect" algorithm; we will only require that it gives the right answer with some probability $1/2 + \epsilon$, where $\epsilon > 0$ (i.e., it is more accurate than a random guess of "1" or "2"). DISTINGUISH is allowed to query the random oracle, and therefore it can compute encryptions of plaintexts. In other words, we are assuming it is a chosen plaintext attack.

As mentioned above, we will prove that Cryptosystem 5.3 is semantically secure in the random oracle model. The main features of this model (which we introduced in Section 4.2.1), and the reduction we describe, are as follows.

1. G is assumed to be a random oracle, so the only way to determine any information about a value $G(r)$ is to call the function G with input r.

2. We construct a new algorithm INVERT, by modifying the algorithm DISTINGUISH, which will invert randomly chosen elements y with probability bounded away from 0 (i.e., given a value $y = f(x)$ where x is chosen randomly, the algorithm INVERT will find x with some specified probability).

3. The algorithm INVERT will replace the random oracle by a specific function that we will describe, SIMG, all of whose outputs are random numbers. SIMG is a perfect simulation of a random oracle.

The algorithm INVERT is presented as Algorithm 5.14.

Algorithm 5.14: INVERT(y)

external f
global $RList, GList, \ell$
procedure SIMG(r)
$i \leftarrow 1$
found $\leftarrow$ **false**
while $i \leq \ell$ **and not** *found*
$\quad$ **do** $\begin{cases} \textbf{if } RList[i] = r \\ \quad \textbf{then } found \leftarrow \textbf{true} \\ \quad \textbf{else } i \leftarrow i + 1 \end{cases}$
if *found*
$\quad$ **then return** ($GList[i]$)
if $f(r) = y$
$\quad$ **then** $\begin{cases} \text{let } j \in \{1, 2\} \text{ be chosen at random} \\ g \leftarrow y_2 \oplus x_j \end{cases}$
$\quad$ **else** let g be chosen at random
$\ell \leftarrow \ell + 1$
$RList[\ell] \leftarrow r$
$GList[\ell] \leftarrow g$
return (g)

main
$y_1 \leftarrow y$
choose y_2 at random
$\ell \leftarrow 0$
insert the code for DISTINGUISH($x_1, x_2, (y_1, y_2)$) here
for $i \leftarrow 1$ **to** ℓ
$\quad$ **do** $\begin{cases} \textbf{if } f(RList[i]) = y \\ \quad \textbf{then return } (RList[i]) \end{cases}$
return ("failure")

Given two plaintexts x_1 and x_2, DISTINGUISH solves the **Ciphertext Distinguishability** problem with probability $1/2+\epsilon$. The input to INVERT is the element y to be inverted; the objective is to output $f^{-1}(y)$. INVERT begins by constructing a ciphertext (y_1, y_2) in which $y_1 = y$ and y_2 is random. INVERT runs the algorithm DISTINGUISH on the ciphertext (y_1, y_2), attempting to determine if it is an encryption of x_1 or of x_2. DISTINGUISH will query the simulated oracle, SIMG, at various points during its execution. The following points summarize the operation of SIMG:

1. SIMG maintains a list, denoted $RList$, of all inputs r for which it is queried during the execution of DISTINGUISH; and the corresponding list, denoted $GList$, of outputs $\text{SIMG}(r)$.

2. If an input r satisfies $f(r) = y$, then $\text{SIMG}(r)$ is defined so that (y_1, y_2) is a valid encryption of one of x_1 or x_2 (chosen at random).

3. If the oracle was previously queried with input r, then $\text{SIMG}(r)$ is already defined.

4. Otherwise, the value for $\text{SIMG}(r)$ is chosen randomly.

Observe that, for any possible plaintext $x_0 \in X$, (y_1, y_2) is a valid encryption of x_0 if and only if

$$\text{SIMG}(f^{-1}(y_1)) = y_2 \oplus x_0.$$

In particular, (y_1, y_2) can be a valid encryption of either of x_1 or x_2, provided that $\text{SIMG}(f^{-1}(y_1))$ is defined appropriately. The description of the algorithm SIMG ensures that (y_1, y_2) is a valid encryption of one of x_1 or x_2.

Eventually, the algorithm DISTINGUISH will terminate with an answer "1" or "2," which may or may not be correct. At this point, the algorithm INVERT examines the list $RList$ to see if any r in the $RList$ satisfies $f(r) = y$. If such a value r is found, then it is the desired value $f^{-1}(y)$, and the algorithm INVERT succeeds (INVERT fails if $f^{-1}(y)$ is not discovered in $RList$).

It is in fact possible to make algorithm INVERT more efficient by observing that the function SIMG checks to see if $y = f(r)$ for every r that it is queried with. Once it is discovered, within the function SIMG, that $y = f(r)$, we can terminate the algorithm INVERT immediately, returning the value r as its output. It is not necessary to keep running the algorithm DISTINGUISH to its conclusion. However, the analysis of the success probability, which we are going to do next, is a bit easier to understand for Algorithm 5.14 as we have presented it. (The reader might want to verify that the abovementioned modification of INVERT will not change its success probability.)

We now proceed to compute a lower bound on the success probability of the algorithm INVERT. We do this by examining the success probability of DISTINGUISH. We are assuming that the success probability of DISTINGUISH is at least $1/2 + \epsilon$ when it interacts with a random oracle. In the algorithm INVERT, DISTINGUISH interacts with the simulated random oracle, SIMG. Clearly SIMG is completely indistinguishable from a true random oracle for all inputs, except possibly for the input $r = f^{-1}(y)$. However, if $f(r) = y$ and (y, y_2) is a valid encryption of x_1 or x_2, then it must be the case that $\text{SIMG}(r) = y_2 \oplus x_1$ or $\text{SIMG}(r) = y_2 \oplus x_2$. SIMG is choosing randomly from these two possible alternatives. Therefore, the output it produces is indistinguishable from a true random oracle for the input $r = f^{-1}(y)$, as well. Consequently, the success probability of DISTINGUISH is at least $1/2 + \epsilon$ when it interacts with the simulated random oracle, SIMG.

We now calculate the success probability of DISTINGUISH, conditioned on whether (or not) $f^{-1}(y) \in RList$:

$$\mathbf{Pr}[\text{DISTINGUISH succeeds}] =$$
$$\mathbf{Pr}[\text{DISTINGUISH succeeds} \mid f^{-1}(y) \in RList] \, \mathbf{Pr}[f^{-1}(y) \in RList] +$$
$$\mathbf{Pr}[\text{DISTINGUISH succeeds} \mid f^{-1}(y) \notin RList] \, \mathbf{Pr}[f^{-1}(y) \notin RList].$$

It is clear that

$$\mathbf{Pr}[\text{DISTINGUISH succeeds} \mid f^{-1}(y) \notin RList] = 1/2,$$

because there is no way to distinguish an encryption of x_1 from an encryption of x_2 if the value of $\text{SIMG}(f^{-1}(y))$ is not determined. Now, using the fact that

$$\mathbf{Pr}[\text{DISTINGUISH succeeds} \mid f^{-1}(y) \in RList] \leq 1,$$

we obtain the following:

$$\frac{1}{2} + \epsilon \leq \mathbf{Pr}[\text{DISTINGUISH succeeds}]$$

$$\leq \mathbf{Pr}[f^{-1}(y) \in RList] + \frac{1}{2} \mathbf{Pr}[f^{-1}(y) \notin RList]$$

$$\leq \mathbf{Pr}[f^{-1}(y) \in RList] + \frac{1}{2}.$$

Therefore, it follows that

$$\mathbf{Pr}[f^{-1}(y) \in RList] \geq \epsilon.$$

Since
$$\mathbf{Pr}[\text{INVERSE succeeds}] = \mathbf{Pr}[f^{-1}(y) \in RList],$$

it follows that
$$\mathbf{Pr}[\text{INVERSE succeeds}] \geq \epsilon.$$

It is straightforward to consider the running time of INVERT as compared to that of DISTINGUISH. Suppose that t_1 is the running time of DISTINGUISH, t_2 is the time required to evaluate the function f, and q denotes the number of oracle queries made by DISTINGUISH. Then it is not difficult to see that the running time of INVERT is $t_1 + O(q^2 + qt_2)$.

We close this section by presenting a more efficient provably secure cryptosystem.

Cryptosystem 5.4: *Optimal Asymmetric Encryption Padding*

Let m, k be positive integers with $m < k$, and let $k_0 = k - m$. Let $\mathcal{F}$ be a family of trapdoor one-way permutations such that $f : \{0,1\}^k \to \{0,1\}^k$ for all $f \in \mathcal{F}$. Let $G : \{0,1\}^{k_0} \to \{0,1\}^m$ and let $H : \{0,1\}^m \to \{0,1\}^{k_0}$ be "random" functions. Define $\mathcal{P} = \{0,1\}^m$, $\mathcal{C} = \{0,1\}^k$, and define

$$\mathcal{K} = \{(f, f^{-1}, G, H) : f \in \mathcal{F}\}.$$

For $K = (f, f^{-1}, G, H)$, let $r \in \{0,1\}^{k_0}$ be chosen randomly, and define

$$e_K(x) = f(y_1 \parallel y_2),$$

where

$$y_1 = x \oplus G(r)$$

and

$$y_2 = r \oplus H(x \oplus G(r)),$$

$x, y_1 \in \{0,1\}^m$, $y_2 \in \{0,1\}^{k_0}$, and "$\parallel$" denotes concatenation of vectors. Further, define

$$f^{-1}(y) = x_1 \parallel x_2,$$

where $x_1 \in \{0,1\}^m$ and $x_2 \in \{0,1\}^{k_0}$. Then define

$$r = x_2 \oplus H(x_1)$$

and

$$d_K(y) = G(r) \oplus x_1.$$

The functions f, G and H are the public key; the function f^{-1} is the private key.

In Cryptosystem 5.4, it suffices to take k_0 to be large enough that 2^{k_0} is an infeasibly large running time; $k_0 = 128$ should suffice for most applications. The length of a ciphertext in Cryptosystem 5.4 exceeds the length of a plaintext by k_0 bits, so the data expansion is considerably less than Cryptosystem 5.3. The security proof for Cryptosystem 5.4 is more complicated, however.

The adjective "optimal" in Cryptosystem 5.4 refers to the message expansion. Observe that each plaintext has 2^{k_0} possible valid encryptions. One way of solving the problem of Ciphertext Distinguishability would be simply to compute all the possible encryptions of one of the two given plaintexts, say x_1, and check to see if the given ciphertext y is obtained. The complexity of this algorithm is 2^{k_0}. It is therefore clear that the message expansion of the cryptosystem must be at least as big as the logarithm to the base 2 of the amount of computation time of an algorithm that solves the Ciphertext Distinguishability problem.

5.10   Notes and References

The idea of public-key cryptography was introduced in the open literature by Diffie and Hellman in 1976. Although [68] is the most cited reference, the conference paper [67] actually appeared a bit earlier. The *RSA Cryptosystem* was discovered by Rivest, Shamir and Adleman [181]. For a general survey article on public-key cryptography, we recommend Diffie [66] (it is now somewhat dated but still worth reading). A specialized survey on *RSA* was written by Boneh [31].

The Solovay-Strassen test was first described in [202]. The Miller-Rabin test was given in [147] and [178]. Our discussion of error probabilities is motivated by observations of Brassard and Bratley [37] (see also [7]). The best current bounds on the error probability of the Miller-Rabin algorithm can be found in [58].

There are many sources of information on factoring algorithms. Lenstra [130] is a good survey on factoring, and Lenstra and Lenstra [132] is a good article on number-theoretic algorithms in general. Bressoud [39] is an elementary textbook on factoring and primality testing. One recommended cryptography textbook that emphasizes number theory is Koblitz [120] (note that Example 5.12 is taken from Koblitz's book). Recommended number theory books that are useful for the study of cryptography include Bach and Shallit [4], von zur Gathen and Gerhard [90] and Yan [219]. Lenstra and Lenstra [131] is a monograph on the number field sieve. The factorization of RSA-155 is reported in [43].

The material in Sections 5.7.2 and 5.9.1 is based on the treatment by Salomaa [184, pp. 143–154] (the factorization of n given the decryption exponent was proved in [60]; the results on partial information revealed by RSA ciphertexts is from [96]). Wiener's attack can be found in [214]; a recent strengthening of the attack, due to Boneh and Durfee, was published in [33].

The *Rabin Cryptosystem* was described in Rabin [177]. Provably secure systems in which decryption is unambiguous have been found by Williams [217] and Kurosawa, Ito and Takeuchi [125].

Partial information leaked by RSA ciphertexts is studied in Alexi, Chor, Goldreich and Schnorr [1]. The concept of semantic security is due to Goldwasser and Micali [95]; the *Blum-Goldwasser Cryptosystem* [29] is an example of an early probabilistic public-key cryptosystem that is provably (semantically) secure.

For a recent survey on the topic of provably secure cryptosystems, see Bellare [9]. The random oracle model was first described by Bellare and Rogaway in [14]; Cryptosystem 5.3 was presented in that paper. *Optimal Asymmetric Encryption Padding* is presented in [15]; it is incorporated into the IEEE P1363 standard specifications for public-key cryptography. There have been several recent

works discussing the security of *Optimal Asymmetric Encryption Padding* and
related cryptosystems against chosen-ciphertext attacks: Shoup [193], Fujisaki,
Okamoto, Pointcheval and Stern [87] and Boneh [32].

Exercises 5.15–5.17 give some examples of protocol failures. For a nice article
on this subject, see Moore [151].

Exercises

5.1 In Algorithm 5.1, prove that

$$\gcd(r_0, r_1) = \gcd(r_1, r_2) = \cdots = \gcd(r_{m-1}, r_m) = r_m$$

and, hence, $r_m = \gcd(a, b)$.

5.2 Suppose that $a > b$ in Algorithm 5.1.
   (a) Prove that $r_i \geq 2r_{i+2}$ for all i such that $0 \leq i \leq m - 2$.
   (b) Prove that m is $O((\log a)^2)$.
   (c) Prove that m is $O((\log b)^2)$.

5.3 Use the EXTENDED EUCLIDEAN ALGORITHM to compute the following multiplicative inverses:
   (a) $17^{-1} \bmod 101$
   (b) $357^{-1} \bmod 1234$
   (c) $3125^{-1} \bmod 9987$.

5.4 Compute $\gcd(57, 93)$, and find integers s and t such that $57s + 93t = \gcd(57, 93)$.

5.5 Suppose $\chi : \mathbb{Z}_{105} \to \mathbb{Z}_3 \times \mathbb{Z}_5 \times \mathbb{Z}_7$ is defined as

$$\chi(x) = (x \bmod 3, x \bmod 5, x \bmod 7).$$

   Give an explicit formula for the function χ^{-1}, and use it to compute $\chi^{-1}(2, 2, 3)$.

5.6 Solve the following system of congruences:

$$x \equiv 12 \ (\bmod \ 25)$$

$$x \equiv 9 \ (\bmod \ 26)$$

$$x \equiv 23 \ (\bmod \ 27).$$

5.7 Solve the following system of congruences:

$$13x \equiv 4 \ (\bmod \ 99)$$

$$15x \equiv 56 \ (\bmod \ 101).$$

   HINT First use the EXTENDED EUCLIDEAN ALGORITHM, and then apply the
   Chinese remainder theorem.

5.8 Use Theorem 5.8 to find the smallest primitive element modulo 97.

5.9 Suppose that $p = 2q + 1$, where p and q are odd primes. Suppose further that
   $\alpha \in \mathbb{Z}_p^*$, $\alpha \not\equiv \pm 1 \ (\bmod \ p)$. Prove that α is a primitive element modulo p if and
   only if $\alpha^q \equiv -1 \ (\bmod \ p)$.

5.10 Suppose that $n = pq$, where p and q are distinct odd primes and $ab \equiv 1 \ (\bmod \ (p-1)(q-1))$. The RSA encryption operation is $e(x) = x^b \bmod n$ and the decryption operation is $d(y) = y^a \bmod n$. We proved that $d(e(x)) = x$ if $x \in \mathbb{Z}_n^*$. Prove that the same statement is true for any $x \in \mathbb{Z}_n$.

HINT   Use the fact that $x_1 \equiv x_2 \pmod{pq}$ if and only if $x_1 \equiv x_2 \pmod{p}$ and $x_1 \equiv x_2 \pmod{q}$. This follows from the Chinese remainder theorem.

5.11 For $n = pq$, where p and q are distinct odd primes, define
$$\lambda(n) = \frac{(p-1)(q-1)}{\gcd(p-1, q-1)}.$$
Suppose that we modify the *RSA Cryptosystem* by requiring that $ab \equiv 1 \pmod{\lambda(n)}$.
   (a) Prove that encryption and decryption are still inverse operations in this modified cryptosystem.
   (b) If $p = 37$, $q = 79$, and $b = 7$, compute a in this modified cryptosystem, as well as in the original *RSA Cryptosystem*.

5.12 Two samples of RSA ciphertext are presented in Tables 5.1 and 5.2. Your task is to decrypt them. The public parameters of the system are $n = 18923$ and $b = 1261$ (for Table 5.1) and $n = 31313$ and $b = 4913$ (for Table 5.2). This can be accomplished as follows. First, factor n (which is easy because it is so small). Then compute the exponent a from $\phi(n)$, and, finally, decrypt the ciphertext. Use the SQUARE-AND-MULTIPLY ALGORITHM to exponentiate modulo n.

   In order to translate the plaintext back into ordinary English text, you need to know how alphabetic characters are "encoded" as elements in $\mathbb{Z}_n$. Each element of $\mathbb{Z}_n$ represents three alphabetic characters as in the following examples:

$$
\begin{aligned}
DOG &\rightarrow & 3 \times 26^2 + 14 \times 26 + 6 &=& 2398 \\
CAT &\rightarrow & 2 \times 26^2 + 0 \times 26 + 19 &=& 1371 \\
ZZZ &\rightarrow & 25 \times 26^2 + 25 \times 26 + 25 &=& 17575.
\end{aligned}
$$

You will have to invert this process as the final step in your program.

   The first plaintext was taken from "The Diary of Samuel Marchbanks," by Robertson Davies, 1947, and the second was taken from "Lake Wobegon Days," by Garrison Keillor, 1985.

5.13 A common way to speed up RSA decryption incorporates the Chinese remainder theorem, as follows. Suppose that $d_K(y) = y^a \bmod n$ and $n = pq$. Define $d_p = d \bmod (p-1)$ and $d_q = d \bmod (q-1)$; and let $M_p = q^{-1} \bmod p$ and $M_q = p^{-1} \bmod q$. Then consider the following algorithm:

Algorithm 5.15: CRT-OPTIMIZED RSA DECRYPTION $(n, d_p, d_q, M_p, M_q, y)$

$x_p \leftarrow y^{d_p} \bmod p$
$x_q \leftarrow y^{d_q} \bmod q$
$x \leftarrow M_p q x_p + M_q p x_q \bmod n$
return (x)

Algorithm 5.15 replaces an exponentiation modulo n by modular exponentiations modulo p and q. If p and q are ℓ-bit integers and exponentiation modulo an ℓ-bit integer takes time $c\ell^3$, then the time to perform the required exponentiation(s) is reduced from $c(2\ell)^3$ to $2c\ell^3$, a savings of 75%. The final step, involving the Chinese remainder theorem, requires time $O(\ell^2)$ if d_p, d_q, M_p and M_q have been pre-computed.

TABLE 5.1
RSA ciphertext

12423	11524	7243	7459	14303	6127	10964	16399
9792	13629	14407	18817	18830	13556	3159	16647
5300	13951	81	8986	8007	13167	10022	17213
2264	961	17459	4101	2999	14569	17183	15827
12693	9553	18194	3830	2664	13998	12501	18873
12161	13071	16900	7233	8270	17086	9792	14266
13236	5300	13951	8850	12129	6091	18110	3332
15061	12347	7817	7946	11675	13924	13892	18031
2620	6276	8500	201	8850	11178	16477	10161
3533	13842	7537	12259	18110	44	2364	15570
3460	9886	8687	4481	11231	7547	11383	17910
12867	13203	5102	4742	5053	15407	2976	9330
12192	56	2471	15334	841	13995	17592	13297
2430	9741	11675	424	6686	738	13874	8168
7913	6246	14301	1144	9056	15967	7328	13203
796	195	9872	16979	15404	14130	9105	2001
9792	14251	1498	11296	1105	4502	16979	1105
56	4118	11302	5988	3363	15827	6928	4191
4277	10617	874	13211	11821	3090	18110	44
2364	15570	3460	9886	9988	3798	1158	9872
16979	15404	6127	9872	3652	14838	7437	2540
1367	2512	14407	5053	1521	297	10935	17137
2186	9433	13293	7555	13618	13000	6490	5310
18676	4782	11374	446	4165	11634	3846	14611
2364	6789	11634	4493	4063	4576	17955	7965
11748	14616	11453	17666	925	56	4118	18031
9522	14838	7437	3880	11476	8305	5102	2999
18628	14326	9175	9061	650	18110	8720	15404
2951	722	15334	841	15610	2443	11056	2186

(a) Prove that the value x returned by Algorithm 5.15 is, in fact, $y^d \bmod n$.

(b) Given that $p = 1511$ and $q = 2003$, compute d_p, d_q, M_p and M_q.

(c) Given the above values of p and q, decrypt the ciphertext $y = 152702$ using Algorithm 5.15.

5.14 Prove that the *RSA Cryptosystem* is insecure against a chosen ciphertext attack. In particular, given a ciphertext y, describe how to choose a ciphertext $\hat{y} \neq y$, such that knowledge of the plaintext $\hat{x} = d_K(\hat{y})$ allows $x = d_K(y)$ to be computed.

HINT   Use the multiplicative property of the *RSA Cryptosystem*, i.e., that

$$e_K(x_1)e_K(x_2) \bmod n = e_K(x_1 x_2 \bmod n).$$

5.15 This exercise exhibits what is called a *protocol failure*. It provides an example where ciphertext can be decrypted by an opponent, without determining the key, if a cryptosystem is used in a careless way. The moral is that it is not sufficient to use a "secure" cryptosystem in order to guarantee "secure" communication.

Suppose Bob has an *RSA Cryptosystem* with a large modulus n for which the

TABLE 5.2
RSA ciphertext

6340	8309	14010	8936	27358	25023	16481	25809
23614	7135	24996	30590	27570	26486	30388	9395
27584	14999	4517	12146	29421	26439	1606	17881
25774	7647	23901	7372	25774	18436	12056	13547
7908	8635	2149	1908	22076	7372	8686	1304
4082	11803	5314	107	7359	22470	7372	22827
15698	30317	4685	14696	30388	8671	29956	15705
1417	26905	25809	28347	26277	7897	20240	21519
12437	1108	27106	18743	24144	10685	25234	30155
23005	8267	9917	7994	9694	2149	10042	27705
15930	29748	8635	23645	11738	24591	20240	27212
27486	9741	2149	29329	2149	5501	14015	30155
18154	22319	27705	20321	23254	13624	3249	5443
2149	16975	16087	14600	27705	19386	7325	26277
19554	23614	7553	4734	8091	23973	14015	107
3183	17347	25234	4595	21498	6360	19837	8463
6000	31280	29413	2066	369	23204	8425	7792
25973	4477	30989					

factorization cannot be found in a reasonable amount of time. Suppose Alice sends a message to Bob by representing each alphabetic character as an integer between 0 and 25 (i.e., $A \leftrightarrow 0$, $B \leftrightarrow 1$, etc.), and then encrypting each residue modulo 26 as a separate plaintext character.

(a) Describe how Oscar can easily decrypt a message which is encrypted in this way.

(b) Illustrate this attack by decrypting the following ciphertext (which was encrypted using an *RSA Cryptosystem* with $n = 18721$ and $b = 25$) without factoring the modulus:

$$365, 0, 4845, 14930, 2608, 2608, 0.$$

5.16 This exercise illustrates another example of a protocol failure (due to Simmons) involving the *RSA Cryptosystem*; it is called the "common modulus protocol failure." Suppose Bob has an *RSA Cryptosystem* with modulus n and encryption exponent b_1, and Charlie has an *RSA Cryptosystem* with (the same) modulus n and encryption exponent b_2. Suppose also that $\gcd(b_1, b_2) = 1$. Now, consider the situation that arises if Alice encrypts the same plaintext x to send to both Bob and Charlie. Thus, she computes $y_1 = x^{b_1} \bmod n$ and $y_2 = x^{b_2} \bmod n$, and then she sends y_1 to Bob and y_2 to Charlie. Suppose Oscar intercepts y_1 and y_2, and performs the computations indicated in Algorithm 5.16.

Algorithm 5.16: RSA COMMON MODULUS DECRYPTION (n, b_1, b_2, y_1, y_2)

$c_1 \leftarrow b_1^{-1} \bmod b_2$
$c_2 \leftarrow (c_1 b_1 - 1)/b_2$
$x_1 \leftarrow y_1^{c_1}(y_2^{c_2})^{-1} \bmod n$
return (x_1)

   (a) Prove that the value x_1 computed in Algorithm 5.16 is in fact Alice's plaintext, x. Thus, Oscar can decrypt the message Alice sent, even though the cryptosystem may be "secure."

   (b) Illustrate the attack by computing x by this method if $n = 18721$, $b_1 = 43$, $b_2 = 7717$, $y_1 = 12677$ and $y_2 = 14702$.

5.17 We give yet another protocol failure involving the *RSA Cryptosystem*. Suppose that three users in a network, say Bob, Bart and Bert, all have public encryption exponents $b = 3$. Let their moduli be denoted by n_1, n_2, n_3, and assume that n_1, n_2 and n_3, are pairwise relatively prime. Now suppose Alice encrypts the same plaintext x to send to Bob, Bart and Bert. That is, Alice computes $y_i = x^3 \bmod n_i$, $1 \le i \le 3$. Describe how Oscar can compute x, given y_1, y_2 and y_3, without factoring any of the moduli.

5.18 A plaintext x is said to be *fixed* if $e_K(x) = x$. Show that, for the *RSA Cryptosystem*, the number of fixed plaintexts $x \in \mathbb{Z}_n^*$ is equal to

$$\gcd(b - 1, p - 1) \times \gcd(b - 1, q - 1).$$

HINT Consider the following system of two congruences:

$$e_K(x) \equiv x \pmod{p},$$

$$e_K(x) \equiv x \pmod{q}.$$

5.19 Suppose **A** is a deterministic algorithm which is given as input an RSA modulus n, an encryption exponent b, and a ciphertext y. **A** will either decrypt y or return no answer. Supposing that there are $\epsilon(n - 1)$ ciphertexts which **A** is able to decrypt, show how to use **A** as an oracle in a Las Vegas decryption algorithm having success probability ϵ.

5.20 Write a program to evaluate Jacobi symbols using the four properties presented in Section 5.4. The program should not do any factoring, other than dividing out powers of two. Test your program by computing the following Jacobi symbols:

$$\left(\frac{610}{987}\right), \left(\frac{20964}{1987}\right), \left(\frac{1234567}{11111111}\right).$$

5.21 For $n = 837$, 851 and 1189, find the number of bases b such that n is an Euler pseudo-prime to the base b.

5.22 The purpose of this question is to prove that the error probability of the Solovay-Strassen primality test is at most $1/2$. Let $\mathbb{Z}_n^*$ denote the group of units modulo n. Define

$$G(n) = \left\{ a : a \in \mathbb{Z}_n^*, \left(\frac{a}{n}\right) \equiv a^{(n-1)/2} \pmod{n} \right\}.$$

   (a) Prove that $G(n)$ is a subgroup of $\mathbb{Z}_n^*$. Hence, by Lagrange's theorem, if $G(n) \ne \mathbb{Z}_n^*$, then

$$|G(n)| \le \frac{|\mathbb{Z}_n^*|}{2} \le \frac{n-1}{2}.$$

(b) Suppose $n = p^k q$, where p and q are odd, p is prime, $k \geq 2$, and $\gcd(p,q) = 1$. Let $a = 1 + p^{k-1}q$. Prove that
$$\left(\frac{a}{n}\right) \not\equiv a^{(n-1)/2} \pmod{n}.$$

HINT   Use the binomial theorem to compute $a^{(n-1)/2}$.

(c) Suppose $n = p_1 \ldots p_s$, where the p_i's are distinct odd primes. Suppose $a \equiv u \pmod{p_1}$ and $a \equiv 1 \pmod{p_2 p_3 \ldots p_s}$, where u is a quadratic non-residue modulo p_1 (note that such an a exists by the Chinese remainder theorem). Prove that
$$\left(\frac{a}{n}\right) \equiv -1 \pmod{n},$$
but
$$a^{(n-1)/2} \equiv 1 \pmod{p_2 p_3 \ldots p_s},$$
so
$$a^{(n-1)/2} \not\equiv -1 \pmod{n}.$$

(d) If n is odd and composite, prove that $|G(n)| \leq (n-1)/2$.

(e) Summarize the above: prove that the error probability of the Solovay-Strassen primality test is at most $1/2$.

5.23 Suppose we have a Las Vegas algorithm with failure probability ϵ.

(a) Prove that the probability of first achieving success on the nth trial is $p_n = \epsilon^{n-1}(1 - \epsilon)$.

(b) The average (expected) number of trials to achieve success is
$$\sum_{n=1}^{\infty}(n \times p_n).$$
Show that this average is equal to $1/(1 - \epsilon)$.

(c) Let δ be a positive real number less than 1. Show that the number of iterations required in order to reduce the probability of failure to at most δ is
$$\left\lfloor \frac{\log_2 \delta}{\log_2 \epsilon} \right\rfloor.$$

5.24 Suppose throughout this question that p is an odd prime and $\gcd(a,p) = 1$.

(a) Suppose that $i \geq 2$ and $b^2 \equiv a \pmod{p^{i-1}}$. Prove that there is a unique $x \in \mathbb{Z}_{p^i}$ such that $x^2 \equiv a \pmod{p^i}$ and $x \equiv b \pmod{p^{i-1}}$. Describe how this x can be computed efficiently.

(b) Illustrate your method in the following situation: starting with the congruence $6^2 \equiv 17 \pmod{19}$, find square roots of 17 modulo 19^2 and modulo 19^3.

(c) For all $i \geq 1$, prove that the number of solutions to the congruence $x^2 \equiv a \pmod{p^i}$ is either 0 or 2.

5.25 Using various choices for the bound, B, attempt to factor 262063 and 9420457 using the $p - 1$ method. How big does B have to be in each case to be successful?

5.26 Factor 262063, 9420457 and 181937053 using the POLLARD RHO ALGORITHM, if the function f is defined to be $f(x) = x^2 + 1$. How many iterations are needed to factor each of these three integers?

5.27 Suppose we want to factor the integer $n = 256961$ using the RANDOM SQUARES ALGORITHM. Using the factor base
$$\{-1, 2, 3, 5, 7, 11, 13, 17, 19, 23, 29, 31\},$$

test the integers $z^2 \bmod n$ for $z = 500, 501, \ldots$, until a congruence of the form $x^2 \equiv y^2 \pmod{n}$ is obtained and the factorization of n is found.

5.28 In the RANDOM SQUARES ALGORITHM, we need to test a positive integer $w \le n - 1$ to see if it factors completely over the factor base $\mathcal{B} = \{p_1, \ldots, p_B\}$ consisting of the B smallest prime numbers. Recall that $p_B = m \approx 2^s$ and $n \approx 2^r$.

    (a) Prove that this can be done using at most $B + r$ divisions of an integer having at most r bits by an integer having at most s bits.

    (b) Assuming that $r < m$, prove that the complexity of this test is $O(rsm)$.

5.29 In this exercise, we show that parameter generation for the *RSA Cryptosystem* should take care to ensure that $q - p$ is not too small, where $n = pq$ and $q > p$.

    (a) Suppose that $q - p = 2d > 0$, and $n = pq$. Prove that $n + d^2$ is a perfect square.

    (b) Given an integer n which is the product of two odd primes, and given a small positive integer d such that $n + d^2$ is a perfect square, show how this information can be used to factor n.

    (c) Use this technique to factor $n = 2189284635403183$.

5.30 Suppose Bob has carelessly revealed his decryption exponent to be $a = 14039$ in an *RSA Cryptosystem* with public key $n = 36581$ and $b = 4679$. Implement the randomized algorithm to factor n given this information. Test your algorithm with the "random" choices $w = 9983$ and $w = 13461$. Show all computations.

5.31 If $q_1, \ldots, q_m$ is the sequence of quotients obtained in the applying the EUCLIDEAN ALGORITHM with input r_0, r_1, prove that the continued fraction $[q_1, \ldots, q_m] = r_0/r_1$.

5.32 Suppose that $n = 317940011$ and $b = 77537081$ in the *RSA Cryptosystem*. Using WIENER'S ALGORITHM, attempt to factor n.

5.33 Consider the modification of the *Rabin Cryptosystem* in which $e_K(x) = x(x + B) \bmod n$, where $B \in \mathbb{Z}_n$ is part of the public key. Supposing that $p = 199$, $q = 211$, $n = pq$ and $B = 1357$, perform the following computations.

    (a) Compute the encryption $y = e_K(32767)$.

    (b) Determine the four possible decryptions of this given ciphertext y.

5.34 Prove Equations (5.3) and (5.4) relating the functions *half* and *parity*.

5.35 Prove that Cryptosystem 5.3 is not semantically secure against a chosen ciphertext attack. Given x_1, x_2, a ciphertext (y_1, y_2) that is an encryption of x_i ($i = 1$ or 2), and given a decryption oracle DECRYPT for Cryptosystem 5.3, describe an algorithm to determine whether $i = 1$ or $i = 2$. You are allowed to call the algorithm DECRYPT with any input except for the given ciphertext (y_1, y_2), and it will output the corresponding plaintext.

6
Public-key Cryptosystems Based on the Discrete Logarithm Problem

The theme of this chapter concerns public-key cryptosystems based on the **Discrete Logarithm** problem. The first and best-known of these is the *ElGamal Cryptosystem*. The **Discrete Logarithm** problem forms the basis of numerous cryptographic protocols that we will study throughout the rest of the text. Thus we devote a considerable amount of time to discussion of this important problem. In later sections of this chapter, we give treatments of some other ElGamal-type systems based on finite fields and elliptic curves.

6.1   The ElGamal Cryptosystem

The *ElGamal Cryptosystem* is based on the **Discrete Logarithm** problem. We begin by describing this problem in the setting of a finite multiplicative group $(G, \cdot)$. For an element $\alpha \in G$ having order n, define

$$\langle \alpha \rangle = \{\alpha^i : 0 \le i \le n - 1\}.$$

It is easy to see that $\langle \alpha \rangle$ is a subgroup of G, and that $\langle \alpha \rangle$ is cyclic of order n.

An often-used example is to take G to be the multiplicative group of a finite field $\mathbb{Z}_p$ (where p is prime), and to let α be a primitive element modulo p. In this situation, we have that $n = |\langle \alpha \rangle| = p - 1$. Another frequently used setting is to take α to be an element having prime order q in the multiplicative group $\mathbb{Z}_p{}^*$ (where p is prime and $p - 1 \equiv 0 \pmod{q}$). Such an α can be obtained by raising a primitive element in $\mathbb{Z}_p{}^*$ to the $(p - 1)/q$th power.

We now define the **Discrete Logarithm** problem in a subgroup $\langle \alpha \rangle$ of a group $(G, \cdot)$.

Problem 6.1: Discrete Logarithm

Instance: A multiplicative group $(G, \cdot)$, an element $\alpha \in G$ having order n, and an element $\beta \in \langle \alpha \rangle$.

Question: Find the unique integer a, $0 \le a \le n - 1$, such that

$$\alpha^a = \beta.$$

We will denote this integer a by $\log_\alpha \beta$.

The utility of the **Discrete Logarithm** problem in a cryptographic setting is that finding discrete logarithms is (probably) difficult, but the inverse operation of exponentiation can be computed efficiently by using the square-and-multiply method (Algorithm 5.5). Stated another way, exponentiation is a one-way function in suitable groups G.

ElGamal proposed a public-key cryptosystem which is based on the **Discrete Logarithm** problem in $(\mathbb{Z}_p^*, \cdot)$. This system is presented as Cryptosystem 6.1.

Cryptosystem 6.1: *ElGamal Public-key Cryptosystem in $\mathbb{Z}_p^*$*

Let p be a prime such that the **Discrete Logarithm** problem in $(\mathbb{Z}_p^*, \cdot)$ is infeasible, and let $\alpha \in \mathbb{Z}_p^*$ be a primitive element. Let $\mathcal{P} = \mathbb{Z}_p^*$, $\mathcal{C} = \mathbb{Z}_p^* \times \mathbb{Z}_p^*$, and define

$$\mathcal{K} = \{(p, \alpha, a, \beta) : \beta \equiv \alpha^a \pmod{p}\}.$$

The values p, α and β are the public key, and a is the private key.

For $K = (p, \alpha, a, \beta)$, and for a (secret) random number $k \in \mathbb{Z}_{p-1}$, define

$$e_K(x, k) = (y_1, y_2),$$

where

$$y_1 = \alpha^k \bmod p$$

and

$$y_2 = x\beta^k \bmod p.$$

For $y_1, y_2 \in \mathbb{Z}_p^*$, define

$$d_K(y_1, y_2) = y_2(y_1{}^a)^{-1} \bmod p.$$

The encryption operation in the *ElGamal Cryptosystem* is randomized, since the ciphertext depends on both the plaintext x and on the random value k chosen by Alice. Hence, there will be many ciphertexts ($p-1$, in fact) that are encryptions of the same plaintext.

Informally, this is how the *ElGamal Cryptosystem* works: The plaintext x is "masked" by multiplying it by β^k, yielding y_2. The value α^k is also transmitted as part of the ciphertext. Bob, who knows the private key, a, can compute β^k from α^k. Then he can "remove the mask" by dividing y_2 by β^k to obtain x.

A small example will illustrate the computations performed in the *ElGamal Cryptosystem*.

Example 6.1   Suppose $p = 2579$ and $\alpha = 2$. α is a primitive element modulo p. Let $a = 765$, so

$$\beta = 2^{765} \bmod 2579 = 949.$$

Now, suppose that Alice wishes to send the message $x = 1299$ to Bob. Say $k = 853$ is the random integer she chooses. Then she computes

$$y_1 = 2^{853} \bmod 2579$$

$$= 435,$$

and

$$y_2 = 1299 \times 949^{853} \bmod 2579$$

$$= 2396.$$

When Bob receives the ciphertext $y = (435, 2396)$, he computes

$$x = 2396 \times (435^{765})^{-1} \bmod 2579$$

$$= 1299,$$

which was the plaintext that Alice encrypted. $\square$

Clearly the *ElGamal Cryptosystem* will be insecure if Oscar can compute the value $a = \log_\alpha \beta$, for then Oscar can decrypt ciphertexts exactly as Bob does. Hence, a necessary condition for the *ElGamal Cryptosystem* to be secure is that the Discrete Logarithm problem in $\mathbb{Z}_p^*$ is infeasible. This is generally regarded as being the case if p is carefully chosen and α is a primitive element modulo p. In particular, there is no known polynomial-time algorithm for this version of the Discrete Logarithm problem. To thwart known attacks, p should have at least 300 digits, and $p - 1$ should have at least one "large" prime factor.

6.2   Algorithms for the Discrete Logarithm Problem

Throughout this section, we assume that $(G, \cdot)$ is a multiplicative group and $\alpha \in G$ has order n. Hence the Discrete Logarithm problem can be phrased in the following form: Given $\beta \in \langle \alpha \rangle$, find the unique exponent a, $0 \leq a \leq n - 1$, such that $\alpha^a = \beta$.

We begin by anaylzing some elementary algorithms which can be used to solve the **Discrete Logarithm** problem. In our analyses, we will assume that computing a product of two elements in the group G requires constant (i.e., $O(1)$) time.

First, we observe that the **Discrete Logarithm** problem can be solved by exhaustive search in $O(n)$ time and $O(1)$ space, simply by computing $\alpha, \alpha^2, \alpha^3, \ldots,$ until $\beta = \alpha^a$ is found. (Each term α^i in the above list is computed by multiplying the previous term α^{i-1} by α, and hence the total time required is $O(n)$.)

Another approach is to precompute all possible values α^i, and then sort the list of ordered pairs (i, α^i) with respect to their second coordinates. Then, given β, we can perform a binary search of the sorted list in order to find the value a such that $\alpha^a = \beta$. This requires precomputation time $O(n)$ to compute the n powers of α, and time $O(n \log n)$ to sort the list of size n. (The sorting step can be done in time $O(n \log n)$ if an efficient sorting algorithm, such as the QUICKSORT algorithm, is used.) If we neglect logarithmic factors, as is usually done in the analysis of these algorithms, the precomputation time is $O(n)$. The time for a binary search of a sorted list of size n is $O(\log n)$. If we (again) ignore the logarithmic term, then we see that we can solve the **Discrete Logarithm** problem in $O(1)$ time with $O(n)$ precomputation and $O(n)$ memory.

6.2.1 Shanks' Algorithm

The first non-trivial algorithm we describe is a time-memory trade-off due to Shanks. SHANKS' ALGORITHM is presented in Algorithm 6.1.

Algorithm 6.1: SHANKS(G, n, α, β)

1. $m \leftarrow \lceil \sqrt{n} \rceil$
2.   **for** $j \leftarrow 0$ **to** $m - 1$
        do compute α^{mj}
3.   Sort the m ordered pairs (j, α^{mj}) with respect to their second coordinates, obtaining a list L_1
4.   **for** $i \leftarrow 0$ **to** $m - 1$
        do compute $\beta \alpha^{-i}$
5.   Sort the m ordered pairs $(i, \beta \alpha^{-i})$ with respect to their second coordinates, obtaining a list L_2
6.   Find a pair $(j, y) \in L_1$ and a pair $(i, y) \in L_2$ (i.e., find two pairs having identical second coordinates)
7. $\log_\alpha \beta \leftarrow (mj + i) \bmod n$

Some comments are in order. First, steps 2 and 3 can be precomputed, if desired (this will not affect the asymptotic running time, however). Next, observe that if

$(j, y) \in L_1$ and $(i, y) \in L_2$, then

$$\alpha^{mj} = y = \beta\alpha^{-i},$$

so

$$\alpha^{mj+i} = \beta,$$

as desired. Conversely, for any $\beta \in \langle\alpha\rangle$, we have that $0 \leq \log_\alpha \beta \leq n - 1$. If we divide $\log_\alpha \beta$ by the integer m, then we can express $\log_\alpha \beta$ in the form

$$\log_\alpha \beta = mj + i,$$

where $0 \leq j, i \leq m - 1$. The fact that $j \leq m - 1$ follows because

$$\log_\alpha \beta \leq n - 1 \leq m^2 - 1 = m(m - 1) + m - 1.$$

Hence, the search in step 6 will be successful. (However, if it happened that $\beta \notin \langle\alpha\rangle$, then step 6 will not be successful.)

It is not difficult to implement the algorithm to run in $O(m)$ time with $O(m)$ memory (neglecting logarithmic factors). Here are a few details: Step 2 can be performed by first computing α^m, and then computing its powers by successively multiplying by α^m. The total time for this step is $O(m)$. In a similar fashion, step 4 also takes time $O(m)$. Steps 3 and 5 can be done in time $O(m \log m)$ using an efficient sorting algorithm. Finally, step 6 can be done with one (simultaneous) pass through each of the two lists L_1 and L_2; so it requires time $O(m)$.

Here is a small example to illustrate SHANKS' ALGORITHM.

Example 6.2   Suppose we wish to find $\log_3 525$ in $(\mathbb{Z}_{809}{}^*, \cdot)$. Note that 809 is prime and 3 is a primitive element in $\mathbb{Z}_{809}{}^*$, so we have $\alpha = 3, n = 808, \beta = 525$ and $m = \lceil\sqrt{808}\rceil = 29$. Then

$$\alpha^{29} \bmod 809 = 99.$$

First, we compute the ordered pairs $(j, 99^j \bmod 809)$ for $0 \leq j \leq 28$. We obtain the list

$(0, 1)$	$(1, 99)$	$(2, 93)$	$(3, 308)$	$(4, 559)$
$(5, 329)$	$(6, 211)$	$(7, 664)$	$(8, 207)$	$(9, 268)$
$(10, 644)$	$(11, 654)$	$(12, 26)$	$(13, 147)$	$(14, 800)$
$(15, 727)$	$(16, 781)$	$(17, 464)$	$(18, 632)$	$(19, 275)$
$(20, 528)$	$(21, 496)$	$(22, 564)$	$(23, 15)$	$(24, 676)$
$(25, 586)$	$(26, 575)$	$(27, 295)$	$(28, 81)$	

which is then sorted to produce L_1.

The second list contains the ordered pairs $(i, 525 \times (3^i)^{-1} \bmod 809)$, $0 \le j \le 28$. It is as follows:

$$
\begin{array}{lllll}
(0, 525) & (1, 175) & (2, 328) & (3, 379) & (4, 396) \\
(5, 132) & (6, 44) & (7, 554) & (8, 724) & (9, 511) \\
(10, 440) & (11, 686) & (12, 768) & (13, 256) & (14, 355) \\
(15, 388) & (16, 399) & (17, 133) & (18, 314) & (19, 644) \\
(20, 754) & (21, 521) & (22, 713) & (23, 777) & (24, 259) \\
(25, 356) & (26, 658) & (27, 489) & (28, 163)
\end{array}
$$

After sorting this list, we get L_2.

Now, if we proceed simultaneously through the two sorted lists, we find that $(10, 644)$ is in L_1 and $(19, 644)$ is in L_2. Hence, we can compute

$$
\log_3 525 = (29 \times 10 + 19) \bmod 808
$$

$$
= 309.
$$

As a check, it can be verified that $3^{309} \equiv 525 \pmod{809}$. $\quad\quad$ ▯

6.2.2 The Pollard Rho Discrete Logarithm Algorithm

We previously discussed the POLLARD RHO ALGORITHM for factoring in Section 5.6.2. There is a corresponding algorithm for finding discrete logarithms, which we describe now. As before, let $(G, \cdot)$ be a group and let $\alpha \in G$ be an element having order n. Let $\beta \in \langle \alpha \rangle$ be the element whose discrete logarithm we want to find. Since $\langle \alpha \rangle$ is cyclic of order n, we can treat $\log_\alpha \beta$ as an element of $\mathbb{Z}_n$.

As with the rho algorithm for factoring, we form a sequence $x_1, x_2, \ldots$ by iteratively applying a random-looking function, f. Once we obtain two elements x_i and x_j in the sequence such that $x_i = x_j$ and $i < j$, we can (hopefully) compute $\log_\alpha \beta$. Just as we did in the case of the factoring algorithm, we will seek a collision of the form $x_i = x_{2i}$, in order to save time and memory.

Let $S_1 \cup S_2 \cup S_3$ be a partition of G into three subsets of roughly equal size. We define a function $f : \langle \alpha \rangle \times \mathbb{Z}_n \times \mathbb{Z}_n \to \langle \alpha \rangle \times \mathbb{Z}_n \times \mathbb{Z}_n$ as follows:

$$
f(x, a, b) = \begin{cases} (\beta x, a, b+1) & \text{if } x \in S_1 \\ (x^2, 2a, 2b) & \text{if } x \in S_2 \\ (\alpha x, a+1, b) & \text{if } x \in S_3. \end{cases}
$$

Further, each of the triples (x, a, b) that we form are required to have the property that $x = \alpha^a \beta^b$. We begin with an initial triple having this property, say $(1, 0, 0)$. Observe that $f(x, a, b)$ satisfies the desired property if (x, a, b) does. So we define

$$
(x_i, a_i, b_i) = \begin{cases} (1, 0, 0) & \text{if } i = 0 \\ f(x_{i-1}, a_{i-1}, b_{i-1}) & \text{if } i \ge 1. \end{cases}
$$

We compare the triples (x_{2i}, a_{2i}, b_{2i}) and (x_i, a_i, b_i) until we find a value of $i \geq 1$ such that $x_{2i} = x_i$. When this occurs, we have that

$$\alpha^{a_{2i}} \beta^{b_{2i}} = \alpha^{a_i} \beta^{b_i}.$$

If we denote $c = \log_\alpha \beta$, then it must be the case that

$$\alpha^{a_{2i} + cb_{2i}} = \alpha^{a_i + cb_i}.$$

Since α has order n, it follows that

$$a_{2i} + cb_{2i} \equiv a_i + cb_i \pmod{n}.$$

This can be rewritten as

$$c(b_{2i} - b_i) \equiv a_i - a_{2i} \pmod{n}.$$

If $\gcd(b_{2i} - b_i, n) = 1$, then we can solve for c as follows:

$$c = (a_i - a_{2i})(b_{2i} - b_i)^{-1} \bmod n.$$

We illustrate the application of the above algorithm with an example. Notice that we take care to ensure that $1 \notin S_2$ (since we would obtain $x_i = (1, 0, 0)$ for all integers $i \geq 0$ if $1 \in S_2$).

Example 6.3 The integer $p = 809$ is prime, and it can be verified that the element $\alpha = 89$ has order $n = 101$ in $\mathbb{Z}_{809}^*$. The element $\beta = 618$ is in the subgroup $\langle \alpha \rangle$; we will compute $\log_\alpha \beta$.

Suppose we define the sets S_1, S_2 and S_3 as follows:

$$S_1 = \{x \in \mathbb{Z}_{809} : x \equiv 1 \pmod 3\}$$

$$S_2 = \{x \in \mathbb{Z}_{809} : x \equiv 0 \pmod 3\}$$

$$S_3 = \{x \in \mathbb{Z}_{809} : x \equiv 2 \pmod 3\}.$$

For $i = 1, 2, \ldots$, we obtain triples (x_{2i}, a_{2i}, b_{2i}) and (x_i, a_i, b_i) as follows:

i	(x_i, a_i, b_i)	(x_{2i}, a_{2i}, b_{2i})
1	$(618, 0, 1)$	$(76, 0, 2)$
2	$(76, 0, 2)$	$(113, 0, 4)$
3	$(46, 0, 3)$	$(488, 1, 5)$
4	$(113, 0, 4)$	$(605, 4, 10)$
5	$(349, 1, 4)$	$(422, 5, 11)$
6	$(488, 1, 5)$	$(683, 7, 11)$
7	$(555, 2, 5)$	$(451, 8, 12)$
8	$(605, 4, 10)$	$(344, 9, 13)$
9	$(451, 5, 10)$	$(112, 11, 13)$
10	$(422, 5, 11)$	$(422, 11, 15)$

The first collision in the above list is $x_{10} = x_{20} = 422$. The equation to be solved is

$$c = (11 - 5)(11 - 15)^{-1} \bmod 101 = (6 \times 25) \bmod 101 = 49.$$

Therefore, $\log_{89} 618 = 49$ in the group $\mathbb{Z}_{809}^*$. ▯

The POLLARD RHO ALGORITHM for discrete logarithms is presented as Algorithm 6.2. In this algorithm, we assume, as usual, that $\alpha \in G$ has order n and $\beta \in \langle \alpha \rangle$.

Algorithm 6.2: POLLARD RHO DISCRETE LOG ALGORITHM(G, n, α, β)

procedure $f(x, a, b)$
  if $x \in S_1$
    then $f \leftarrow (\beta \cdot x, a, (b + 1) \bmod n)$
    else if $x \in S_2$
    then $f \leftarrow (x^2, 2a \bmod n, 2b \bmod n)$
    else $f \leftarrow (\alpha \cdot x, (a + 1) \bmod n, b)$
  return (f)

main
  define the partition $G = S_1 \cup S_2 \cup S_3$
  $(x, a, b) \leftarrow f(1, 0, 0)$
  $(x', a', b') \leftarrow f(x, a, b)$
  while $x \neq x'$
  do $\begin{cases} (x, a, b) \leftarrow f(x, a, b) \\ (x', a', b') \leftarrow f(x', a', b') \\ (x', a', b') \leftarrow f(x', a', b') \end{cases}$
  if $\gcd(b' - b, n) \neq 1$
    then return ("failure")
    else return $((a - a')(b' - b)^{-1} \bmod n)$

In the situation where $\gcd(b' - b, n) > 1$, Algorithm 6.2 terminates with the output "failure." The situation may not be so bleak, however. If $\gcd(b' - b, n) = d$, then it is not hard to show that the congruence $c(b' - b) \equiv a - a' \pmod{n}$ has exactly d possible solutions. If d is not too large, then it is relatively straightforward to find the d solutions to the congruence and check to see which one is the correct one.

Algorithm 6.2 can be analyzed in a similar fashion as the Pollard rho factoring algorithm. Under reasonable assumptions concerning the randomness of the function f, we expect to be able to compute discrete logarithms in cyclic groups of order n in $O(\sqrt{n})$ iterations of the algorithm.

6.2.3   The Pohlig-Hellman Algorithm

The next algorithm we study is the POHLIG-HELLMAN ALGORITHM. Suppose that

$$n = \prod_{i=1}^{k} p_i^{c_i},$$

where the p_i's are distinct primes. The value $a = \log_\alpha \beta$ is determined (uniquely) modulo n. We first observe that if we can compute $a \bmod p_i^{c_i}$ for each i, $1 \leq i \leq k$, then we can compute $a \bmod n$ by the Chinese remainder theorem. So, let's suppose that q is prime,

$$n \equiv 0 \pmod{q^c}$$

and

$$n \not\equiv 0 \pmod{q^{c+1}}.$$

We will show how to compute the value

$$x = a \bmod q^c,$$

where $0 \leq x \leq q^c - 1$. We can express x in radix q representation as

$$x = \sum_{i=0}^{c-1} a_i q^i,$$

where $0 \leq a_i \leq q - 1$ for $0 \leq i \leq c - 1$. Also, observe that we can express a as

$$a = x + sq^c$$

for some integer s. Hence, we have that

$$a = \sum_{i=0}^{c-1} a_i q^i + sq^c.$$

The first step of the algorithm is to compute a_0. The main observation used in the algorithm is the following:

$$\beta^{n/q} = \alpha^{a_0 n/q}. \tag{6.1}$$

We prove that equation (6.1) holds as follows:

$$\begin{aligned}
\beta^{n/q} &= (\alpha^a)^{n/q} \\
&= (\alpha^{a_0 + a_1 q + \cdots + a_{c-1} q^{c-1} + sq^c})^{n/q} \\
&= (\alpha^{a_0 + Kq})^{n/q} \qquad \text{where } K \text{ is an integer} \\
&= \alpha^{a_0 n/q} \alpha^{Kn} \\
&= \alpha^{a_0 n/q}.
\end{aligned}$$

Using equation (6.1), it is a simple matter to determine a_0. This can be done, for example, by computing

$$\gamma = \alpha^{n/q}, \gamma^2, \ldots,$$

until

$$\gamma^i = \beta^{n/q}$$

for some $i \leq q - 1$. When this happens, we know that $a_0 = i$.

Now, if $c = 1$, we're done. Otherwise $c > 1$, and we proceed to determine $a_1, \ldots, a_{c-1}$. This is done in a similar fashion as the computation of a_0. Denote $\beta_0 = \beta$, and define

$$\beta_j = \beta \alpha^{-(a_0 + a_1 q + \cdots + a_{j-1} q^{j-1})}$$

for $1 \leq j \leq c - 1$. We make use of the following generalization of equation (6.1):

$$\beta_j^{n/q^{j+1}} = \alpha^{a_j n/q}. \tag{6.2}$$

Observe that equation (6.2) reduces to equation (6.1) when $j = 0$.

The proof of equation (6.2) is similar to that of equation (6.1):

$$\begin{aligned}
\beta_j^{n/q^{j+1}} &= (\alpha^{a - (a_0 + a_1 q + \cdots + a_{j-1} q^{j-1})})^{n/q^{j+1}} \\
&= (\alpha^{a_j q^j + \cdots + a_{c-1} q^{c-1} + s q^c})^{n/q^{j+1}} \\
&= (\alpha^{a_j q^j + K_j q^{j+1}})^{n/q^{j+1}} \quad \text{where } K_j \text{ is an integer} \\
&= \alpha^{a_j n/q} \alpha^{K_j n} \\
&= \alpha^{a_j n/q}.
\end{aligned}$$

Hence, given β_j, it is straightforward to compute a_j from equation (6.2).

To complete the description of the algorithm, it suffices to observe that β_{j+1} can be computed from β_j by means of a simple recurrence relation, once a_j is known:

$$\beta_{j+1} = \beta_j \alpha^{-a_j q^j}. \tag{6.3}$$

Therefore, we can compute $a_0, \beta_1, a_1, \beta_2, \ldots, \beta_{c-1}, a_{c-1}$ by alternately applying equations (6.2) and (6.3).

A pseudo-code description of the POHLIG-HELLMAN ALGORITHM is given as Algorithm 6.3. To summarize the operation of this algorithm, α is an element of order n in a multiplicative group G, q is prime,

$$n \equiv 0 \pmod{q^c}$$

and

$$n \not\equiv 0 \pmod{q^{c+1}}.$$

The algorithm calculates $a_0, \ldots, a_{c-1}$, where

$$\log_\alpha \beta \bmod q^c = \sum_{i=0}^{c-1} a_i q^i.$$

Algorithm 6.3: POHLIG-HELLMAN$(G, n, \alpha, \beta, q, c)$

$j \leftarrow 0$
$\beta_j \leftarrow \beta$
while $j \leq c - 1$

$$\textbf{do} \begin{cases} \delta \leftarrow \beta_j^{n/q^{j+1}} \\ \text{find } i \text{ such that } \delta = \alpha^{in/q} \\ a_j \leftarrow i \\ \beta_{j+1} \leftarrow \beta_j \alpha^{-a_j q^j} \\ j \leftarrow j + 1 \end{cases}$$

return $(a_0, \ldots, a_{c-1})$

We illustrate the Pohlig-Hellman algorithm with a small example.

Example 6.4 Suppose $p = 29$ and $\alpha = 2$. p is prime and α is a primitive element modulo p, and we have that

$$n = p - 1 = 28 = 2^2 7^1.$$

Suppose $\beta = 18$, so we want to determine $a = \log_2 18$. We proceed by first computing $a \bmod 4$ and then computing $a \bmod 7$.

We start by setting $q = 2$ and $c = 2$ and applying Algorithm 6.3. We find that $a_0 = 1$ and $a_1 = 1$. Hence, $a \equiv 3 \pmod 4$.

Next, we apply Algorithm 6.3 with $q = 7$ and $c = 1$. We find that $a_0 = 4$, so $a \equiv 4 \pmod 7$.

Finally, solving the system

$$a \equiv 3 \pmod 4$$

$$a \equiv 4 \pmod 7$$

using the Chinese remainder theorem, we get $a \equiv 11 \pmod{28}$. That is, we have computed $\log_2 18 = 11$ in $\mathbb{Z}_{29}$. $\quad\square$

Let's consider the complexity of Algorithm 6.3. It is not difficult to see that a straightforward implementation of this algorithm runs in time $O(cq)$. This can be improved, however, by observing that each computation of a value i such that $\delta = \alpha^{in/q}$ can be viewed as the solution of a particular instance of the **Discrete Logarithm** problem. To be specific, we have that $\delta = \alpha^{in/q}$ if and only if

$$i = \log_{\alpha^{n/q}} \delta.$$

The element $\alpha^{n/q}$ has order q, and therefore each i can be computed (using SHANKS' ALGORITHM, for example) in time $O(\sqrt{q})$. The complexity of Algorithm 6.3 can therefore be reduced to $O(c\sqrt{q})$.

6.2.4 The Index Calculus Method

The algorithms in the three previous sections can be applied to any group. The algorithm we describe in this section, the INDEX CALCULUS ALGORITHM, is more specialized: it applies to the particular situation of finding discrete logarithms in $\mathbb{Z}_p^*$ when p is prime and α is a primitive element modulo p. In this situation, the INDEX CALCULUS ALGORITHM is faster than the algorithms previously considered.

The INDEX CALCULUS ALGORITHM for computing discrete logarithms bears considerable resemblance to many of the best factoring algorithms. The method uses a *factor base*, which, as in Section 5.6.3, is a set $\mathcal{B}$ of "small" primes. Suppose $\mathcal{B} = \{p_1, p_2, \ldots, p_B\}$. The first step (a preprocessing step) is to find the logarithms of the B primes in the factor base. The second step is to compute the discrete logarithm of a desired element β, using the knowledge of the discrete logarithms of the elements in the factor base.

Let C be a bit bigger than B; say $C = B + 10$. In the precomputation phase, we will construct C congruences modulo p, which have the following form:

$$\alpha^{x_j} \equiv p_1^{a_{1j}} p_2^{a_{2j}} \ldots p_B^{a_{Bj}} \pmod{p},$$

for $1 \leq j \leq C$. Notice that these congruences can be written equivalently as

$$x_j \equiv a_{1j} \log_\alpha p_1 + \cdots + a_{Bj} \log_\alpha p_B \pmod{p-1},$$

$1 \leq j \leq C$. Given C congruences in the B "unknowns" $\log_\alpha p_i$ ($1 \leq i \leq B$), we hope that there is a unique solution modulo $p - 1$. If this is the case, then we can compute the logarithms of the elements in the factor base.

How do we generate the C congruences of the desired form? One elementary way is to take a random value x, compute $\alpha^x \bmod p$, and then determine if $\alpha^x \bmod p$ has all its factors in $\mathcal{B}$ (using trial division, for example).

Now, supposing that we have already successfully carried out the precomputation step, we compute a desired logarithm $\log_\alpha \beta$ by means of a Las Vegas type randomized algorithm. Choose a random integer s ($1 \leq s \leq p - 2$) and compute

$$\gamma = \beta \alpha^s \bmod p.$$

Now attempt to factor γ over the factor base $\mathcal{B}$. If this can be done, then we obtain a congruence of the form

$$\beta \alpha^s \equiv p_1^{c_1} p_2^{c_2} \ldots p_B^{c_B} \pmod{p}.$$

This can be written equivalently as

$$\log_\alpha \beta + s \equiv c_1 \log_\alpha p_1 + \cdots + c_B \log_\alpha p_B \pmod{p-1}.$$

Since all terms in the above congruence are now known, except for $\log_\alpha \beta$, we can easily solve for $\log_\alpha \beta$.

Here is a small, very artificial, example to illustrate the two steps in the algorithm.

Example 6.5   The integer $p = 10007$ is prime. Suppose that $\alpha = 5$ is the primitive element used as the base of logarithms modulo p. Suppose we take $\mathcal{B} = \{2, 3, 5, 7\}$ as the factor base. Of course $\log_5 5 = 1$, so there are three logs of factor base elements to be determined.

Some examples of "lucky" exponents that might be chosen are 4063, 5136 and 9865.

With $x = 4063$, we compute

$$5^{4063} \bmod 10007 = 42 = 2 \times 3 \times 7.$$

This yields the congruence

$$\log_5 2 + \log_5 3 + \log_5 7 \equiv 4063 \pmod{10006}.$$

Similarly, since

$$5^{5136} \bmod 10007 = 54 = 2 \times 3^3$$

and

$$5^{9865} \bmod 10007 = 189 = 3^3 \times 7,$$

we obtain two more congruences:

$$\log_5 2 + 3 \log_5 3 \equiv 5136 \pmod{10006}$$

and

$$3 \log_5 3 + \log_5 7 \equiv 9865 \pmod{10006}.$$

We now have three congruences in three unknowns, and there happens to be a unique solution modulo 10006, namely $\log_5 2 = 6578$, $\log_5 3 = 6190$ and $\log_5 7 = 1301$.

Now, let's suppose that we wish to find $\log_5 9451$. Suppose we choose the "random" exponent $s = 7736$, and compute

$$9451 \times 5^{7736} \bmod 10007 = 8400.$$

Since $8400 = 2^4 3^1 5^2 7^1$ factors over $\mathcal{B}$, we obtain

$$\log_5 9451 = (4 \log_5 2 + \log_5 3 + 2 \log_5 5 + \log_5 7 - s) \bmod 10006$$

$$= (4 \times 6578 + 6190 + 2 \times 1 + 1301 - 7736) \bmod 10006$$

$$= 6057.$$

To verify, we can check that $5^{6057} \equiv 9451 \pmod{10007}$.   □

Heuristic analyses of various versions of the INDEX CALCULUS ALGORITHM have been done. Under reasonable assumptions, such as those considered in the analysis of DIXON'S ALGORITHM in Section 5.6.3, the asymptotic running time of the precomputation phase is

$$O\left(e^{(1+o(1))\sqrt{\ln p \ln \ln p}}\right),$$

and the time to find a particular discrete logarithm is

$$O\left(e^{(1/2+o(1))\sqrt{\ln p \ln \ln p}}\right).$$

6.3 Lower Bounds on the Complexity of Generic Algorithms

In this section, we turn our attention to an interesting lower bound on the complexity of the Discrete Logarithm problem. Several of the algorithms we have described for the discrete logarithm problem can be applied in any group. An algorithm of this type is called a *generic algorithm*, because it does not depend on any property of the representation of the group. Examples of generic algorithms for the discrete logarithm problem include SHANKS' ALGORITHM, the POLLARD RHO ALGORITHM and the POHLIG-HELLMAN ALGORITHM. On the other hand, the INDEX CALCULUS ALGORITHM studied in the previous section is not generic. This algorithm involves treating elements of $\mathbb{Z}_p^*$ as integers, and then computing their factorizations into primes. Clearly this is something that cannot be done in an arbitrary group.

Another example of a non-generic algorithm for a particular group is provided by studying the Discrete Logarithm problem in the additive group $(\mathbb{Z}_n, +)$. (We defined the Discrete Logarithm problem in a multiplicative group, but this was done solely to establish a consistent notation for the algorithms we presented.) Suppose that $\gcd(\alpha, n) = 1$, so α is a generator of $\mathbb{Z}_n$. Since the group operation is addition modulo n, an "exponentiation" operation, α^a, corresponds to multiplication by a modulo n. Hence, in this setting, the Discrete Logarithm problem is to find the integer a such that

$$\alpha a \equiv \beta \pmod{n}.$$

Since $\gcd(\alpha, n) = 1$, α has a multiplicative inverse modulo n, and we can compute $\alpha^{-1} \bmod n$ easily using the EXTENDED EUCLIDEAN ALGORITHM. Then we can solve for a, obtaining

$$\log_\alpha \beta = \beta \alpha^{-1} \bmod n.$$

This algorithm is of course very fast; its complexity is polynomial in $\log n$.

An even more trivial algorithm can be used to solve the DISCRETE LOGARITHM problem in $(\mathbb{Z}_n, +)$ when $\alpha = 1$. In this situation, we have that $\log_1 \beta = \beta$ for all $\beta \in \mathbb{Z}_n$.

The DISCRETE LOGARITHM problem, by definition, takes place in a cyclic (sub)group of order n. It is well known, and almost trivial to prove, that all cyclic groups of order n are isomorphic. By the discussion above, we know how to compute discrete logarithms quickly in the additive group $(\mathbb{Z}_n, +)$. This suggests that we might be able to solve the Discrete Logarithm problem in any subgroup $\langle \alpha \rangle$ of order n of any group G by "reducing" the problem to the the easily solved formulation in $(\mathbb{Z}_n, +)$.

Let us think about how (in theory, at least) this could be done. The statement that $\langle \alpha \rangle$ is isomorphic to $(\mathbb{Z}_n, +)$ means that there is a bijection

$$\phi : \langle \alpha \rangle \to \mathbb{Z}_n$$

such that

$$\phi(xy) = (\phi(x) + \phi(y)) \bmod n$$

for all $x, y \in \langle \alpha \rangle$. It follows easily that

$$\phi(\alpha^a) = a\phi(\alpha) \bmod n,$$

so we have that

$$\beta = \alpha^a \Leftrightarrow a\phi(\alpha) \equiv \phi(\beta) \pmod{n}.$$

Hence, solving for a as described above (using the EXTENDED EUCLIDEAN ALGORITHM), we have that

$$\log_\alpha \beta = \phi(\beta)(\phi(\alpha))^{-1} \bmod n.$$

Consequently, if we have an efficient method of computing the isomorphism ϕ, then we would have an efficient algorithm to compute discrete logarithms in $\langle \alpha \rangle$. The catch is that there is no known general method to efficiently compute the isomorphism ϕ for an arbitrary subgroup $\langle \alpha \rangle$ of an arbitrary group G, even though we know the two groups in question are isomorphic. In fact, it is not hard to see that computing discrete logarithms in $\langle \alpha \rangle$ is equivalent to finding an explicit isomorphism between $\langle \alpha \rangle$ and $(\mathbb{Z}_n, +)$. Hence, this approach seems to lead to a dead end.

The fact that an extremely efficient algorithm exists for the DISCRETE LOGARITHM problem in $(\mathbb{Z}_n, +)$ makes it seem unlikely that there are any interesting lower bounds on the complexity of the general problem. However, this is not the case. A result of Shoup provides a lower bound on the complexity of generic algorithms for the DISCRETE LOGARITHM problem. Recall that Shanks' and the rho algorithms have the property that their complexity (in terms of the number of group operations required to run the algorithm) is roughly $\sqrt{n}$, where n is the order of the (sub)group in which the discrete logarithm is being sought. Shoup's result establishes that these algorithms are essentially optimal within the class of generic algorithms.

We begin by giving a precise description of what we mean by a generic algorithm. We consider a cyclic group or subgroup of order n, which is therefore isomorphic to $(\mathbb{Z}_n, +)$. We will study generic algorithms for the DISCRETE LOG-ARITHM problem in $(\mathbb{Z}_n, +)$. (As we shall see, the particular group that is used is irrelevant in the context of generic algorithms; the choice of $(\mathbb{Z}_n, +)$ is arbitrary.)

An *encoding* of $(\mathbb{Z}_n, +)$ is any injective mapping $\sigma : \mathbb{Z}_n \to S$, where S is a finite set. The encoding function specifies how group elements are represented. Any discrete logarithm problem in a (sub)group of cardinality n of an arbitrary group G can be specified by defining a suitable encoding function. For example, consider the multiplicative group $(\mathbb{Z}_p^*, \cdot)$, and let α be a primitive element in $\mathbb{Z}_p^*$. Let $n = p - 1$, and define the encoding function σ as follows: $\sigma(i) = \alpha^i \bmod p$, $0 \leq i \leq n - 1$. Then it should be clear that solving the Discrete Logarithm problem in $(\mathbb{Z}_p^*, \cdot)$ with respect to the primitive element α is equivalent to solving the Discrete Logarithm problem in $(\mathbb{Z}_n, +)$ with generator 1 under the encoding σ.

A generic algorithm is one that works for any encoding. In particular, a generic algorithm must work correctly when the encoding function σ is a random injective function; for example, when $S = \mathbb{Z}_n$ and σ is a random permutation of $\mathbb{Z}_n$. This is very similar to the random oracle model, in which a hash function is regarded as a random function in order to define an idealized model in which a formal security proof can be given.

We suppose that we have a random encoding, σ, for the group $(\mathbb{Z}_n, +)$. In this group, the discrete logarithm of any element a to the base 1 is just a, of course. Given the encoding function σ, the encoding $\sigma(1)$ of the generator, and an encoding of an arbitrary group element $\sigma(a)$, a generic algorithm is trying to compute the value of a. In order to perform operations in this group when group elements are encoded using the function σ, we hypothesize the existence of an oracle (or subroutine) to perform this task.

Given encodings of two group elements, say $\sigma(i)$ and $\sigma(j)$, it should be possible to compute the encodings $\sigma((i + j) \bmod n)$ and $\sigma((i - j) \bmod n)$. This is necessary if we are going to add and subtract group elements, and we assume that our oracle will do this for us. By combining operations of the above type, it is possible to compute arbitrary linear combinations of the form $\sigma((ci \pm dj) \bmod n)$, where $c, d \in \mathbb{Z}_n$. However, using the fact that $-j \equiv n - j \pmod{n}$, we observe that we only need to be able to compute linear combinations of the form $\sigma((ci + dj) \bmod n)$. We will assume that the oracle can directly compute linear combinations of this form in one unit of time.

Group operations of the type described above are the only ones allowed in a generic algorithm. That is, we assume that we have some method of performing group operations on encoded elements, but we cannot do any more than that. Now let us consider how a generic algorithm, say GENLOG, can go about trying to compute a discrete logarithm. The input to the algorithm GENLOG consists of $\sigma_1 = \sigma(1)$ and $\sigma_2 = \sigma(a)$, where $a \in \mathbb{Z}_n$ is chosen randomly. GENLOG will be successful if and only if it outputs the value a. (We will assume that n is prime, in order to simplify the analysis.)

GENLOG will use the oracle to generate a sequence of m, say, encodings of linear combinations of 1 and a. The execution of GENLOG can be specified by a list of ordered pairs $(c_i, d_i) \in \mathbb{Z}_n \times \mathbb{Z}_n$, $1 \leq m$. (We can assume that these m ordered pairs are distinct.) For each ordered pair (c_i, d_i), the oracle computes the encoding $\sigma_i = \sigma((c_i + d_i a) \bmod n)$. Note that we can define $(c_1, d_1) = (1, 0)$ and $(c_2, d_2) = (0, 1)$ so our notation is consistent with the input to the algorithm.

In this way, the algorithm GENLOG obtains a list of encoded group elements, $(\sigma_1, \ldots, \sigma_m)$. Because the encoding function σ is injective, it follows immediately that $c_i + d_i a \equiv c_j + d_j a \pmod{n}$ if and only if $\sigma_i = \sigma_j$. This provides a method to possibly compute the value of the unknown a: Suppose that $\sigma_i = \sigma_j$ for two integers $i \neq j$. If $d_i = d_j$, then $c_i = c_j$ and the two ordered pairs (c_i, d_i) and (c_j, d_j) are the same. Since we are assuming the ordered pairs are distinct, it follows that $d_i \neq d_j$. Because n is prime, we can compute a as follows:

$$a = (c_i - c_j)(d_j - d_i)^{-1} \bmod n.$$

(Recall that we used a similar method of computing the value of a discrete logarithm in the POLLARD RHO ALGORITHM.)

Suppose first that the algorithm GENLOG chooses a set

$$\mathcal{C} = \{(c_i, d_i) : 1 \leq i \leq m\} \subseteq \mathbb{Z}_n \times \mathbb{Z}_n$$

of m distinct ordered pairs all at once, at the beginning of the algorithm. Such an algorithm is called a *non-adaptive algorithm* (SHANKS' ALGORITHM is an example of a non-adaptive algorithm). Then the list of m corresponding encodings is obtained from the oracle. Define $\mathsf{Good}(\mathcal{C})$ to consist of all elements $a \in \mathbb{Z}_n$ that are the solution of an equation $a = (c_i - c_j)(d_j - d_i)^{-1} \bmod n$ with $i \neq j$, $i, j \in \{1, \ldots, m\}$. By what we have said above, we know that the value of a can be computed by GENLOG if and only if $a \in \mathsf{Good}(\mathcal{C})$. It is clear that $|\mathsf{Good}(\mathcal{C})| \leq \binom{m}{2}$, so there are at most $\binom{m}{2}$ elements for which GENLOG can compute the discrete logarithm after having obtained a sequence of m encoded group elements corresponding to the ordered pairs in $\mathcal{C}$. The probability that $a \in \mathsf{Good}(\mathcal{C})$ is at most $\binom{m}{2}/n$. If $a \notin \mathsf{Good}(\mathcal{C})$, then the best strategy for the algorithm GENLOG is to guess the value of a by choosing a random value in $\mathbb{Z}_n \backslash \mathsf{Good}(\mathcal{C})$. Denote $g = |\mathsf{Good}(\mathcal{C})|$. Then, by conditioning on whether or not $a \in \mathsf{Good}(\mathcal{C})$, we can compute a bound on the success probability of the algorithm. Suppose we define the following random variables: **a** denotes the event $a \in \mathsf{Good}(\mathcal{C})$; and **b** denotes the event "the algorithm returns the correct value of a." Then we have that

$$\mathbf{Pr[b]} = \mathbf{Pr[b|a]Pr[a]} + \mathbf{Pr[b|\bar{a}]Pr[\bar{a}]}$$

$$= 1 \times \frac{g}{n} + \frac{1}{n-g} \times \frac{n-g}{n}$$

$$= \frac{g+1}{n}$$

$$\leq \frac{\binom{m}{2} + 1}{n}.$$

A generic discrete logarithm algorithm is not required to choose all the ordered pairs in $\mathcal{C}$ at the beginning of the algorithm, of course. It can choose later pairs after seeing what encodings of previous linear combinations look like (i.e., we allow the algorithm to be an *adaptive algorithm*). However, we will show, using an inductive argument, that this does not improve the success probability of the algorithm.

Let GENLOG be an adaptive generic algorithm for the discrete logarithm problem. For $1 \leq i \leq m$, let $\mathcal{C}_i$ consist of the first i ordered pairs, for which the oracle computes the corresponding encodings $\sigma_1, \ldots, \sigma_i$. The set $\mathcal{C}_i$ and the list $\sigma_1, \ldots, \sigma_i$ represent all the information available to GENLOG at time i of its execution.

We claim that the value of a can be computed at time i if $a \in \mathsf{Good}(\mathcal{C}_i)$; and if $a \notin \mathsf{Good}(\mathcal{C}_i)$, then the probability that a has any given value in the set $\mathbb{Z}_n \backslash \mathsf{Good}(\mathcal{C}_i)$ is exactly $1/(n - |\mathsf{Good}(\mathcal{C}_i)|)$. (This probability is the conditional probability of a value of a at time i, assuming that a is chosen randomly from $\mathbb{Z}_n$.)

This claim is true for $i = 1$ (which forms the base case for an inductive proof) because $\mathcal{C}_1 = \{(1, 0)\}$ by definition, and $\mathsf{Good}(\mathcal{C}_1) = \emptyset$. If the claim is true for $i = j - 1$, then it is easy to see that it is also true for $i = j$. Then the claimed result follows by induction. When $i = m$, the claim establishes that the success probability of the algorithm is at most $(\binom{m}{2} + 1)/n$, exactly as in the case of a non-adaptive algorithm.

If GENLOG is guaranteed to compute the desired discrete logarithm, then its success probability is equal to 1. Then the result proved above says that $(\binom{m}{2} + 1)/n \geq 1$, or $m^2 + m \geq 2n$. Hence, m is $\Omega(\sqrt{n})$. This provides a lower bound on the complexity of any generic algorithm for the **Discrete Logarithm** problem in a (sub)group of prime order n.

6.4 Finite Fields

The *ElGamal Cryptosystem* can be implemented in any group where the **Discrete Logarithm** problem is infeasible. We used the multiplicative group $\mathbb{Z}_p^*$ in the description of Cryptosystem 6.1, but other groups are also suitable candidates. Two such classes of groups are

1. the multiplicative group of the finite field $\mathbb{F}_{p^n}$
2. the group of an elliptic curve defined over a finite field.

We will discuss these two classes of groups in the next sections.

We have already discussed the fact that $\mathbb{Z}_p$ is a field if p is prime. However, there are other examples of finite fields not of this form. In fact, there is a finite field with q elements if $q = p^n$ where p is prime and $n \geq 1$ is an integer. We will now describe very briefly how to construct such a field. First, we need several definitions.

Definition 6.1:   Suppose p is prime. Define $\mathbb{Z}_p[x]$ to be the set of all polynomials in the indeterminate x. By defining addition and multiplication of polynomials in the usual way (and reducing coefficients modulo p), we construct a ring.

For $f(x), g(x) \in \mathbb{Z}_p[x]$, we say that $f(x)$ *divides* $g(x)$ (notation: $f(x) \mid g(x)$) if there exists $q(x) \in \mathbb{Z}_p[x]$ such that

$$g(x) = q(x)f(x).$$

For $f(x) \in \mathbb{Z}_p[x]$, define $\deg(f)$, the *degree* of f, to be the highest exponent in a term of f.

Suppose $f(x), g(x), h(x) \in \mathbb{Z}_p[x]$, and $\deg(f) = n \geq 1$. We define

$$g(x) \equiv h(x) \pmod{f(x)}$$

if

$$f(x) \mid (g(x) - h(x)).$$

Notice the resemblance of the definition of congruence of polynomials to that of congruence of integers.

We are now going to define a ring of polynomials "modulo $f(x)$" which we denote by $\mathbb{Z}_p[x]/(f(x))$. The construction of $\mathbb{Z}_p[x]/(f(x))$ from $\mathbb{Z}_p[x]$ is based on the idea of congruences modulo $f(x)$ and is analogous to the construction of $\mathbb{Z}_m$ from $\mathbb{Z}$.

Suppose $\deg(f) = n$. If we divide $g(x)$ by $f(x)$, we obtain a (unique) *quotient* $q(x)$ and *remainder* $r(x)$, where

$$g(x) = q(x)f(x) + r(x)$$

and

$$\deg(r) < n.$$

This can be done by usual long division of polynomials. Hence any polynomial in $\mathbb{Z}_p[x]$ is congruent modulo $f(x)$ to a unique polynomial of degree at most $n - 1$.

Now we define the elements of $\mathbb{Z}_p[x]/(f(x))$ to be the p^n polynomials in $\mathbb{Z}_p[x]$ of degree at most $n - 1$. Addition and multiplication in $\mathbb{Z}_p[x]/(f(x))$ is defined as in $\mathbb{Z}_p[x]$, followed by a reduction modulo $f(x)$. Equipped with these operations, $\mathbb{Z}_p[x]/(f(x))$ is a ring.

Recall that $\mathbb{Z}_m$ is a field if and only if m is prime, and multiplicative inverses can be found using the Euclidean algorithm. A similar situation holds for $\mathbb{Z}_p[x]/(f(x))$. The analog of primality for polynomials is irreducibility, which we define as follows:

> **Definition 6.2:** A polynomial $f(x) \in \mathbb{Z}_p[x]$ is said to be *irreducible* if there do not exist polynomials $f_1(x), f_2(x) \in \mathbb{Z}_p[x]$ such that
>
> $$f(x) = f_1(x)f_2(x),$$
>
> where $\deg(f_1) > 0$ and $\deg(f_2) > 0$.

A very important fact is that $\mathbb{Z}_p[x]/(f(x))$ is a field if and only if $f(x)$ is irreducible. Further, multiplicative inverses in $\mathbb{Z}_p[x]/(f(x))$ can be computed using a straightforward modification of the (extended) Euclidean algorithm.

Here is an example to illustrate the concepts described above.

Example 6.6 Let's attempt to construct a field having eight elements. This can be done by finding an irreducible polynomial of degree three in $\mathbb{Z}_2[x]$. It is sufficient to consider the polynomials having constant term equal to 1, since any polynomial with constant term 0 is divisible by x and hence is reducible. There are four such polynomials:

$$f_1(x) = x^3 + 1$$
$$f_2(x) = x^3 + x + 1$$
$$f_3(x) = x^3 + x^2 + 1$$
$$f_4(x) = x^3 + x^2 + x + 1.$$

Now, $f_1(x)$ is reducible because

$$x^3 + 1 = (x + 1)(x^2 + x + 1)$$

(remember that all coefficients are to be reduced modulo 2). Also, f_4 is reducible because

$$x^3 + x^2 + x + 1 = (x + 1)(x^2 + 1).$$

However, $f_2(x)$ and $f_3(x)$ are both irreducible, and either one can be used to construct a field having eight elements.

Let us use $f_2(x)$, and thus construct the field $\mathbb{Z}_2[x]/(x^3 + x + 1)$. The eight field elements are the eight polynomials $0, 1, x, x + 1, x^2, x^2 + 1, x^2 + x$ and $x^2 + x + 1$.

To compute a product of two field elements, we multiply the two polynomials together, and reduce modulo $x^3 + x + 1$ (i.e., divide by $x^3 + x + 1$ and find the remainder polynomial). Since we are dividing by a polynomial of degree three, the remainder will have degree at most two and hence is an element of the field.

For example, to compute $(x^2 + 1)(x^2 + x + 1)$ in $\mathbb{Z}_2[x]/(x^3 + x + 1)$, we first compute the product in $\mathbb{Z}_2[x]$, which is $x^4 + x^3 + x + 1$. Then we divide by $x^3 + x + 1$, obtaining the expression

$$x^4 + x^3 + x + 1 = (x + 1)(x^3 + x + 1) + x^2 + x.$$

Hence, in the field $\mathbb{Z}_2[x]/(x^3 + x + 1)$, we have that

$$(x^2 + 1)(x^2 + x + 1) = x^2 + x.$$

Below, we present a complete multiplication table for the non-zero field elements. To save space, we write a polynomial $a_2x^2 + a_1x + a_0$ as the ordered triple $a_2a_1a_0$.

	001	010	011	100	101	110	111
001	001	010	011	100	101	110	111
010	010	100	110	011	001	111	101
011	011	110	101	111	100	001	010
100	100	011	111	110	010	101	001
101	101	001	100	010	111	011	110
110	110	111	001	101	011	010	100
111	111	101	010	001	110	100	011

Computation of inverses can be done by using a straightforward adaptation of the extended Euclidean algorithm.

Finally, the multiplicative group of the non-zero polynomials in the field is a cyclic group of order seven. Since 7 is prime, it follows that any non-zero field element is a generator of this group, i.e., a primitive element of the field.

For example, if we compute the powers of x, we obtain

$$x^1 = x$$
$$x^2 = x^2$$
$$x^3 = x + 1$$
$$x^4 = x^2 + x$$
$$x^5 = x^2 + x + 1$$
$$x^6 = x^2 + 1$$
$$x^7 = 1,$$

which comprise all the non-zero field elements.   ⬚

It remains to discuss existence and uniqueness of fields of this type. It can be shown that there is at least one irreducible polynomial of any given degree $n \geq 1$ in $\mathbb{Z}_p[x]$. Hence, there is a finite field with p^n elements for all primes p and all integers $n \geq 1$. There are usually many irreducible polynomials of degree n in $\mathbb{Z}_p[x]$. But the finite fields constructed from any two irreducible polynomials of degree n can be shown to be isomorphic. Thus there is a unique finite field of any size p^n (p prime, $n \geq 1$), which is denoted by $\mathbb{F}_{p^n}$. In the case $n = 1$, the resulting field $\mathbb{F}_p$ is the same thing as $\mathbb{Z}_p$. Finally, it can be shown that there does not exist a finite field with r elements unless $r = p^n$ for some prime p and some integer $n \geq 1$.

We have already noted that the multiplicative group $\mathbb{Z}_p^*$ (p prime) is a cyclic group of order $p - 1$. In fact, the multiplicative group of any finite field is cyclic: $\mathbb{F}_{p^n} \backslash \{0\}$ is a cyclic group of order $p^n - 1$. This provides further examples of cyclic groups in which the Discrete Logarithm problem can be studied.

In practice, the finite fields $\mathbb{F}_{2^n}$ have been most studied. Any generic algorithm works in a field $\mathbb{F}_{2^n}$, of course. More importantly, however, the INDEX CALCULUS ALGORITHM can be modified in a straightforward manner to work in these fields. Recall that the main steps in the INDEX CALCULUS ALGORITHM involve trying to factor elements in $\mathbb{Z}_p$ over a given factor base which consists of small primes. The analog of a factor base in $\mathbb{Z}_2[x]$ is a set of irreducible polynomials of low degree. The idea then is to try to factor elements in $\mathbb{F}_{2^n}$ into polynomials in the given factor base. The reader can easily fill in the details.

Once the appropriate modifications have been made, the precomputation time of the INDEX CALCULUS ALGORITHM in $\mathbb{F}_{2^n}$ turns out to be

$$O\left(e^{(1.405 + o(1))n^{1/3}(\ln n)^{2/3}}\right),$$

and the time to find an individual discrete logarithm is

$$O\left(e^{(1.098 + o(1))n^{1/3}(\ln n)^{2/3}}\right).$$

For large values of n (say $n > 1024$), the Discrete Logarithm problem in $\mathbb{F}_{2^n}$ is thought to be infeasible at present, provided that $2^n - 1$ has at least one "large" prime factor (in order to thwart a Pohlig-Hellman attack).

6.5 Elliptic Curves

Elliptic curves are described by the set of solutions to certain equations in two variables. Elliptic curves defined modulo a prime p are of central importance in public-key cryptography. We begin by looking briefly at elliptic curves defined over the real numbers, because some of the basic concepts are easier to motivate in this setting.

6.5.1 Elliptic Curves over the Reals

Definition 6.3: Let $a, b \in \mathbb{R}$ be constants such that $4a^3 + 27b^2 \neq 0$. A *non-singular elliptic curve* is the set E of solutions $(x, y) \in \mathbb{R} \times \mathbb{R}$ to the equation

$$y^2 = x^3 + ax + b, \tag{6.4}$$

together with a special point $\mathcal{O}$ called the *point at infinity*.

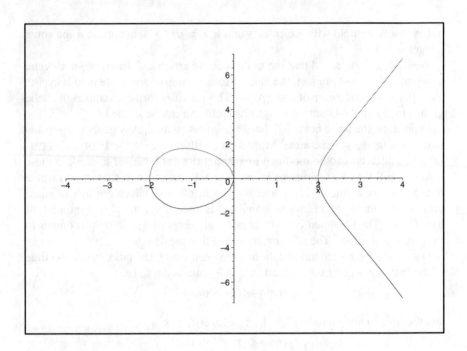

FIGURE 6.1
An elliptic curve over the reals

It can be shown that the condition $4a^3 + 27b^2 \neq 0$ is necessary and sufficient to ensure that the equation $x^3 + ax + b = 0$ has three distinct roots (which may be real or complex numbers). If $4a^3 + 27b^2 = 0$, then the corresponding elliptic curve is called a *singular elliptic curve*.

In Figure 6.1, we depict the elliptic curve $y^2 = x^3 - 4x$.

Suppose E is a non-singular elliptic curve. We will define a binary operation over E which makes E into an abelian group. This operation is usually denoted by addition. The point at infinity, $\mathcal{O}$, will be the identity element, so $P + \mathcal{O} = \mathcal{O} + P = P$ for all $P \in E$.

Suppose $P, Q \in E$, where $P = (x_1, y_1)$ and $Q = (x_2, y_2)$. We consider three cases:

1. $x_1 \neq x_2$
2. $x_1 = x_2$ and $y_1 = -y_2$
3. $x_1 = x_2$ and $y_1 = y_2$

In case 1, we define L to be the line through P and Q. L intersects E in the two points P and Q, and it is easy to see that L will intersect E in one further point, which we call R'. If we reflect R in the x-axis, then we get a point which we name R. We define $P + Q = R$.

Let's work out an algebraic formula to compute R. First, the equation of L is $y = \lambda x + \nu$, where the slope of L is

$$\lambda = \frac{y_2 - y_1}{x_2 - x_1},$$

and

$$\nu = y_1 - \lambda x_1 = y_2 - \lambda x_2.$$

In order to find the points in $E \cap L$, we substitute $y = \lambda x + \nu$ into the equation for E, obtaining the following:

$$(\lambda x + \nu)^2 = x^3 + ax + b,$$

which is the same as

$$x^3 - \lambda^2 x^2 + (a - 2\lambda\nu)x + b - \nu^2 = 0. \tag{6.5}$$

The roots of equation (6.5) are the x-co-ordinates of the points in $E \cap L$. We already know two points in $E \cap L$, namely, P and Q. Hence x_1 and x_2 are two roots of equation (6.5).

Since equation (6.5) is a cubic equation over the reals having two real roots, the third root, say x_3, must also be real. The sum of the three roots must be the negative of the coefficient of the quadratic term, or λ^2. Therefore

$$x_3 = \lambda^2 - x_1 - x_2.$$

x_3 is the x-co-ordinate of the point R'. We will denote the y-co-ordinate of R' by $-y_3$, so the y-co-ordinate of R will be y_3. An easy way to compute y_3 is to use the fact that the slope of L, namely λ, is determined by any two points on L. If we use the points (x_1, y_1) and $(x_3, -y_3)$ to compute this slope, we get

$$\lambda = \frac{-y_3 - y_1}{x_3 - x_1},$$

or

$$y_3 = \lambda(x_1 - x_3) - y_1.$$

Therefore we have derived a formula for $P + Q$ in case 1: if $x_1 \neq x_2$, then $(x_1, y_1) + (x_2, y_2) = (x_3, y_3)$, where

$$x_3 = \lambda^2 - x_1 - x_2,$$

$$y_3 = \lambda(x_1 - x_3) - y_1, \quad \text{and}$$

$$\lambda = \frac{y_2 - y_1}{x_2 - x_1}.$$

Case 2, where $x_1 = x_2$ and $y_1 = -y_2$, is simple: we define $(x, y) + (x, -y) = \mathcal{O}$ for all $(x, y) \in E$. Therefore (x, y) and $(x, -y)$ are inverses with respect to the elliptic curve addition operation.

Case 3 remains to be considered. Here we are adding a point $P = (x_1, y_1)$ to itself. We can assume that $y_1 \neq 0$, for then we would be in case 2. Case 3 is handled much like case 1, except that we define L to be the tangent to E at the point P. A little bit of calculus makes the computation quite simple. The slope of L can be computed using implicit differentiation of the equation of E:

$$2y\frac{dy}{dx} = 3x^2 + a.$$

Substituting $x = x_1, y = y_1$, we see that the slope of the tangent is

$$\lambda = \frac{3x_1^2 + a}{2y_1}.$$

The rest of the analysis in this case is the same as in case 1. The formula obtained is identical, except that λ is computed differently.

At this point the following properties of the addition operation, as defined above, should be clear:

1. addition is closed on the set E,

2. addition is commutative,

3. $\mathcal{O}$ is an identity with respect to addition, and

4. every point on E has an inverse with respect to addition.

In order to show that $(E, +)$ is an abelian group, it still must be proven that addition is associative. This is quite messy to prove by algebraic methods. The proof of associativity can be made simpler by using some results from geometry; however, we will not discuss the proof here.

6.5.2   Elliptic Curves Modulo a Prime

Let $p > 3$ be prime. Elliptic curves over $\mathbb{Z}_p$ can be defined exactly as they were over the reals (and the addition operation is also defined in an identical fashion) provided that all operations over $\mathbb{R}$ are replaced by analogous operations in $\mathbb{Z}_p$.

Definition 6.4:   Let $p > 3$ be prime. The *elliptic curve* $y^2 = x^3 + ax + b$ over $\mathbb{Z}_p$ is the set of solutions $(x, y) \in \mathbb{Z}_p \times \mathbb{Z}_p$ to the congruence

$$y^2 \equiv x^3 + ax + b \pmod{p}, \qquad (6.6)$$

where $a, b \in \mathbb{Z}_p$ are constants such that $4a^3 + 27b^2 \not\equiv 0 \pmod{p}$, together with a special point $\mathcal{O}$ called the *point at infinity*.

The addition operation on E is defined as follows (where all arithmetic operations are performed in $\mathbb{Z}_p$): Suppose

$$P = (x_1, y_1)$$

and

$$Q = (x_2, y_2)$$

are points on E. If $x_2 = x_1$ and $y_2 = -y_1$, then $P + Q = \mathcal{O}$; otherwise $P + Q = (x_3, y_3)$, where

$$x_3 = \lambda^2 - x_1 - x_2$$

$$y_3 = \lambda(x_1 - x_3) - y_1,$$

and

$$\lambda = \begin{cases} (y_2 - y_1)(x_2 - x_1)^{-1}, & \text{if } P \neq Q \\ (3x_1{}^2 + a)(2y_1)^{-1}, & \text{if } P = Q. \end{cases}$$

Finally, define

$$P + \mathcal{O} = \mathcal{O} + P = P$$

for all $P \in E$.

Note that the addition of points on an elliptic curve over $\mathbb{Z}_p$ does not have the nice geometric interpretation that it does on an elliptic curve over the reals. However, the same formulas can be used to define addition, and the resulting pair $(E, +)$ still forms an abelian group.

Let us look at a small example.

Example 6.7 Let E be the elliptic curve $y^2 = x^3 + x + 6$ over $\mathbb{Z}_{11}$. Let's first determine the points on E. This can be done by looking at each possible $x \in \mathbb{Z}_{11}$, computing $x^3 + x + 6 \bmod 11$, and then trying to solve equation (6.6) for y. For a given x, we can test to see if $z = x^3 + x + 6 \bmod 11$ is a quadratic residue by applying Euler's criterion. Recall that there is an explicit formula to compute square roots of quadratic residues modulo p for primes $p \equiv 3 \pmod 4$. Applying this formula, we have that the square roots of a quadratic residue z are

$$\pm z^{(11+1)/4} \bmod 11 = \pm z^3 \bmod 11.$$

The results of these computations are tabulated in Table 6.1.

E has 13 points on it. Since any group of prime order is cyclic, it follows that E is isomorphic to $\mathbb{Z}_{13}$, and any point other than the point at infinity is a generator of E. Suppose we take the generator $\alpha = (2, 7)$. Then we can compute the "powers" of α (which we will write as multiples of α, since the group operation is additive). To compute $2\alpha = (2, 7) + (2, 7)$, we first compute

$$\lambda = (3 \times 2^2 + 1)(2 \times 7)^{-1} \bmod 11$$

$$= 2 \times 3^{-1} \bmod 11$$

$$= 2 \times 4 \bmod 11$$

$$= 8.$$

TABLE 6.1
Points on the elliptic curve $y^2 = x^3 + x + 6$ over $\mathbb{Z}_{11}$

x	$x^3 + x + 6 \bmod 11$	quadratic residue?	y
0	6	no	
1	8	no	
2	5	yes	4, 7
3	3	yes	5, 6
4	8	no	
5	4	yes	2, 9
6	8	no	
7	4	yes	2, 9
8	9	yes	3, 8
9	7	no	
10	4	yes	2, 9

Then we have

$$x_3 = 8^2 - 2 - 2 \bmod 11$$
$$= 5$$

and

$$y_3 = 8(2 - 5) - 7 \bmod 11$$
$$= 2,$$

so $2\alpha = (5, 2)$.

The next multiple would be $3\alpha = 2\alpha + \alpha = (5, 2) + (2, 7)$. Again, we begin by computing λ, which in this situation is done as follows:

$$\lambda = (7 - 2)(2 - 5)^{-1} \bmod 11$$
$$= 5 \times 8^{-1} \bmod 11$$
$$= 5 \times 7 \bmod 11$$
$$= 2.$$

Then we have

$$x_3 = 2^2 - 5 - 2 \bmod 11$$
$$= 8$$

and

$$y_3 = 2(5 - 8) - 2 \bmod 11$$
$$= 3,$$

so $3\alpha = (8, 3)$.

Continuing in this fashion, the remaining multiples can be computed to be the following:

$$
\begin{array}{llllll}
\alpha & = & (2, 7) & 2\alpha & = & (5, 2) & 3\alpha & = & (8, 3) \\
4\alpha & = & (10, 2) & 5\alpha & = & (3, 6) & 6\alpha & = & (7, 9) \\
7\alpha & = & (7, 2) & 8\alpha & = & (3, 5) & 9\alpha & = & (10, 9) \\
10\alpha & = & (8, 8) & 11\alpha & = & (5, 9) & 12\alpha & = & (2, 4)
\end{array}
$$

Hence, as we already knew, $\alpha = (2, 7)$ is indeed a primitive element. ▯

Let's now look at an example of ElGamal encryption and decryption using the elliptic curve of Example 6.7. Since the group operation of an elliptic curve is written additively, we will need to translate the operations in Cryptosystem 6.1 into additive notation.

Example 6.8 Suppose that $\alpha = (2, 7)$ and Bob's private key is 7, so

$$\beta = 7\alpha = (7, 2).$$

Thus the encryption operation is

$$e_K(x, k) = (k(2, 7), x + k(7, 2)),$$

where $x \in E$ and $0 \le k \le 12$, and the decryption operation is

$$d_K(y_1, y_2) = y_2 - 7y_1.$$

Suppose that Alice wishes to encrypt the plaintext $x = (10, 9)$ (which is a point on E). If she chooses the random value $k = 3$, then she will compute

$$
\begin{aligned}
y_1 &= 3\,(2, 7) \\
&= (8, 3)
\end{aligned}
$$

and

$$
\begin{aligned}
y_2 &= (10, 9) + 3\,(7, 2) \\
&= (10, 9) + (3, 5) \\
&= (10, 2).
\end{aligned}
$$

Hence, $y = ((8, 3), (10, 2))$. Now, if Bob receives the ciphertext y, he decrypts it as follows:

$$
\begin{aligned}
x &= (10, 2) - 7(8, 3) \\
&= (10, 2) - (3, 5) \\
&= (10, 2) + (3, 6) \\
&= (10, 9).
\end{aligned}
$$

Hence, the decryption yields the correct plaintext.　　　　　　　　　　　　□

6.5.3　Properties of Elliptic Curves

An elliptic curve E defined over $\mathbb{Z}_p$ (p prime, $p > 3$) will have roughly p points on it. More precisely, a well-known theorem due to Hasse asserts that the number of points on E, which we denote by $\#E$, satisfies the following inequality

$$p + 1 - 2\sqrt{p} \le \#E \le p + 1 + 2\sqrt{p}.$$

Computing the exact value of $\#E$ is more difficult, but there is an efficient algorithm to do this, due to Schoof. (By "efficient" we mean that it has a running time that is polynomial in $\log p$. Schoof's algorithm has a running time of $O((\log p)^8)$ bit operations ($O((\log p)^6)$ operations in $\mathbb{Z}_p$) and is practical for primes p having several hundred digits.)

Now, given that we can compute $\#E$, we further want to find a cyclic subgroup of E in which the **Discrete Logarithm** problem is intractible. So we would like to know something about the structure of the group E. The following theorem gives a considerable amount of information on the group structure of E.

THEOREM 6.1　*Let E be an elliptic curve defined over $\mathbb{Z}_p$, where p is prime and $p > 3$. Then there exist positive integers n_1 and n_2 such that $(E, +)$ is isomorphic to $\mathbb{Z}_{n_1} \times \mathbb{Z}_{n_2}$. Further, $n_2 \mid n_1$ and $n_2 \mid (p - 1)$.*

Note that $n_2 = 1$ is allowed in the above theorem. In fact, $n_2 = 1$ if and only if E is a cyclic group. Also, if $\#E$ is a prime, or the product of distinct primes, then E must be a cyclic group.

In any event, if the integers n_1 and n_2 are computed, then we know that $(E, +)$ has a cyclic subgroup isomorphic to $\mathbb{Z}_{n_1}$ that can potentially be used as a setting for an *ElGamal Cryptosystem*.

Generic algorithms apply to the elliptic curve **Discrete Logarithm** problem, but there is no known adaptation of the INDEX CALCULUS ALGORITHM to the setting of elliptic curves. However, there is a method of exploiting an explicit isomorphism between elliptic curves and finite fields that leads to efficient algorithms for certain classes of elliptic curves. This technique, due to Menezes, Okamoto and Vanstone, can be applied to some particular examples within a special class of elliptic curves called supersingular curves that were suggested for use in cryptosystems.

Another class of weak elliptic curves are the so-called curves of "trace one." These are elliptic curves defined over $\mathbb{Z}_p$ (where p is prime) having exactly p points on them. The elliptic curve **Discrete Logarithm** problem can easily be solved on these elliptic curves.

If the classes of curves described above are avoided, however, then it appears that an elliptic curve having a cyclic subgroup of size about 2^{160} will provide a secure setting for a cryptosystem, provided that the order of the subgroup is divisible by at least one large prime factor (again, to guard against a Pohlig-Hellman attack).

6.5.4 Point Compression and the ECIES

There are some practical difficulties in implementing an *ElGamal Cryptosystem* on an elliptic curve. When implemented in $\mathbb{Z}_p$, the *ElGamal Cryptosystem* has a message expansion factor of two. An elliptic curve implementation has a message expansion factor of (about) four. This happens since there are approximately p plaintexts, but each ciphertext consists of four field elements. However, a more serious problem is that the plaintext space consists of the points on the curve E, and there is no convenient method known of deterministically generating points on E.

A more efficient ElGamal-type system is used in the so-called *ECIES* (*Elliptic Curve Integrated Encryption Scheme*). *ECIES* has a fairly complicated description that incorporates a symmetric-key encryption as well as a message authentication code; we present a simplification which basically implements the elliptic curve based ElGamal public-key encryption scheme used in *ECIES*. In this variation, the x-co-ordinate of a point on an elliptic curve is used for "masking," and a plaintext is allowed to be an arbitrary (nonzero) field element (i.e., it is not required to be a point on E).

We also use one other standard trick, called *point compression*, which reduces the storage requirement for points on elliptic curves. A (non-infinite) point on an elliptic curve E is a pair (x, y), where $y^2 \equiv x^3 + ax + b \pmod{p}$. Given a value for x, there are two possible values for y (unless $x^3 + ax + b \equiv 0 \pmod{p}$). These two possible y-values are negatives of each other modulo p. Since p is odd, one of the two possible values of $y \bmod p$ is even and the other is odd. Therefore we can determine a unique point $P = (x, y)$ on E by specifying the value of x, together with the single bit $y \bmod 2$. This reduces the storage by (almost) 50%, at the expense of requiring additional computations to reconstruct the y-co-ordinate of P.

The operation of point compression can be expressed as a function

$$\text{POINTCOMPRESS} : E \backslash \{\mathcal{O}\} \to \mathbb{Z}_p \times \mathbb{Z}_2,$$

which is defined as follows:

$$\text{POINTCOMPRESS}(P) = (x, y \bmod 2), \text{ where } P = (x, y) \in E.$$

The inverse operation, POINTDECOMPRESS, reconstructs the elliptic curve point $P = (x, y)$ from $(x, y \bmod 2)$. It can be implemented as follows:

Algorithm 6.4: POINTDECOMPRESS(x, i)

$z \leftarrow x^3 + ax + b \bmod p$
if z is a quadratic non-residue modulo p
   then return ("failure")
   else $\begin{cases} y \leftarrow \sqrt{z} \bmod p \\ \textbf{if } y \equiv i \ (\mathrm{mod}\ 2) \\ \quad \textbf{then return } (x, y) \\ \quad \textbf{else return } (x, p - y) \end{cases}$

As previously mentioned, $\sqrt{z}$ can be computed as $z^{(p+1)/4} \bmod p$ provided that $p \equiv 3 \pmod 4$ and z is a quadratic residue modulo p (or $z = 0$).

The cryptosystem that we call *Simplified ECIES* is presented as Cryptosystem 6.2.

Cryptosystem 6.2: *Simplified ECIES*

Let E be an elliptic curve defined over $\mathbb{Z}_p$ ($p > 3$ prime) such that E contains a cyclic subgroup $H = \langle P \rangle$ of prime order n in which the **Discrete Logarithm** problem is infeasible.
Let $\mathcal{P} = \mathbb{Z}_p^*$, $\mathcal{C} = (\mathbb{Z}_p \times \mathbb{Z}_2) \times \mathbb{Z}_p^*$, and define

$$\mathcal{K} = \{(E, P, m, Q, n) : Q = mP\}.$$

The values P, Q and n are the public key, and $m \in \mathbb{Z}_n^*$ is the private key.

For $K = (E, P, m, Q, n)$, for a (secret) random number $k \in \mathbb{Z}_n^*$, and for $x \in \mathbb{Z}_p^*$, define

$$e_K(x, k) = (\text{POINTCOMPRESS}(kP), xx_0 \bmod p),$$

where $kQ = (x_0, y_0)$ and $x_0 \neq 0$.

For a ciphertext $y = (y_1, y_2)$, where $y_1 \in \mathbb{Z}_p \times \mathbb{Z}_2$ and $y_2 \in \mathbb{Z}_p^*$, define

$$d_K(y) = y_2(x_0)^{-1} \bmod p,$$

where

$$(x_0, y_0) = m\, \text{POINTDECOMPRESS}(y_1).$$

Simplified ECIES has a message expansion (approximately) equal to two, which is similar to the *ElGamal Cryptosystem* over $\mathbb{Z}_p^*$. We illustrate encryption and decryption in *Simplified ECIES* using the curve $y^2 = x^3 + x + 6$ defined over $\mathbb{Z}_{11}$.

Example 6.9 As in the previous example, suppose that $P = (2, 7)$ and Bob's private key is 7, so

$$Q = 7P = (7, 2).$$

Suppose Alice wants to encrypt the plaintext $x = 9$, and she chooses the random value $k = 6$. First, she computes

$$kP = 6(2, 7) = (7, 9)$$

and

$$kQ = 6(7, 2) = (8, 3),$$

so $x_0 = 8$.

Next, she calculates

$$y_1 = \text{POINTCOMPRESS}(7, 9) = (7, 1)$$

and

$$y_2 = 8 \times 9 \bmod 11 = 6.$$

The ciphertext she sends to Bob is

$$y = (y_1, y_2) = ((7, 1), 6).$$

When Bob receives the ciphertext y, he computes

$$\text{POINTDECOMPRESS}(7, 1) = (7, 9),$$

$$7(7, 9) = (8, 3), \quad \text{and}$$

$$6 \times 8^{-1} \bmod 11 = 9.$$

Hence, the decryption yields the correct plaintext, 9. □

6.5.5 Computing Point Multiples on Elliptic Curves

We can compute powers α^a in a multiplicative group efficiently using the SQUARE-AND-MULTIPLY ALGORITHM (Algorithm 5.5). In an elliptic curve setting, where the group operation is written additively, we would compute a multiple aP of an elliptic curve point P using an analogous DOUBLE-AND-ADD ALGORITHM. (The squaring operation $\alpha \mapsto \alpha^2$ would be replaced by the doubling operation $P \mapsto 2P$, and the multiplication of two group elements would be replaced by the addition of two elliptic curve points.)

The addition operation on an elliptic curve has the property that additive inverses are very easy to compute. This fact can be exploited in a generalization of the DOUBLE-AND-ADD ALGORITHM which we might call the DOUBLE-AND-(ADD OR SUBTRACT) ALGORITHM. We describe this technique now.

Let c be an integer. A *signed binary representation* of c is an equation of the form

$$c = \sum_{i=0}^{\ell-1} c_i 2^i,$$

where $c_i \in \{-1, 0, 1\}$ for all i. In general, there will be more than one signed binary representation of an integer c. For example, we have that

$$11 = 8 + 2 + 1 = 16 - 4 - 1,$$

so

$$(c_4, c_3, c_2, c_1, c_0) = (0, 1, 0, 1, 1) \quad \text{or} \quad (1, 0, -1, 0, -1)$$

are both signed binary representations of 11.

Let P be a point of order n on an elliptic curve. Given any signed binary representation $(c_{\ell-1}, \dots, c_0)$ of an integer c, where $0 \le c \le n - 1$, it is possible to compute the multiple cP of the elliptic curve point P by a series of doublings, additions and subtractions, using the following algorithm.

Algorithm 6.5: DOUBLE-AND-(ADD OR SUBTRACT)$(P, (c_{\ell-1}, \dots, c_0))$

$Q \leftarrow \mathcal{O}$
for $i \leftarrow \ell - 1$ **downto** 0

do $\begin{cases} Q \leftarrow 2Q \\ \textbf{if } c_i = 1 \\ \quad \textbf{then } Q \leftarrow Q + P \\ \quad \textbf{else if } c_i = -1 \\ \quad \textbf{then } Q \leftarrow Q - P \end{cases}$

return (Q)

In Algorithm 6.5, the subtraction operation $Q - P$ would be performed by first computing the additive inverse $-P$ of P, and then adding the result to Q.

A signed binary representation $(c_{\ell-1}, \dots, c_0)$ of an integer c is said to be in *non-adjacent form* provided that no two consecutive c_i's are non-zero. Such a representation is denoted as a *NAF* representation. It is a simple matter to transform a binary representation of a positive integer c into a NAF representation. The basis of this transformation is to replace substrings of the form $(0, 1, \dots, 1, 1)$ in the binary representation by $(1, 0, \dots, 0, -1)$. Substitutions of this type do not change the value of c, due to the identity

$$2^i + 2^{i-1} + \dots + 2^j = 2^{i+1} - 2^j,$$

where $i > j$. This process is repeated as often as needed, starting with the rightmost (i.e., low-order) bits and proceeding to the left.

We illustrate the above-described process with an example:

$$1 \quad 1 \quad 1 \quad 1 \quad 0 \quad 0 \quad 1 \quad 1 \quad 0 \quad 1 \quad 1 \quad 1$$
$$\downarrow$$
$$1 \quad 1 \quad 1 \quad 1 \quad 0 \quad 0 \quad 1 \quad 1 \quad 1 \quad 0 \quad 0 \quad -1$$
$$\downarrow$$
$$1 \quad 1 \quad 1 \quad 1 \quad 0 \quad 1 \quad 0 \quad 0 \quad -1 \quad 0 \quad 0 \quad -1$$
$$\downarrow$$
$$1 \quad 0 \quad 0 \quad 0 \quad -1 \quad 0 \quad 1 \quad 0 \quad 0 \quad -1 \quad 0 \quad 0 \quad -1$$

Hence the NAF representation of

$$(1, 1, 1, 1, 0, 0, 1, 1, 0, 1, 1, 1)$$

is

$$(1, 0, 0, 0, -1, 0, 1, 0, 0, -1, 0, 0, -1).$$

This discussion establishes that every non-negative integer has a NAF representation. It is also possible to prove that the NAF representation of an integer is unique (see the Exercises). Therefore we can speak of *the* NAF representation of an integer without ambiguity.

In a NAF representation, there do not exist two consecutive non-zero coefficients. We might expect that, on average, a NAF representation contains more zeroes than the traditional binary representation of a positive integer. This is indeed the case: it can be shown that, on average, an ℓ-bit integer contains $\ell/2$ zeroes in its binary representation and $2\ell/3$ zeroes in its NAF representation.

These results make it easy to compare the average efficiency of the DOUBLE-AND-ADD ALGORITHM using a binary representation to the DOUBLE-AND-(ADD OR SUBTRACT) ALGORITHM using the NAF representation. Each algorithm requires ℓ doublings, but the number of additions (or subtractions) is $\ell/2$ in the first case, and $2\ell/3$ in the second case. If we assume that a doubling takes roughly the same amount of time as an addition (or subtraction), then the ratio of the average times required by the two algorithms is approximately

$$\frac{\ell + \frac{\ell}{2}}{\ell + \frac{\ell}{3}} = \frac{9}{8}.$$

We have therefore obtained a (roughly) 11% speedup, on average, by this simple technique.

6.6 Discrete Logarithm Algorithms in Practice

The most important settings (G, α) for the Discrete Logarithm problem in cryptographic applications are the following:

1. $G = (\mathbb{Z}_p^*, \cdot)$, p prime, α a primitive element modulo p
2. $G = (\mathbb{Z}_p^*, \cdot)$, p, q prime, $p \equiv 1 \bmod q$, α an element in $\mathbb{Z}_p$ having order q
3. $G = (\mathbb{F}_{2^n}^*, \cdot)$, α a primitive element in $\mathbb{F}_{2^n}^*$
4. $G = (E, +)$, where E is an elliptic curve modulo a prime p, $\alpha \in E$ is a point having prime order $q = \#E/h$, where (typically) $h = 1, 2$ or 4
5. $G = (E, +)$, where E is an elliptic curve over a finite field $\mathbb{F}_{2^n}$, $\alpha \in E$ is a point having prime order $q = \#E/h$, where (typically) $h = 2$ or 4

Cases 1, 2 and 3 can be attacked using the appropriate form of the INDEX CALCULUS ALGORITHM in $(\mathbb{Z}_p^*, \cdot)$ or $(\mathbb{F}_{2^n}^*, \cdot)$. Cases 2, 4 and 5 can be attacked using POLLARD RHO ALGORITHMS in subgroups of order q.

We briefly report some comparative security estimates due to Lenstra and Verheul. In order for an elliptic curve discrete logarithm based cryptosystem to be secure until the year 2020, it has been suggested that one should take $p \approx 2^{160}$ in case 4 (or $n \approx 160$ in case 5). In contrast, p needs to be at least 2^{1880} in cases 1 and 2 to achieve the same (predicted) level of security. The reason for this significant difference is the lack of a known index calculus attack on elliptic curve discrete logarithms. As a consequence, elliptic curve cryptography has become increasingly popular for practical applications, especially for applications on constrained platforms such as wireless devices and smart cards. On platforms such as these, available memory is very small, and a secure implementation of discrete logarithm based cryptography in $(\mathbb{Z}_p^*, \cdot)$, for example, would require too much space to be practical. The smaller space required for elliptic curve based cryptography is therefore very desirable.

The above estimates are conjectures that are based on the best currently implemented algorithms, as well as some reasonable hypotheses concerning possible progress in algorithm development and computing speed in the coming years. It is also of interest to look at the current state-of-the-art of algorithms for the Discrete Logarithm problem. Due to the practical importance of elliptic curve cryptography, algorithms for the Discrete Logarithm problem on elliptic curves have received the most attention in recent years. *Certicom Corporation* has issued a series of "challenges" to encourage the development of efficient implementations of discrete logarithm algorithms. The most recent and difficult challenge to be solved so far was a discrete logarithm in an elliptic curve defined over the field $\mathbb{F}_{2^{109}}$. This challenge, known as ECC2K-108, was solved in April 2000, by the combined effort of 9500 computers in 40 countries. The computing time required was approximately 50 times that required to factor the RSA challenge known as RSA-512.

6.7 Security of ElGamal Systems

In this section, we study several aspects of the security of ElGamal-type cryptosystems. First, we look at the bit security of discrete logarithms. Then we consider the semantic security of ElGamal-type cryptosystems, and introduce the Diffie-Hellman problems.

6.7.1 Bit Security of Discrete Logarithms

In this section, we consider whether individual bits of a discrete logarithm are easy or hard to compute. To be precise, consider Problem 6.2, which we call the Discrete Logarithm ith Bit problem (the setting for the discrete logarithms considered in this section is $(\mathbb{Z}_p^*, \cdot)$, where p is prime).

Problem 6.2:	Discrete Logarithm ith Bit
Instance:	$I = (p, \alpha, \beta, i)$, where p is prime, $\alpha \in \mathbb{Z}_p^*$ is a primitive element, $\beta \in \mathbb{Z}_p^*$, and i is an integer such that $1 \leq i \leq \lceil \log_2(p-1) \rceil$.
Question:	Compute $L_i(\beta)$, which (for the specified α and p) denotes the ith least significant bit in the binary representation of $\log_\alpha \beta$.

We will first show that computing the least significant bit of a discrete logarithm is easy. In other words, if $i = 1$, then the Discrete Logarithm ith Bit problem can be solved efficiently. This follows from Euler's criterion concerning quadratic residues modulo p, where p is prime.

Consider the mapping $f : \mathbb{Z}_p^* \to \mathbb{Z}_p^*$ defined by

$$f(x) = x^2 \bmod p.$$

Denote by $\mathsf{QR}(p)$ the set of quadratic residues modulo p; thus

$$\mathsf{QR}(p) = \{x^2 \bmod p : x \in \mathbb{Z}_p^*\}.$$

First, observe that $f(x) = f(p - x)$. Next, note that

$$w^2 \equiv x^2 \pmod{p}$$

if and only if

$$p \mid (w - x)(w + x),$$

which happens if and only if

$$w \equiv \pm x \pmod{p}.$$

It follows that

$$|f^{-1}(y)| = 2$$

for every $y \in \mathsf{QR}(p)$, and hence

$$|\mathsf{QR}(p)| = \frac{p-1}{2}.$$

That is, exactly half the residues in $\mathbb{Z}_p{}^*$ are quadratic residues and half are not.

Now, suppose α is a primitive element of $\mathbb{Z}_p$. Then $\alpha^a \in \mathsf{QR}(p)$ if a is even. Since the $(p-1)/2$ elements $\alpha^0 \bmod p, \alpha^2 \bmod p, \ldots, \alpha^{p-3} \bmod p$ are all distinct, it follows that

$$\mathsf{QR}(p) = \{\alpha^{2i} \bmod p : 0 \le i \le (p-3)/2\}.$$

Hence, β is a quadratic residue if and only if $\log_\alpha \beta$ is even, that is, if and only if $L_1(\beta) = 0$. But we already know, by Euler's criterion, that β is a quadratic residue if and only if

$$\beta^{(p-1)/2} \equiv 1 \pmod{p}.$$

So we have the following efficient formula to calculate $L_1(\beta)$:

$$L_1(\beta) = \begin{cases} 0 & \text{if } \beta^{(p-1)/2} \equiv 1 \pmod{p} \\ 1 & \text{otherwise.} \end{cases}$$

Let's now consider the computation of $L_i(\beta)$ for values of i exceeding 1. Suppose $p - 1 = 2^s t$, where t is odd. It can be shown that it is easy to compute $L_i(\beta)$ if $i \le s$. On the other hand, computing $L_{s+1}(\beta)$ is (probably) difficult, in the sense that any hypothetical algorithm (or oracle) to compute $L_{s+1}(\beta)$ could be used to find discrete logarithms in $\mathbb{Z}_p$.

We shall prove this result in the case $s = 1$. More precisely, if $p \equiv 3 \pmod{4}$ is prime, then we show how any oracle for computing $L_2(\beta)$ can be used to solve the Discrete Logarithm problem in $\mathbb{Z}_p$.

Recall that, if β is a quadratic residue in $\mathbb{Z}_p$ and $p \equiv 3 \pmod{4}$, then the two square roots of β modulo p are $\pm\beta^{(p+1)/4} \bmod p$. It is also important that, for any $\beta \ne 0$,

$$L_1(\beta) \ne L_1(p - \beta)$$

if $p \equiv 3 \pmod{4}$. We see this as follows. Suppose

$$\alpha^a \equiv \beta \pmod{p};$$

then

$$\alpha^{a+(p-1)/2} \equiv -\beta \pmod{p}.$$

Since $p \equiv 3 \pmod{4}$, the integer $(p-1)/2$ is odd, and the result follows.

Now, suppose that $\beta = \alpha^a$ for some (unknown) even exponent a. Then either

$$\beta^{(p+1)/4} \equiv \alpha^{a/2} \pmod{p}$$

TABLE 6.2
Values of L_1 and L_2 for $p = 19, \alpha = 2$

γ	$L_1(\gamma)$	$L_2(\gamma)$	γ	$L_1(\gamma)$	$L_2(\gamma)$	γ	$L_1(\gamma)$	$L_2(\gamma)$
1	0	0	7	0	1	13	1	0
2	1	0	8	1	1	14	1	1
3	1	0	9	0	0	15	1	1
4	0	1	10	1	0	16	0	0
5	0	0	11	0	0	17	0	1
6	0	1	12	1	1	18	1	0

or

$$-\beta^{(p+1)/4} \equiv \alpha^{a/2} \pmod{p}.$$

We can determine which of these two possibilities is correct if we know the value $L_2(\beta)$, since

$$L_2(\beta) = L_1(\alpha^{a/2}).$$

This fact is exploited in our algorithm, which we present as Algorithm 6.6.

Algorithm 6.6: $L_2\text{ORACLEDISCRETELOGARITHM}(p, \alpha, \beta)$

external $L_1, \text{ORACLEL}_2$
$x_0 \leftarrow L_1(\beta)$
$\beta \leftarrow \beta/\alpha^{x_0} \bmod p$
$i \leftarrow 1$
while $\beta \neq 1$

$\mathbf{do} \begin{cases} x_i \leftarrow \text{ORACLEL}_2(\beta) \\ \gamma \leftarrow \beta^{(p+1)/4} \bmod p \\ \mathbf{if}\ L_1(\gamma) = x_i \\ \quad \mathbf{then}\ \beta \leftarrow \gamma \\ \quad \mathbf{else}\ \beta \leftarrow p - \gamma \\ \beta \leftarrow \beta/\alpha^{x_i} \bmod p \\ i \leftarrow i + 1 \end{cases}$

return $(x_{i-1}, x_{i-2}, \ldots, x_0)$

At the end of Algorithm 6.6, the x_i's comprise the bits in the binary representation of $\log_\alpha \beta$; that is,

$$\log_\alpha \beta = \sum_{i \geq 0} x_i 2^i.$$

We will work out a small example to illustrate the algorithm.

Example 6.10 Suppose $p = 19$, $\alpha = 2$ and $\beta = 6$. Since the example is so small, we can tabulate the values of $L_1(\gamma)$ and $L_2(\gamma)$ for all $\gamma \in \mathbb{Z}_{19}^*$. (In

---

$x_0 \leftarrow 0, \beta \leftarrow 6, i \leftarrow 1$
$x_1 \leftarrow L_2(6) = 1, \gamma \leftarrow 5, L_1(5) = 0 \neq x_1, \beta \leftarrow 14, \beta \leftarrow 7, i \leftarrow 2$
$x_2 \leftarrow L_2(7) = 1, \gamma \leftarrow 11, L_1(11) = 0 \neq x_2, \beta \leftarrow 8, \beta \leftarrow 4, i \leftarrow 3$
$x_3 \leftarrow L_2(4) = 1, \gamma \leftarrow 17, L_1(17) = 0 \neq x_3, \beta \leftarrow 2, \beta \leftarrow 1, i \leftarrow 4$
return $(1, 1, 1, 0)$

FIGURE 6.2
Computation of $\log_2 6$ in $\mathbb{Z}_{19}^*$ using an oracle for L_2

general, L_1 can be computed efficiently using Euler's criterion, and L_2 is is computed using the hypothetical algorithm ORACLEL2.) These values are given in Table 6.2. Algorithm 6.6 then proceeds as shown in Figure 6.2.

The result is that $\log_2 6 = 1110_2 = 14$, as can easily be verified. ⬚

It is possible to give formal proof of the algorithm's correctness using mathematical induction. Denote

$$x = \log_\alpha \beta = \sum_{i \geq 0} x_i 2^i.$$

For $i \geq 0$, define

$$Y_i = \left\lfloor \frac{x}{2^{i+1}} \right\rfloor.$$

Also, define β_0 to be the value of β just before the start of the **while** loop; and, for $i \geq 1$, define β_i to be the value of β at the end of the ith iteration of the **while** loop. It can be proved by induction that

$$\beta_i \equiv \alpha^{2Y_i} \pmod{p}$$

for all $i \geq 0$. Now, with the observation that

$$2Y_i = Y_{i-1} - x_i,$$

it follows that

$$x_{i+1} = L_2(\beta_i),$$

$i \geq 0$. Since

$$x_0 = L_1(\beta),$$

the algorithm is correct. The details are left to the reader.

6.7.2 Semantic Security of ElGamal Systems

We first observe that the basic *ElGamal Cryptosystem*, as described in Cryptosystem 6.1, is not semantically secure. Recall that $\alpha \in \mathbb{Z}_p^*$ is a primitive element and $\beta = \alpha^a \bmod p$ where a is the private key. Given a plaintext element x, a random number k is chosen, and then $e_K(x, k) = (y_1, y_2)$ is computed, where $y_1 = \alpha^k \bmod p$ and $y_2 = x\beta^k \bmod p$.

We make use of the fact that it is easy, using Euler's criterion, to test elements of $\mathbb{Z}_p$ to see if they are quadratic residues modulo p. Recall from Section 6.7.1 that β is a quadratic residue modulo p if and only if a is even. Similarly, y_1 is a quadratic residue modulo p if and only if k is even. We can determine the parity of both a and k, and hence we can compute the parity of ak. Therefore, we can determine if $\beta^k \ (= \alpha^{ak})$ is a quadratic residue.

Now, suppose that we wish to distinguish encryptions of x_1 from encryptions of x_2, where x_1 is a quadratic residue and x_2 is a quadratic non-residue modulo p. It is a simple matter to determine the quadratic residuosity of y_2, and we have already discussed how the quadratic residuosity of β^k can be determined. It follows that (y_1, y_2) is an encryption of x_1 if and only if β^k and y_2 are both quadratic residues or both quadratic non-residues.

The above attack does not work if β is a quadratic residue and every plaintext x is required to be a quadratic residue. In fact, if $p = 2q + 1$ where q is prime, then it can be shown that restricting β, y_1 and x to be quadratic residues is equivalent to implementing the *ElGamal Cryptosystem* in the subgroup of quadratic residues modulo p (which is a cyclic subgroup of $\mathbb{Z}_p^*$ of order q). This version of the *ElGamal Cryptosystem* is conjectured to be semantically secure if the Discrete Logarithm problem in $\mathbb{Z}_p^*$ is infeasible.

6.7.3 The Diffie-Hellman Problems

We introduce two variants of the so-called Diffie-Hellman problems, a computational version and a decision version. The reason for calling them "Diffie-Hellman problems" comes from the origin of these two problems in connection with Diffie-Hellman key agreement protocols. In this section, we discuss some interesting connections between these problems and security of ElGamal-type cryptosystems.

Here are descriptions of the two problems.

Problem 6.3:   Computational Diffie-Hellman

Instance: A multiplicative group $(G, \cdot)$, an element $\alpha \in G$ having order n, and two elements $\beta, \gamma \in \langle \alpha \rangle$.

Question: Find $\delta \in \langle \alpha \rangle$ such that $\log_\alpha \delta \equiv \log_\alpha \beta \times \log_\alpha \gamma \pmod{n}$. (Equivalently, given α^b and α^c, find α^{bc}.)

Problem 6.4:　Decision Diffie-Hellman

Instance:　　A multiplicative group $(G, \cdot)$, an element $\alpha \in G$ having order n, and three elements $\beta, \gamma, \delta \in \langle \alpha \rangle$.

Question:　　Is it the case that $\log_\alpha \delta \equiv \log_\alpha \beta \times \log_\alpha \gamma \pmod{n}$? (Equivalently, given α^b, α^c and α^d, determine if $d \equiv bc \pmod{n}$.)

It is easy to see that there exist Turing reductions

$$\text{Decision Diffie-Hellman} \propto_T \text{Computational Diffie-Hellman}$$

and

$$\text{Computational Diffie-Hellman} \propto_T \text{Discrete Logarithm.}$$

The first reduction is proven as follows: Let $\alpha, \beta, \gamma, \delta$ be given. Use an algorithm that solves **Computational Diffie-Hellman** to find the value δ' such that

$$\log_\alpha \delta' \equiv \log_\alpha \beta \times \log_\alpha \gamma \pmod{n}.$$

Then check to see if $\delta' = \delta$.

The second reduction is also very simple. Let α, β, γ be given. Use an algorithm that solves **Discrete Logarithm** to find $b = \log_\alpha \beta$ and $c = \log_\alpha \gamma$. Then compute $d = bc \bmod n$ and $\delta = \alpha^d$.

These reductions show that the assumption that **Decision Diffie-Hellman** is infeasible is at least as strong as the assumption that **Computational Diffie-Hellman** is infeasible, which in turn is at least as strong as the assumption that **Discrete Logarithm** is infeasible.

It is not hard to show that the semantic security of the *ElGamal Cryptosystem* is equivalent to the infeasibility of **Decision Diffie-Hellman**; and ElGamal decryption (without knowing the public key) is equivalent to solving **Computational Diffie-Hellman**. The assumptions necessary to prove the security of the *El-Gamal Cryptosystem* are therefore (potentially) stronger than assuming just that **Discrete Logarithm** is infeasible. Indeed, we already showed that the *ElGamal Cryptosystem* in $\mathbb{Z}_p^*$ is not semantically secure, whereas the **Discrete Logarithm** problem is conjectured to be infeasible in $\mathbb{Z}_p^*$ for appropriately chosen primes p. This suggests that the security of the three problems may not be equivalent.

Here, we give a proof that any algorithm that solves **Computational Diffie-Hellman** can be used to decrypt ElGamal ciphertexts, and vice versa. Suppose first that ORACLECDH is an algorithm for **Computational Diffie-Hellman**, and let (y_1, y_2) be a ciphertext for the *ElGamal Cryptosystem* with public key α and β. Compute

$$\delta = \text{ORACLECDH}(\alpha, \beta, y_1),$$

and then define

$$x = y_2 \delta^{-1}.$$

It is easy to see that x is the decryption of the ciphertext (y_1, y_2).

Conversely, suppose that ORACLEELGAMALDECRYPT is an algorithm that decrypts ElGamal ciphertexts. Let α, β, γ be given as in **Computational Diffie-Hellman**. Define α and β to be the public key for the *ElGamal Cryptosystem*. Then define $y_1 = \gamma$ and let $y_2 \in \langle \alpha \rangle$ be chosen randomly. Compute

$$x = \text{ORACLEELGAMALDECRYPT}(\alpha, \beta, (y_1, y_2)),$$

which is the decryption of the ciphertext (y_1, y_2). Finally, compute

$$\delta = y_2 x^{-1}.$$

δ is the solution to the given instance of **Computational Diffie-Hellman**.

6.8 Notes and References

The *ElGamal Cryptosystem* was presented in [70]. The POHLIG-HELLMAN ALGORITHM was published in [168]. For further information on the **Discrete Logarithm** problem in general, we recommend the recent survey article by Odlyzko [159].

The POLLARD RHO ALGORITHM was first described in [170]. Brent [38] described a more efficient method to detect cycles (and, therefore, collisions), which can also be used in the corresponding factoring algorithm. There are many ways of defining the "random walks" used in the algorithm; for a thorough treatment of these topics, see Teske [208].

The lower bound on generic algorithms for the **Discrete Logarithm** problem was proven independently by Nechaev [152] and Shoup [192]. Our discussion is based on the treatment of Chateauneuf, Ling and Stinson [46].

The main reference book for finite fields is Lidl and Niederreiter [134]. McEliece [141] is a good elementary textbook on this subject.

An informative article on the **Discrete Logarithm** problem in $\mathbb{F}_{2^n}$ is Gordon and McCurley [98]. Some more recent results are found in the article by Thomé [209]. Lenstra and Verheul [133] provide predictions of the feasibility of the **Discrete Logarithm** problem in coming years, and the ramifications for choosing key sizes for cryptosystems based on discrete logarithms.

The idea of using elliptic curves for public-key cryptosystems is due to Koblitz [119] and Miller [148]. Koblitz, Menezes and Vanstone [122] have written a recent survey article on this topic. Two recent monographs on elliptic curve cryptography are Blake, Seroussi and Smart [26] and Enge [71]. For a more elementary treatment of elliptic curves, see Silverman and Tate [194].

For a recent article that presents a thorough treatment of fast arithmetic on elliptic curves, see Solinas [201]. The Menezes-Okamoto-Vanstone reduction of discrete logarithms from elliptic curves to finite fields is given in [143] (see also [142]). The attack on "trace one" curves is due to Smart, Satoh, Araki and Semaev; see, for example, Smart [198].

The material we presented concerning the Discrete Logarithm ith Bit problem is based on Peralta [163]. Other papers that discuss related questions are Håstad, Schrift and Shamir [101] and Long and Wigderson [135].

Boneh [30] is an interesting survey article on the Decision Diffie-Hellman problem. Maurer and Wolf [140] is an article which further develops topics considered in Section 6.7.

Exercises

6.1 Implement SHANKS' ALGORITHM for finding discrete logarithms in $\mathbb{Z}_p^*$, where p is prime and α is a primitive element modulo p. Use your program to find $\log_{106} 12375$ in $\mathbb{Z}_{24691}^*$ and $\log_6 248388$ in $\mathbb{Z}_{458009}^*$.

6.2 Describe how to modify SHANKS' ALGORITHM to compute the logarithm of β to the base α in a group G if it is specified ahead of time that this logarithm lies in the interval $[s, t]$, where s and t are integers such that $0 \leq s < t < n$, where n is the order of α. Prove that your algorithm is correct, and show that its complexity is $O(\sqrt{t - s})$.

6.3 The integer $p = 458009$ is prime and $\alpha = 2$ has order 57251 in $\mathbb{Z}_p^*$. Use the POLLARD RHO ALGORITHM to compute the discrete logarithm in $\mathbb{Z}_p^*$ of $\beta = 56851$ to the base α. Take the initial value $x_0 = 1$, and define the partition (S_1, S_2, S_3) as in Example 6.3. Find the smallest integer i such that $x_i = x_{2i}$, and then compute the desired discrete logarithm.

6.4 Suppose that p is an odd prime and k is a positive integer. The multiplicative group $\mathbb{Z}_{p^k}^*$ has order $p^{k-1}(p-1)$, and is known to be cyclic. A generator for this group is called a *primitive element modulo p^k*.

   (a) Suppose that α is a primitive element modulo p. Prove that at least one of α or $\alpha + p$ is a primitive element modulo p^2.

   (b) Describe how to efficiently verify that 3 is a primitive root modulo 29 and modulo 29^2. Note: It can be shown that if α is a primitive root modulo p and modulo p^2, then it is a primitive root modulo p^k for all positive integers k (you do not have to prove this fact). Therefore, it follows that 3 is a primitive root modulo 29^k for all positive integers k.

   (c) Find an integer α that is a primitive root modulo 29 but not a primitive root modulo 29^2.

   (d) Use the POHLIG-HELLMAN ALGORITHM to compute the discrete logarithm of 3344 to the base 3 in the multiplicative group $\mathbb{Z}_{24389}^*$.

6.5 Implement the POHLIG-HELLMAN ALGORITHM for finding discrete logarithms in $\mathbb{Z}_p$, where p is prime and α is a primitive element. Use your program to find $\log_5 8563$ in $\mathbb{Z}_{28703}$ and $\log_{10} 12611$ in $\mathbb{Z}_{31153}$.

6.6 Let $p = 227$. The element $\alpha = 2$ is primitive in $\mathbb{Z}_p^*$.

   (a) Compute α^{32}, α^{40}, α^{59} and α^{156} modulo p, and factor them over the factor base $\{2, 3, 5, 7, 11\}$.

(b) Using the fact that log 2 = 1, compute log 3, log 5, log 7 and log 11 from the factorizations obtained above (all logarithms are discrete logarithms in $\mathbb{Z}_p^*$ to the base α).

(c) Now suppose we wish to compute log 173. Multiply 173 by the "random" value $2^{177} \bmod p$. Factor the result over the factor base, and proceed to compute log 173 using the previously computed logarithms of the numbers in the factor base.

6.7 Suppose that $n = pq$ is an RSA modulus (i.e., p and q are distinct odd primes), and let $\alpha \in \mathbb{Z}_n^*$. For a positive integer m and for any $\alpha \in \mathbb{Z}_m^*$, define $\text{ord}_m(\alpha)$ to be the order of α in the group $\mathbb{Z}_m^*$.

(a) Prove that
$$\text{ord}_n(\alpha) = \text{lcm}(\text{ord}_p(\alpha), \text{ord}_q(\alpha)).$$

(b) Suppose that $\gcd(p - 1, q - 1) = d$. Show that there exists an element $\alpha \in \mathbb{Z}_n^*$ such that
$$\text{ord}_n(\alpha) = \frac{\phi(n)}{d}.$$

(c) Suppose that $\gcd(p - 1, q - 1) = 2$, and we have an oracle that solves the **Discrete Logarithm** problem in the subgroup $\langle \alpha \rangle$, where $\alpha \in \mathbb{Z}_n^*$ has order $\phi(n)/2$. That is, given any $\beta \in \langle \alpha \rangle$, the oracle will find the discrete logarithm $a = \log_\alpha \beta$, where $0 \leq a \leq \phi(n)/2 - 1$. (The value $\phi(n)/2$ is secret however.) Suppose we compute the value $\beta = \alpha^n \bmod n$ and then we use the oracle to find $a = \log_\alpha \beta$. Assuming that $p > 3$ and $q > 3$, prove that $n - a = \phi(n)$.

(d) Describe how n can easily be factored, given the discrete logarithm $a = \log_\alpha \beta$ from (c).

6.8 In this question, we consider a generic algorithm for the **Discrete Logarithm** problem in $(\mathbb{Z}_{19}, +)$.

(a) Suppose that the set C is defined as follows:
$$C = \{(1 - i^2 \bmod 19, i \bmod 19) : i = 0, 1, 2, 4, 7, 12\}.$$

Compute $\text{Good}(C)$.

(b) Suppose that the output of the group oracle, given the ordered pairs in C, is as follows:

$$(0, 1) \mapsto 10111$$
$$(1, 0) \mapsto 01100$$
$$(16, 2) \mapsto 00110$$
$$(4, 4) \mapsto 01010$$
$$(9, 7) \mapsto 00100$$
$$(9, 12) \mapsto 11001,$$

where group elements are encoded as (random) binary 5-tuples. What can you say about the value of "a"?

6.9 Decrypt the ElGamal ciphertext presented in Table 6.3. The parameters of the system are $p = 31847$, $\alpha = 5$, $a = 7899$ and $\beta = 18074$. Each element of $\mathbb{Z}_n$ represents three alphabetic characters as in Exercise 5.6.

TABLE 6.3
ElGamal Ciphertext

(3781, 14409)	(31552, 3930)	(27214, 15442)	(5809, 30274)
(5400, 31486)	(19936, 721)	(27765, 29284)	(29820, 7710)
(31590, 26470)	(3781, 14409)	(15898, 30844)	(19048, 12914)
(16160, 3129)	(301, 17252)	(24689, 7776)	(28856, 15720)
(30555, 24611)	(20501, 2922)	(13659, 5015)	(5740, 31233)
(1616, 14170)	(4294, 2307)	(2320, 29174)	(3036, 20132)
(14130, 22010)	(25910, 19663)	(19557, 10145)	(18899, 27609)
(26004, 25056)	(5400, 31486)	(9526, 3019)	(12962, 15189)
(29538, 5408)	(3149, 7400)	(9396, 3058)	(27149, 20535)
(1777, 8737)	(26117, 14251)	(7129, 18195)	(25302, 10248)
(23258, 3468)	(26052, 20545)	(21958, 5713)	(346, 31194)
(8836, 25898)	(8794, 17358)	(1777, 8737)	(25038, 12483)
(10422, 5552)	(1777, 8737)	(3780, 16360)	(11685, 133)
(25115, 10840)	(14130, 22010)	(16081, 16414)	(28580, 20845)
(23418, 22058)	(24139, 9580)	(173, 17075)	(2016, 18131)
(19886, 22344)	(21600, 25505)	(27119, 19921)	(23312, 16906)
(21563, 7891)	(28250, 21321)	(28327, 19237)	(15313, 28649)
(24271, 8480)	(26592, 25457)	(9660, 7939)	(10267, 20623)
(30499, 14423)	(5839, 24179)	(12846, 6598)	(9284, 27858)
(24875, 17641)	(1777, 8737)	(18825, 19671)	(31306, 11929)
(3576, 4630)	(26664, 27572)	(27011, 29164)	(22763, 8992)
(3149, 7400)	(8951, 29435)	(2059, 3977)	(16258, 30341)
(21541, 19004)	(5865, 29526)	(10536, 6941)	(1777, 8737)
(17561, 11884)	(2209, 6107)	(10422, 5552)	(19371, 21005)
(26521, 5803)	(14884, 14280)	(4328, 8635)	(28250, 21321)
(28327, 19237)	(15313, 28649)		

The plaintext was taken from "The English Patient," by Michael Ondaatje, Alfred A. Knopf, Inc., New York, 1992.

6.10 Determine which of the following polynomials are irreducible over $\mathbb{Z}_2[x]$: $x^5 + x^4 + 1$, $x^5 + x^3 + 1$, $x^5 + x^4 + x^2 + 1$.

6.11 The field $\mathbb{F}_{2^5}$ can be constructed as $\mathbb{Z}_2[x]/(x^5 + x^2 + 1)$. Perform the following computations in this field.

(a) Compute $(x^4 + x^2) \times (x^3 + x + 1)$.

(b) Using the extended Euclidean algorithm, compute $(x^3 + x^2)^{-1}$.

(c) Using the square-and-multiply algorithm, compute x^{25}.

6.12 We give an example of the *ElGamal Cryptosystem* implemented in $\mathbb{F}_{3^3}$. The polynomial $x^3 + 2x^2 + 1$ is irreducible over $\mathbb{Z}_3[x]$ and hence $\mathbb{Z}_3[x]/(x^3 + 2x^2 + 1)$ is the field $\mathbb{F}_{3^3}$. We can associate the 26 letters of the alphabet with the 26 nonzero field elements, and thus encrypt ordinary text in a convenient way. We will use a lexicographic ordering of the (nonzero) polynomials to set up the correspondence.

This correspondence is as follows:

A	$\leftrightarrow$	1	B	$\leftrightarrow$	2	C	$\leftrightarrow$	x

$$
\begin{array}{lllllll}
A \leftrightarrow 1 & & B \leftrightarrow 2 & & C \leftrightarrow x \\
D \leftrightarrow x+1 & & E \leftrightarrow x+2 & & F \leftrightarrow 2x \\
G \leftrightarrow 2x+1 & & H \leftrightarrow 2x+2 & & I \leftrightarrow x^2 \\
J \leftrightarrow x^2+1 & & K \leftrightarrow x^2+2 & & L \leftrightarrow x^2+x \\
M \leftrightarrow x^2+x+1 & & N \leftrightarrow x^2+x+2 & & O \leftrightarrow x^2+2x \\
P \leftrightarrow x^2+2x+1 & & Q \leftrightarrow x^2+2x+2 & & R \leftrightarrow 2x^2 \\
S \leftrightarrow 2x^2+1 & & T \leftrightarrow 2x^2+2 & & U \leftrightarrow 2x^2+x \\
V \leftrightarrow 2x^2+x+1 & & W \leftrightarrow 2x^2+x+2 & & X \leftrightarrow 2x^2+2x \\
Y \leftrightarrow 2x^x+2x+1 & & Z \leftrightarrow 2x^x+2x+2 & &
\end{array}
$$

Suppose Bob uses $\alpha = x$ and $a = 11$ in an *ElGamal Cryptosystem*; then $\beta = x+2$. Show how Bob will decrypt the following string of ciphertext:

(K,H) (P,X) (N,K) (H,R) (T,F) (V,Y) (E,H) (F,A) (T,W) (J,D) (U,J)

6.13 Let E be the elliptic curve $y^2 = x^3 + x + 28$ defined over $\mathbb{Z}_{71}$.
  (a) Determine the number of points on E.
  (b) Show that E is not a cyclic group.
  (c) What is the maximum order of an element in E? Find an element having this order.

6.14 Suppose that $p > 3$ is an odd prime, and $a, b \in \mathbb{Z}_p$. Further, suppose that the equation $x^3 + ax + b \equiv 0 \pmod{p}$ has three distinct roots in $\mathbb{Z}_p$. Prove that the corresponding elliptic curve group $(E, +)$ is not cyclic.

  HINT Show that the points of order two generate a subgroup of $(E, +)$ that is isomorphic to $\mathbb{Z}_2 \times \mathbb{Z}_2$.

6.15 Consider an elliptic curve E described by the formula $y^2 \equiv x^3 + ax + b \pmod{p}$, where $4a^3 + 27b^2 \not\equiv 0 \pmod{p}$ and $p > 3$ is prime.
  (a) It is clear that a point $P = (x_1, y_1) \in E$ has order 3 if and only if $2P = -P$. Use this fact to prove that, if $P = (x_1, y_1) \in E$ has order 3, then

$$3x_1^{\,4} + 6ax_1^{\,2} + 12x_1 b - a^2 \equiv 0 \pmod{p}. \qquad (6.7)$$

  (b) Conclude from equation (6.7) that there are at most 8 points of order 3 on the elliptic curve E.
  (c) Using equation (6.7), determine all points of order 3 on the elliptic curve $y^2 \equiv x^3 + 34x \pmod{73}$.

6.16 Suppose that E is an elliptic curve defined over $\mathbb{Z}_p$, where $p > 3$ is prime. Suppose that $\#E$ is prime, $P \in E$, and $P \neq \mathcal{O}$.
  (a) Prove that the discrete logarithm $\log_P(-P) = \#E - 1$.
  (b) Describe how to compute $\#E$ in time $O(p^{1/4})$ by using Hasse's bound on $\#E$, together with a modification of SHANKS' ALGORITHM. Give a pseudocode description of the algorithm.

6.17 Let E be the elliptic curve $y^2 = x^3 + 2x + 7$ defined over $\mathbb{Z}_{31}$. It can be shown that $\#E = 39$ and $P = (2, 9)$ is an element of order 39 in E. The *Simplified ECIES* defined on E has $\mathbb{Z}_{31}^*$ as its plaintext space. Suppose the private key is $m = 8$.
  (a) Compute $Q = mP$.
  (b) Decrypt the following string of ciphertext:

$$((18, 1), 21), ((3, 1), 18), ((17, 0), 19), ((28, 0), 8).$$

(c) Assuming that each plaintext represents one alphabetic character, convert the plaintext into an English word. (Here we will use the correspondence $A \leftrightarrow 1$, ..., $Z \leftrightarrow 26$, because 0 is not allowed in a (plaintext) ordered pair.)

6.18   (a) Determine the NAF representation of the integer 87.

         (b) Using the NAF representation of 87, use Algorithm 6.5 to compute $87P$, where $P = (2, 6)$ is a point on the elliptic curve $y^2 = x^3 + x + 26$ defined over $\mathbb{Z}_{127}$. Show the partial results during each iteration of the algorithm.

6.19 Let $\mathcal{L}_i$ denote the set of positive integers that have exactly i coefficients in their NAF representation, such that the leading coefficient is 1. Denote $k_i = |\mathcal{L}_i|$.

         (a) By means of a suitable decomposition of $\mathcal{L}_i$, prove that the k_i's satisfy the following recurrence relation:

$$k_1 = 1$$

$$k_2 = 1$$

$$k_{i+1} = 2(k_1 + k_2 + ... + k_{i-1}) + 1 \quad \text{(for } i \geq 2\text{)}.$$

         (b) Derive a second degree recurrence relation for the k_i's, and obtain an explicit solution of the recurrence relation.

6.20 Find $\log_5 896$ in $\mathbb{Z}_{1103}$ using Algorithm 6.6, given that $L_2(\beta) = 1$ for $\beta = 25, 219$ and 841, and $L_2(\beta) = 0$ for $\beta = 163, 532, 625$ and 656.

6.21 Throughout this question, suppose that $p \equiv 5 \pmod 8$ is prime and suppose that a is a quadratic residue modulo p.

         (a) Prove that $a^{(p-1)/4} \equiv \pm 1 \pmod p$.

         (b) If $a^{(p-1)/4} \equiv 1 \pmod p$, prove that $a^{(p+3)/8} \bmod p$ is a square root of a modulo p.

         (c) If $a^{(p-1)/4} \equiv -1 \pmod p$, prove that $2^{-1}(4a)^{(p+3)/8} \bmod p$ is a square root of a modulo p.

         HINT   Use the fact that $\left(\frac{2}{p}\right) = -1$ when $p \equiv 5 \pmod 8$ is prime.

         (d) Given a primitive element $\alpha \in \mathbb{Z}_p^*$, and given any $\beta \in \mathbb{Z}_p^*$, show that $L_2(\beta)$ can be computed efficiently.

         HINT   Use the fact that it is possible to compute square roots modulo p, as well as the fact that $L_1(\beta) = L_1(p - \beta)$ for all $\beta \in \mathbb{Z}_p^*$, when $p \equiv 5 \pmod 8$ is prime.

6.22 The *ElGamal Cryptosystem* can be implemented in any subgroup $\langle \alpha \rangle$ of a finite multiplicative group $(G, \cdot)$, as follows: Let $\beta \in \langle \alpha \rangle$ and define (α, β) to be the public key. The plaintext space is $\mathcal{P} = \langle \alpha \rangle$, and the encryption operation is $e_K(x) = (y_1, y_2) = (\alpha^k, x \cdot \beta^k)$, where k is random.

Here we show that distinguishing ElGamal encryptions of two plaintexts can be Turing reduced to **Decision Diffie-Hellman**, and vice versa.

         (a) Assume that ORACLEDDH is an oracle that solves **Decision Diffie-Hellman** in $(G, \cdot)$. Prove that ORACLEDDH can be used as a subroutine in an algorithm that distinguishes ElGamal encryptions of two given plaintexts, say x_1 and x_2. (That is, given $x_1, x_2 \in \mathcal{P}$, and given a ciphertext (y_1, y_2) which is an encryption of x_i for some $i \in \{1, 2\}$, the distinguishing algorithm will determine if $i = 1$ or $i = 2$.)

(b) Assume that ORACLEDISTINGUISH is an oracle that distinguishes ElGamal encryptions of any two given plaintexts x_1 and x_2, for any *ElGamal Cryptosystem* implemented in the group $(G, \cdot)$ as described above. Suppose further that ORACLEDISTINGUISH will determine if a ciphertext (y_1, y_2) is not a valid encryption of either of x_1 or x_2. Prove that ORACLEDISTINGUISH can be used as a subroutine in an algorithm that solves Decision Diffie-Hellman in $(G, \cdot)$.

7

Signature Schemes

7.1 Introduction

In this chapter, we study signature schemes, which are also called digital signatures. A "conventional" handwritten signature attached to a document is used to specify the person responsible for it. A signature is used in everyday situations such as writing a letter, withdrawing money from a bank, signing a contract, etc.

A signature scheme is a method of signing a message stored in electronic form. As such, a signed message can be transmitted over a computer network. In this chapter, we will study several signature schemes, but first we discuss some fundamental differences between conventional and digital signatures.

First is the question of signing a document. With a conventional signature, a signature is part of the physical document being signed. However, a digital signature is not attached physically to the message that is signed, so the algorithm that is used must somehow "bind" the signature to the message.

Second is the question of verification. A conventional signature is verified by comparing it to other, authentic signatures. For example, when someone signs a credit card purchase, the salesperson is supposed to compare the signature on the sales slip to the signature on the back of the credit card in order to verify the signature. Of course, this is not a very secure method as it is relatively easy to forge someone else's signature. Digital signatures, on the other hand, can be verified using a publicly known verification algorithm. Thus, "anyone" can verify a digital signature. The use of a secure signature scheme will prevent the possibility of forgeries.

Another fundamental difference between conventional and digital signatures is that a "copy" of a signed digital message is identical to the original. On the other hand, a copy of a signed paper document can usually be distinguished from an original. This feature means that care must be taken to prevent a signed digital message from being reused. For example, if Alice signs a digital message authorizing Bob to withdraw $100 from her bank account (i.e., a check), she only wants Bob to be able to do so once. So the message itself should contain information, such as a date, that prevents it from being reused.

A signature scheme consists of two components: a signing algorithm and a verification algorithm. Alice can sign a message x using a (private) signing algorithm sig. The resulting signature $\text{sig}(x)$ can subsequently be verified using a public verification algorithm ver. Given a pair (x, y), the verification algorithm returns an answer "true" or "false" depending on whether the signature is valid.

Here is a formal definition of a signature scheme.

Definition 7.1: A *signature scheme* is a five-tuple $(\mathcal{P}, \mathcal{A}, \mathcal{K}, \mathcal{S}, \mathcal{V})$, where the following conditions are satisfied:

1. $\mathcal{P}$ is a finite set of possible *messages*
2. $\mathcal{A}$ is a finite set of possible *signatures*
3. $\mathcal{K}$, the *keyspace*, is a finite set of possible *keys*
4. For each $K \in \mathcal{K}$, there is a *signing algorithm* $\text{sig}_K \in \mathcal{S}$ and a corresponding *verification algorithm* $\text{ver}_K \in \mathcal{V}$. Each $\text{sig}_K : \mathcal{P} \to \mathcal{A}$ and $\text{ver}_K : \mathcal{P} \times \mathcal{A} \to \{true, false\}$ are functions such that the following equation is satisfied for every message $x \in \mathcal{P}$ and for every signature $y \in \mathcal{A}$:
$$\text{ver}(x, y) = \begin{cases} true & \text{if } y = \text{sig}(x) \\ false & \text{if } y \neq \text{sig}(x). \end{cases}$$
A pair (x, y) with $x \in \mathcal{P}$ and $y \in \mathcal{A}$ is called a *signed message*.

For every $K \in \mathcal{K}$, the functions sig_K and ver_K should be polynomial-time functions. ver_K will be a public function and sig_K will be private. Given a message x, it should be computationally infeasible for anyone other than Alice to compute a signature y such that $\text{ver}_K(x, y) = true$ (and note that there might be more than one such y for a given x, depending on how the function ver is defined). If Oscar can compute a pair (x, y) such that $\text{ver}_K(x, y) = true$ and x was not previously signed by Alice, then the signature y is called a *forgery*. Informally, a forged signature is a valid signature produced by someone other than Alice.

As our first example of a signature scheme, we observe that the *RSA Cryptosystem* can be used to provide digital signatures; in this context, it is known as the *RSA Signature Scheme*. See Cryptosystem 7.1.

Cryptosystem 7.1: *RSA Signature Scheme*

Let $n = pq$, where p and q are primes. Let $\mathcal{P} = \mathcal{A} = \mathbb{Z}_n$, and define

$$\mathcal{K} = \{(n, p, q, a, b) : n = pq, p, q \text{ prime}, ab \equiv 1 \pmod{\phi(n)}\}.$$

The values n and b are the public key, and the values p, q, a are the private key.

For $K = (n, p, q, a, b)$, define

$$\text{sig}_K(x) = x^a \bmod n$$

and

$$\text{ver}_K(x, y) = \text{true} \Leftrightarrow x \equiv y^b \pmod{n}$$

$(x, y \in \mathbb{Z}_n)$.

Thus, Alice signs a message x using the RSA decryption rule d_K. Alice is the only person who can create the signature because $d_K = \text{sig}_K$ is private. The verification algorithm uses the RSA encryption rule e_K. Anyone can verify a signature because e_K is public.

Note that anyone can forge Alice's RSA signature by choosing a random y and computing $x = e_K(y)$; then $y = \text{sig}_K(x)$ is a valid signature on the message x. (Note, however, that there does not seem to be an obvious way to first choose x and then compute the corresponding signature y; if this could be done, then the *RSA Cryptosystem* would be insecure.) One way to prevent this attack is to require that messages contain sufficient redundancy that a forged signature of this type does not correspond to a "meaningful" message x except with a very small probability. Alternatively, the use of hash functions in conjunction with signature schemes will eliminate this method of forging (cryptographic hash functions were discussed in Chapter 4). We pursue this approach further in the next section.

Finally, let's look briefly at how we would combine signing and public-key encryption. Suppose Alice wishes to send a signed, encrypted message to Bob. Given a plaintext x, Alice would compute her signature $y = \text{sig}_{\text{Alice}}(x)$, and then encrypt both x and y using Bob's public encryption function e_{Bob}, obtaining $z = e_{\text{Bob}}(x, y)$. The ciphertext z would be transmitted to Bob. When Bob receives z, he first decrypts it with his decryption function d_{Bob} to get (x, y). Then he uses Alice's public verification function to check that $\text{ver}_{\text{Alice}}(x, y) = \text{true}$.

What if Alice first encrypted x, and then signed the result? Then she would compute

$$z = e_{\text{Bob}}(x) \text{ and } y = \text{sig}_{\text{Alice}}(z).$$

Alice would transmit the pair (z, y) to Bob. Bob would decrypt z, obtaining x, and then verify the signature y on x using $\text{ver}_{\text{Alice}}$. One potential problem with this approach is that if Oscar obtains a pair (z, y) of this type, he could replace

Alice's signature y by his own signature

$$y' = \text{sig}_{\text{Oscar}}(z).$$

(Note that Oscar can sign the ciphertext $z = e_{\text{Bob}}(x)$ even though he doesn't know the plaintext x.) Then, if Oscar transmits (z, y') to Bob, Oscar's signature will be verified by Bob using $\text{ver}_{\text{Oscar}}$, and Bob may infer that the plaintext x originated with Oscar. Because of this potential difficulty, most people recommend signing before encrypting.

The rest of this chapter is organized as follows. Section 7.2 introduces the notion of security for signature schemes and how hash functions are used in conjunction with signature schemes. Section 7.3 introduces the *ElGamal Signature Scheme* and discusses its security. Section 7.4 deals with three important schemes that evolved from the *ElGamal Signature Scheme*, namely, the *Schnorr Signature Scheme*, the *Digital Signature Algorithm* and the *Elliptic Curve Digital Signature Algorithm*. Provably secure signature schemes are the topic of Section 7.5. Finally, in Sections 7.6 and 7.7, we consider certain signature schemes with additional properties.

7.2 Security Requirements for Signature Schemes

In this section, we discuss what it means for a signature scheme to be "secure." As was the case with a cryptosystem, we need to specify an attack model, the goal of the adversary, and the type of security provided by the scheme.

Recall from Section 1.2 that the attack model defines the information available to the adversary. In the case of signature schemes, the following types of attack models are commonly considered:

key-only attack

Oscar possesses Alice's public key, i.e., the verification function, ver_K.

known message attack

Oscar possesses a list of messages previously signed by Alice, say

$$(x_1, y_1), (x_2, y_2), \ldots,$$

where the x_i's are messages and the y_i's are Alice's signatures on these messages (so $y_i = \text{sig}_K(x_i)$, $i = 1, 2, \ldots$).

chosen message attack

Oscar requests Alice's signatures on a list of messages. Therefore he chooses messages $x_1, x_2, \ldots$, and Alice supplies her signatures on these messages, namely, $y_i = \text{sig}_K(x_i)$, $i = 1, 2, \ldots$.

We consider several possible adversarial goals:

total break

The adversary is able to determine Alice's private key, i.e., the signing function sig_K. Therefore he can create valid signatures on any message.

selective forgery

With some non-negligible probability, the adversary is able to create a valid signature on a message chosen by someone else. In other words, if the adversary is given a message x, then he can determine (with some probability) the signature y such that $\text{ver}_K(x, y) = \text{true}$. The message x should not be one that has previously been signed by Alice.

existential forgery

The adversary is able to create a valid signature for at least one message. In other words, the adversary can create a pair (x, y) where x is a message and $\text{ver}_K(x, y) = \text{true}$. The message x should not be one that has previously been signed by Alice.

A signature scheme cannot be unconditionally secure, since Oscar can test all possible signatures $y \in \mathcal{A}$ for a given message x, using the public algorithm ver_K, until he finds a valid signature. So, given sufficient time, Oscar can always forge Alice's signature on any message. Thus, as was the case with public-key cryptosystems, our goal is to find signature schemes that are computationally or provably secure.

Notice that the above definitions have some similarity to the attacks on MACs that we considered in Section 4.4. In the MAC setting, there is no such thing as a public key; so it does not make sense to speak of a key-only attack (and a MAC does not have separate signing and verifying functions, of course). The attacks in Section 4.4 were existential forgeries using chosen message attacks.

We illustrate the concepts described above with a couple of examples based on the *RSA Signature Scheme*. In Section 7.1, we observed that Oscar can construct a valid signed message by choosing a signature y and then computing x such that $\text{ver}_K(x, y) = \text{true}$. This would be an existential forgery using a key-only attack.

Another type of attack is based on the multiplicative property of RSA. Suppose that $y_1 = \text{sig}_K(x_1)$ and $y_2 = \text{sig}_K(x_2)$ are any two messages previously signed by Alice. Then

$$\text{ver}_K(x_1 x_2 \bmod n, y_1 y_2 \bmod n) = \text{true},$$

FIGURE 7.1
Signing a message digest

message	x	$x \in \{0, 1\}^*$
	$\downarrow$	
message digest	$z = h(x)$	$z \in \mathcal{Z}$
	$\downarrow$	
signature	$y = sig_K(z)$	$y \in \mathcal{Y}$

and therefore Oscar can create the valid signature $y_1 y_2 \bmod n$ on the message $x_1 x_2 \bmod n$. This is an example of an existential forgery using a known message attack.

Here is one more variation. Suppose Oscar wants to forge a signature on the message x, where x was possibly chosen by someone else. It is a simple matter for him to find $x_1, x_2 \in \mathbb{Z}_n$ such that $x \equiv x_1 x_2 \pmod{n}$. Now suppose he asks Alice for her signatures on messages x_1 and x_2, which we denote by y_1 and y_2 respectively. Then, as in the previous attack, $y_1 y_2 \bmod n$ is the signature for the message $x = x_1 x_2 \bmod n$. This is a selective forgery using a chosen message attack.

7.2.1 Signatures and Hash Functions

Signature schemes are almost always used with in conjunction with a very fast public cryptographic hash function. The hash function $h : \{0, 1\}^* \rightarrow \mathcal{Z}$ will take a message of arbitrary length and produce a message digest of a specified size (160 bits is a popular choice). The message digest will then be signed using a signature scheme $(\mathcal{P}, \mathcal{A}, \mathcal{K}, \mathcal{S}, \mathcal{V})$, where $\mathcal{Z} \subseteq \mathcal{P}$. This use of a hash function and signature scheme is depicted diagramatically in Figure 7.1.

Suppose Alice wants to sign a message x, which is a bitstring of arbitrary length. She first constructs the message digest $z = h(x)$, and then computes the signature on z, namely, $y = sig_K(z)$. Then she transmits the ordered pair (x, y) over the channel. Now the verification can be performed (by anyone) by first reconstructing the message digest $z = h(x)$ using the public hash function h, and then checking that $ver_K(z, y) = true$.

We have to be careful that the use of a hash function h does not weaken the security of the signature scheme, for it is the message digest that is signed, not the message. It will be necessary for h to satisfy certain properties in order to prevent various attacks. The desired properties of hash functions were the ones that were already discussed in Section 4.2.

The most obvious type of attack is for Oscar to start with a valid signed message (x, y), where $y = sig_K(h(x))$. (The pair (x, y) could be any message previously signed by Alice.) Then he computes $z = h(x)$ and attempts to find $x' \neq x$ such that $h(x') = h(x)$. If Oscar can do this, (x', y) would be a valid signed message; so y is a forged signature for the message x'. This is an existential forgery using a known message attack. In order to prevent this type of attack, we require that h be second preimage resistant.

Another possible attack is the following: Oscar first finds two messages $x \neq x'$ such that $h(x) = h(x')$. Oscar then gives x to Alice and persuades her to sign the message digest $h(x)$, obtaining y. Then (x', y) is a valid signed message and y is a forged signature for the message x'. This is an existential forgery using a chosen message attack; it can be prevented if h is collision resistant.

Here is a third variety of attack. It is often possible with certain signature schemes to forge signatures on random message digests z (we observed already that this could be done with the *RSA Signature Scheme*). That is, we assume that the signature scheme (without the hash function) is subject to existential forgery using a key-only attack. Now, suppose Oscar computes a signature on some message digest z, and then he finds a message x such that $z = h(x)$. If he can do this, then (x, y) is a valid signed message and y is a forged signature for the message x. This is an existential forgery on the signature scheme using a key-only attack. In order to prevent this attack, we desire that h be a preimage resistant hash function.

7.3 The ElGamal Signature Scheme

In this section, we present the *ElGamal Signature Scheme*, which was described in a 1985 paper. A modification of this scheme has been adopted as the *Digital Signature Algorithm* (or *DSA*) by the National Institute of Standards and Technology. The *DSA* also incorporates some ideas used in a scheme known as the *Schnorr Signature Scheme*. All of these schemes are designed specifically for the purpose of signatures, as opposed to the *RSA Cryptosystem*, which can be used both as a public-key cryptosystem and a signature scheme.

The *ElGamal Signature Scheme* is non-deterministic (recall that the *ElGamal Public-key Cryptosystem* is also non-deterministic). This means that there are many valid signatures for any given message, and the verification algorithm must be able to accept any of these valid signatures as authentic. The description of the *ElGamal Signature Scheme* is given as Cryptosystem 7.2.

Cryptosystem 7.2: *ElGamal Signature Scheme*

Let p be a prime such that the discrete log problem in $\mathbb{Z}_p$ is intractable, and let $\alpha \in \mathbb{Z}_p^*$ be a primitive element. Let $\mathcal{P} = \mathbb{Z}_p^*$, $\mathcal{A} = \mathbb{Z}_p^* \times \mathbb{Z}_{p-1}$, and define

$$\mathcal{K} = \{(p, \alpha, a, \beta) : \beta \equiv \alpha^a \pmod{p}\}.$$

The values p, α and β are the public key, and a is the private key.

For $K = (p, \alpha, a, \beta)$, and for a (secret) random number $k \in \mathbb{Z}_{p-1}^*$, define

$$\text{sig}_K(x, k) = (\gamma, \delta),$$

where

$$\gamma = \alpha^k \bmod p$$

and

$$\delta = (x - a\gamma)k^{-1} \bmod (p-1).$$

For $x, \gamma \in \mathbb{Z}_p^*$ and $\delta \in \mathbb{Z}_{p-1}$, define

$$\text{ver}_K(x, (\gamma, \delta)) = \text{true} \Leftrightarrow \beta^\gamma \gamma^\delta \equiv \alpha^x \pmod{p}.$$

If the signature was constructed correctly, then the verification will succeed, because

$$\beta^\gamma \gamma^\delta \equiv \alpha^{a\gamma} \alpha^{k\delta} \pmod{p}$$
$$\equiv \alpha^x \pmod{p},$$

where we use the fact that

$$a\gamma + k\delta \equiv x \pmod{p-1}.$$

Actually, it is probably less mysterious to begin with the verification equation, and then derive the signing function. Suppose we start with the congruence

$$\alpha^x \equiv \beta^\gamma \gamma^\delta \pmod{p}. \tag{7.1}$$

Then we make the substitutions

$$\gamma \equiv \alpha^k \pmod{p}$$

and

$$\beta \equiv \alpha^a \pmod{p},$$

but we do not substitute for γ in the exponent of (7.1). We obtain the following:

$$\alpha^x \equiv \alpha^{a\gamma + k\delta} \pmod{p}.$$

Now, α is a primitive element modulo p; so this congruence is true if and only if the exponents are congruent modulo $p - 1$, i.e., if and only if

$$x \equiv a\gamma + k\delta \pmod{p-1}.$$

Given x, a, γ and k, this congruence can be solved for δ, yielding the formula used in the signing function of Cryptosystem 7.2.

Alice computes a signature using both the private key, a, and the secret random number, k (which is used to sign one message, x). The verification can be accomplished using only public information.

Let's do a small example to illustrate the arithmetic.

Example 7.1 Suppose we take $p = 467, \alpha = 2, a = 127$; then

$$\begin{aligned} \beta &= \alpha^a \bmod p \\ &= 2^{127} \bmod 467 \\ &= 132. \end{aligned}$$

Suppose Alice wants to sign the message $x = 100$ and she chooses the random value $k = 213$ (note that $\gcd(213, 466) = 1$ and $213^{-1} \bmod 466 = 431$). Then

$$\gamma = 2^{213} \bmod 467 = 29$$

and

$$\delta = (100 - 127 \times 29)431 \bmod 466 = 51.$$

Anyone can verify this signature by checking that

$$132^{29}29^{51} \equiv 189 \pmod{467}$$

and

$$2^{100} \equiv 189 \pmod{467}.$$

Hence, the signature is valid.   ◻

7.3.1 Security of the ElGamal Signature Scheme

Let's look at the security of the *ElGamal Signature Scheme*. Suppose Oscar tries to forge a signature for a given message x, without knowing a. If Oscar chooses a value γ and then tries to find the corresponding δ, he must compute the discrete logarithm $\log_\gamma \alpha^x \beta^{-\gamma}$. On the other hand, if he first chooses δ and then tries to find γ, he is trying to "solve" the equation

$$\beta^\gamma \gamma^\delta \equiv \alpha^x \pmod{p}$$

for the "unknown" γ. This is a problem for which no feasible solution is known; however, it does not seem to be related to any well-studied problem such as the Discrete Logarithm problem. There also remains the possibility that there might be some way to compute γ and δ simultaneously in such a way that (γ, δ) will be a signature. No one has discovered a way to do this, but conversely, no one has proved that it cannot be done.

If Oscar chooses γ and δ and then tries to solve for x, he is again faced with an instance of the Discrete Logarithm problem, namely the computation of $\log_\alpha \beta^\gamma \gamma^\delta$. Hence, Oscar cannot sign a given message x using this approach.

However, there is a method by which Oscar can sign a random message by choosing γ, δ and x simultaneously. Thus an existential forgery is possible under a key-only attack (assuming a hash function is not used). We describe how to do this now.

Suppose i and j are integers such that $0 \leq i \leq p - 2$, $0 \leq j \leq p - 2$, and suppose we express γ in the form $\gamma = \alpha^i \beta^j \bmod p$. Then the verification condition is

$$\alpha^x \equiv \beta^\gamma (\alpha^i \beta^j)^\delta \pmod{p}.$$

This is equivalent to

$$\alpha^{x - i\delta} \equiv \beta^{\gamma + j\delta} \pmod{p}.$$

This latter congruence will be satisfied if

$$x - i\delta \equiv 0 \pmod{p - 1}$$

and

$$\gamma + j\delta \equiv 0 \pmod{p - 1}.$$

Given i and j, we can easily solve these two congruences modulo $p - 1$ for δ and x, provided that $\gcd(j, p - 1) = 1$. We obtain the following:

$$\gamma = \alpha^i \beta^j \bmod p,$$
$$\delta = -\gamma j^{-1} \bmod (p - 1), \quad \text{and}$$
$$x = -\gamma i j^{-1} \bmod (p - 1).$$

By the way in which we constructed (γ, δ), it is clear that it is a valid signature for the message x.

We illustrate with an example.

Example 7.2 As in the previous example, suppose $p = 467$, $\alpha = 2$ and $\beta = 132$. Suppose Oscar chooses $i = 99$ and $j = 179$; then $j^{-1} \bmod (p - 1) = 151$. He would compute the following:

$$
\begin{aligned}
\gamma &= 2^{99} 132^{179} \bmod 467 &= 117 \\
\delta &= -117 \times 151 \bmod 466 &= 41 \\
x &= 99 \times 41 \bmod 466 &= 331.
\end{aligned}
$$

Then $(117, 41)$ is a valid signature for the message 331, as may be verified by checking that

$$132^{117}117^{41} \equiv 303 \pmod{467}$$

and

$$2^{331} \equiv 303 \pmod{467}.$$

$\square$

Here is a second type of forgery, in which Oscar begins with a message previously signed by Alice. This is an existential forgery under a known message attack. Suppose (γ, δ) is a valid signature for a message x. Then it is possible for Oscar to sign various other messages. Suppose h, i and j are integers, $0 \le h, i, j \le p - 2$, and $\gcd(h\gamma - j\delta, p - 1) = 1$. Compute the following:

$$\lambda = \gamma^h \alpha^i \beta^j \bmod p$$

$$\mu = \delta\lambda(h\gamma - j\delta)^{-1} \bmod (p - 1), \quad \text{and}$$

$$x' = \lambda(hx + i\delta)(h\gamma - j\delta)^{-1} \bmod (p - 1).$$

Then, it is tedious but straightforward to check that the verification condition

$$\beta^\lambda \lambda^\mu \equiv \alpha^{x'} \pmod{p}$$

holds. Hence (λ, μ) is a valid signature for x'.

Both of these methods are existential forgeries, but it does not appear that they can be modified to yield selective forgeries. Hence, they do not seem to represent a threat to the security of the *ElGamal Signature Scheme*, provided that a secure hash function is used as described in Section 7.2.1.

We also mention a couple of ways in which the *ElGamal Signature Scheme* can be broken if it is used carelessly (these are further examples of protocol failures, some of which were discussed in the Exercises of Chapter 5). First, the random value k used in computing a signature should not be revealed. For, if k is known, then it is a simple matter to compute

$$a = (x - k\delta)\gamma^{-1} \bmod (p - 1).$$

Once a is known, then the system is completely broken and Oscar can forge signatures at will.

Another misuse of the system is to use the same value k in signing two different messages. This also makes it easy for Oscar to compute a and hence break the system. This can be done as follows. Suppose (γ, δ_1) is a signature on x_1 and (γ, δ_2) is a signature on x_2. Then we have

$$\beta^\gamma \gamma^{\delta_1} \equiv \alpha^{x_1} \pmod{p}$$

and

$$\beta^\gamma \gamma^{\delta_2} \equiv \alpha^{x_2} \pmod{p}.$$

Thus

$$\alpha^{x_1 - x_2} \equiv \gamma^{\delta_1 - \delta_2} \pmod{p}.$$

Writing $\gamma = \alpha^k$, we obtain the following equation in the unknown k:

$$\alpha^{x_1 - x_2} \equiv \alpha^{k(\delta_1 - \delta_2)} \pmod{p},$$

which is equivalent to

$$x_1 - x_2 \equiv k(\delta_1 - \delta_2) \pmod{p-1}.$$

Now let $d = \gcd(\delta_1 - \delta_2, p - 1)$. Since $d \mid (p - 1)$ and $d \mid (\delta_1 - \delta_2)$, it follows that $d \mid (x_1 - x_2)$. Define

$$x' = \frac{x_1 - x_2}{d}$$

$$\delta' = \frac{\delta_1 - \delta_2}{d}$$

$$p' = \frac{p - 1}{d}.$$

Then the congruence becomes:

$$x' \equiv k\delta' \pmod{p'}.$$

Since $\gcd(\delta', p') = 1$, we can compute

$$\epsilon = (\delta')^{-1} \bmod p'.$$

Then value of k is determined modulo p' to be

$$k = x'\epsilon \bmod p'.$$

This yields d candidate values for k:

$$k = x'\epsilon + ip' \bmod (p-1)$$

for some i, $0 \le i \le d - 1$. Of these d candidate values, the (unique) correct one can be determined by testing the condition

$$\gamma \equiv \alpha^k \pmod{p}.$$

7.4   Variants of the ElGamal Signature Scheme

In many situations, a message might be encrypted and decrypted only once, so it suffices to use any cryptosystem which is known to be secure at the time the message is encrypted. On the other hand, a signed message could function as a legal document such as a contract or will; so it is very likely that it would be necessary to verify a signature many years after the message is signed. It is therefore important to take even more precautions regarding the security of a signature scheme as opposed to a cryptosystem. Since the *ElGamal Signature Scheme* is no more secure than the Discrete Logarithm problem, this necessitates the use of a large modulus p. Most people would argue that the length of p should be at least 1024 bits in order to provide present-day security, and even larger to provide security into the foreseeable future (this was already mentioned in Section 6.6).

A 1024 bit modulus leads to an ElGamal signature having 2048 bits. For potential applications, many of which involve the use of smart cards, a shorter signature is desirable. In 1989, Schnorr proposed a signature scheme that can be viewed as a variant of the *ElGamal Signature Scheme* in which the signature size is greatly reduced. The *Digital Signature Algorithm* (or *DSA*) is another modification of the *ElGamal Signature Scheme*, which incorporates some of the ideas used in the *Schnorr Signature Scheme*. The *DSA* was published in the Federal Register on May 19, 1994 and was adopted as a standard on December 1, 1994 (however, it was first proposed in August 1991). We describe the *Schnorr Signature Scheme*, the *DSA*, and a modification of the *DSA* to elliptic curves (called the *ECDSA*) in the next subsections.

7.4.1   The Schnorr Signature Scheme

Suppose that p and q are primes such that $p - 1 \equiv 0 \pmod{q}$. Typically we will take $p \approx 2^{1024}$ and $q \approx 2^{160}$. The *Schnorr Signature Scheme* modifies the *ElGamal Signature Scheme* in an ingenious way so that a $\log_2 q$-bit message digest is signed using a $2 \log_2 q$-bit signature, but the computations are done in $\mathbb{Z}_p$. The way that this is accomplished is to work in a subgroup of $\mathbb{Z}_p^*$ of size q. The assumed security of the scheme is based on the belief that finding discrete logarithms in this specified subgroup of $\mathbb{Z}_p^*$ is secure. (This setting for the Discrete Logarithm problem was previously discussed in Section 6.6.)

We will take α to be a qth root of 1 modulo p. (It is easy to construct such an α: Let α_0 be a primitive element of $\mathbb{Z}_p$, and define $\alpha = \alpha_0^{(p-1)/q} \bmod p$.) The key in the *Schnorr Signature Scheme* is similar to the key in the *ElGamal Signature Scheme* in other respects. However, the *Schnorr Signature Scheme* integrates a hash function directly into the signing algorithm (as opposed to the usual hash-then-sign method that we discussed in Section 7.2.1). We will assume that $h : \{0, 1\}^* \to \mathbb{Z}_q$ is a secure hash function. Here is a complete description of the *Schnorr Signature Scheme*.

Cryptosystem 7.3: *Schnorr Signature Scheme*

Let p be a prime such that the discrete log problem in $\mathbb{Z}_p^*$ is intractable, and let q be a prime that divides $p - 1$. Let $\alpha \in \mathbb{Z}_p^*$ be a qth root of 1 modulo p. Let $\mathcal{P} = \{0, 1\}^*$, $\mathcal{A} = \mathbb{Z}_q \times \mathbb{Z}_q$, and define

$$\mathcal{K} = \{(p, q, \alpha, a, \beta) : \beta \equiv \alpha^a \pmod{p}\},$$

where $0 \leq a \leq q - 1$. The values p, q, α and β are the public key, and a is the private key. Finally, let $h : \{0, 1\}^* \to \mathbb{Z}_q$ be a secure hash function.

For $K = (p, q, \alpha, a, \beta)$, and for a (secret) random number k, $1 \leq k \leq q - 1$, define

$$\mathrm{sig}_K(x, k) = (\gamma, \delta),$$

where

$$\gamma = h(x \parallel \alpha^k)$$

and

$$\delta = k + a\gamma \bmod q.$$

For $x \in \{0, 1\}^*$ and $\gamma, \delta \in \mathbb{Z}_q$, verification is done by performing the following computations:

$$\mathrm{ver}_K(x, (\gamma, \delta)) = \text{true} \Leftrightarrow h(x \parallel \alpha^\delta \beta^{-\gamma}) = \gamma.$$

It is easy to check that $\alpha^\delta \beta^{-\gamma} \equiv \alpha^k \pmod{p}$, and hence a Schnorr signature will be verified. Here is a small example to illustrate.

Example 7.3 Suppose we take $q = 101$ and $p = 78q + 1 = 7879$. 3 is a primitive element in $\mathbb{Z}_{7879}^*$, so we can take

$$\alpha = 3^{78} \bmod 7879 = 170.$$

α is a qth root of 1 modulo p. Suppose $a = 75$; then

$$\beta = \alpha^a \bmod 7879 = 4567.$$

Now, suppose Alice wants to sign the message x, and she chooses the random value $k = 50$. She first computes

$$\alpha^k \bmod p = 170^{50} \bmod 7879 = 2518.$$

The next step is to compute $h(x \parallel 2518)$, where h is a given hash function and 2518 is represented in binary (as a bitstring). Suppose for purposes of illustration that $h(x \parallel 2518) = 96$. Then δ is computed as

$$\delta = 50 + 75 \times 96 \bmod 101 = 79,$$

and the signature is $(96, 79)$.

This signature is verified by computing

$$170^{79}4567^{-96} \bmod 7879 = 2518,$$

and then checking that $h(x \,\|\, 2518) = 96$.   ▯

7.4.2   The Digital Signature Algorithm

We will outline the changes that are made to the verification function of the *ElGamal Signature Scheme* in the specification of the *DSA*. The *DSA* uses an order q subgroup of $\mathbb{Z}_p^*$, as does the *Schnorr Signature Scheme*. In the *DSA*, it is required that q is a 160-bit prime and p is an L-bit prime, where $L \equiv 0 \pmod{64}$ and $512 \le L \le 1024$. The key in the *DSA* has the same form as in the *Schnorr Signature Scheme*. It is also specified in the *DSA* that the message will be hashed using *SHA-1* before it is signed. The result is that a 160-bit message digest is signed with a 320-bit signature, and the computations are done in $\mathbb{Z}_p$ and $\mathbb{Z}_q$.

In the *ElGamal Signature Scheme*, suppose we change the "$-$" to a "$+$" in the definition of δ, so

$$\delta = (x + a\gamma)k^{-1} \bmod (p - 1).$$

It is easy to see that this changes the verification condition to the following:

$$\alpha^x \beta^\gamma \equiv \gamma^\delta \pmod{p}. \tag{7.2}$$

Now α has order q, and β and γ are powers of α, so they also have order q. This means that all exponents in (7.2) can be reduced modulo q without affecting the validity of the congruence. Since x will be replaced by a 160-bit message digest in the *DSA*, we will assume that $x \in \mathbb{Z}_q$. Further, we will alter the definition of δ, so that $\delta \in \mathbb{Z}_q$, as follows:

$$\delta = (x + a\gamma)k^{-1} \bmod q.$$

It remains to consider $\gamma = \alpha^k \bmod p$. Suppose we temporarily define

$$\gamma' = \gamma \bmod q = (\alpha^k \bmod p) \bmod q.$$

Note that

$$\delta = (x + a\gamma')k^{-1} \bmod q,$$

so δ is unchanged. We can write the verification equation as

$$\alpha^x \beta^{\gamma'} \equiv \gamma^\delta \pmod{p}. \tag{7.3}$$

Notice that we cannot replace the remaining occurrence of γ by γ'.

Now we proceed to rewrite (7.3), by raising both sides to the power $\delta^{-1} \bmod q$ (this requires that $\delta \neq 0$). We obtain the following:

$$\alpha^{x\delta^{-1}} \beta^{\gamma'\delta^{-1}} \bmod p = \gamma. \tag{7.4}$$

Now we can reduce both sides of (7.4) modulo q, which produces the following:

$$(\alpha^{x\delta^{-1}}\beta^{\gamma'\delta^{-1}} \bmod p) \bmod q = \gamma'. \tag{7.5}$$

The complete description of the *DSA* is given as Cryptosystem 7.4, in which we rename γ' as γ and replace x by SHA-1(x).

Cryptosystem 7.4: *Digital Signature Algorithm*

Let p be a L-bit prime such that the discrete log problem in $\mathbb{Z}_p$ is intractable, where $L \equiv 0 \pmod{64}$ and $512 \leq L \leq 1024$, and let q be a 160-bit prime that divides $p - 1$. Let $\alpha \in \mathbb{Z}_p^*$ be a qth root of 1 modulo p. Let $\mathcal{P} = \{0, 1\}^*$, $\mathcal{A} = \mathbb{Z}_q^* \times \mathbb{Z}_q^*$, and define

$$\mathcal{K} = \{(p, q, \alpha, a, \beta) : \beta \equiv \alpha^a \pmod{p}\},$$

where $0 \leq a \leq q - 1$. The values p, q, α and β are the public key, and a is the private key.

For $K = (p, q, \alpha, a, \beta)$, and for a (secret) random number k, $1 \leq k \leq q - 1$, define

$$\text{sig}_K(x, k) = (\gamma, \delta),$$

where

$$\gamma = (\alpha^k \bmod p) \bmod q \quad \text{and}$$

$$\delta = (\text{SHA-1}(x) + a\gamma)k^{-1} \bmod q.$$

(If $\gamma = 0$ or $\delta = 0$, a new random value of k should be chosen.)

For $x \in \{0, 1\}^*$ and $\gamma, \delta \in \mathbb{Z}_q^*$, verification is done by performing the following computations:

$$e_1 = \text{SHA-1}(x)\,\delta^{-1} \bmod q$$

$$e_2 = \gamma\,\delta^{-1} \bmod q$$

$$\text{ver}_K(x, (\gamma, \delta)) = \text{true} \Leftrightarrow (\alpha^{e_1}\beta^{e_2} \bmod p) \bmod q = \gamma.$$

In October 2001, NIST recommended that p be chosen to be a 1024-bit prime (i.e., 1024 is the only value allowed for L). This is "neither a standard nor a guideline," but it does indicate some concern about the security of the discrete logarithm problem.

Notice that if Alice computes a value $\delta \equiv 0 \pmod{q}$ in the DSA signing algorithm, she should reject it and construct a new signature with a new random k. We should point out that this is not likely to cause a problem in practice: the probability that $\delta \equiv 0 \pmod{q}$ is likely to be on the order of 2^{-160}; so for all intents and purposes it will almost never happen.

Here is an example (with p and q much smaller than they are required to be in the *DSA*) to illustrate.

Example 7.4 Suppose we take the same values of p, q, α, a, β and k as in Example 7.3, and suppose Alice wants to sign the message digest SHA-1$(x) = 22$. Then she computes

$$k^{-1} \bmod 101 = 50^{-1} \bmod 101 = 99,$$

$$\gamma = (170^{50} \bmod 7879) \bmod 101$$

$$= 2518 \bmod 101$$

$$= 94,$$

and

$$\delta = (22 + 75 \times 94)99 \bmod 101$$

$$= 97.$$

The signature $(94, 97)$ on the message digest 22 is verified by the following computations:

$$\delta^{-1} = 97^{-1} \bmod 101 = 25$$
$$e_1 = 22 \times 25 \bmod 101 = 45$$
$$e_2 = 94 \times 25 \bmod 101 = 27$$
$$(170^{45} 4567^{27} \bmod 7879) \bmod 101 = 2518 \bmod 101 = 94.$$

$\square$

When the *DSA* was proposed in 1991, there were several criticisms put forward. One complaint was that the selection process by NIST was not public. The standard was developed by the National Security Agency (NSA) without the input of U.S. industry. Regardless of the merits of the resulting scheme, many people resented the "closed-door" approach.

Of the technical criticisms put forward, the most serious was that the size of the modulus p was fixed initially at 512 bits. Many people suggested that the modulus size not be fixed, so that larger modulus sizes could be used if desired. In reponse to these comments, NIST altered the description of the standard so that a variety of modulus sizes were allowed.

7.4.3 The Elliptic Curve DSA

In 2000, the *Elliptic Curve Digital Signature Algorithm* (*ECDSA*) was approved as FIPS 186-2. We describe the *ECDSA* in this section.

Cryptosystem 7.5: *Elliptic Curve Digital Signature Algorithm*

Let p be a prime or a power of two, and let E be an elliptic curve defined over $\mathbb{F}_p$. Let A be a point on E having prime order q, such that the Discrete Logarithm problem in $\langle A \rangle$ is infeasible. Let $\mathcal{P} = \{0, 1\}^*$, $\mathcal{A} = \mathbb{Z}_q^* \times \mathbb{Z}_q^*$, and define

$$\mathcal{K} = \{(p, q, E, A, m, B) : B = mA\},$$

where $0 \leq m \leq q - 1$. The values p, q, E, A and B are the public key, and m is the private key.

For $K = (p, q, E, A, m, B)$, and for a (secret) random number k, $1 \leq k \leq q - 1$, define

$$\mathsf{sig}_K(x, k) = (r, s),$$

where

$$kA = (u, v)$$

$$r = u \bmod q, \quad \text{and}$$

$$s = k^{-1}(\text{SHA-1}(x) + mr) \bmod q.$$

(If either $r = 0$ or $s = 0$, a new random value of k should be chosen.)

For $x \in \{0, 1\}^*$ and $r, s \in \mathbb{Z}_q^*$, verification is done by performing the following computations:

$$w = s^{-1} \bmod q$$

$$i = w\,\text{SHA-1}(x) \bmod q$$

$$j = wr \bmod q$$

$$(u, v) = iA + jB$$

$$\mathsf{ver}_K(x, (r, s)) = \text{true} \Leftrightarrow u \bmod q = r.$$

We work through a tiny example to illustrate the computations in the *ECDSA*.

Example 7.5 We will base our example on the same elliptic curve that was used in Section 6.5.2, namely, $y^2 = x^3 + x + 6$, defined over $\mathbb{Z}_{11}$. The parameters of the signature scheme are $p = 11$, $q = 13$, $A = (2, 7)$, $m = 7$ and $B = (7, 2)$.

Suppose we have a message x with SHA-1$(x) = 4$, and Alice wants to sign the message x using the random value $k = 3$. She will compute

$$(u, v) = 3\,(2, 7) = (8, 3)$$

$$r = u \bmod 13 = 8, \quad \text{and}$$

$$s = 3^{-1}(4 + 7 \times 8) \bmod 13 = 7.$$

Therefore $(8, 7)$ is the signature.

Bob verifies the signature by performing the following computations:

$$w = 7^{-1} \bmod 13 = 2$$

$$i = 2 \times 4 \bmod 13 = 8$$

$$j = 2 \times 8 \bmod 13 = 3$$

$$(u, v) = 8A + 3B = (8, 3), \quad \text{and}$$

$$u \bmod 13 = 8 = r.$$

Hence, the signature is verified.   ▯

7.5 Provably Secure Signature Schemes

We present some examples of provably secure signature schemes in this section. First, we describe a construction for a one-time signature scheme based on an arbitrary one-way (i.e., preimage resistant) function, say f. This scheme can be proven secure against a key-only attack provided that f is a bijective function. The second construction is for a signature scheme known as *Full Domain Hash*. This signature scheme is provably secure in the random oracle model provided that it is constructed from a trap-door one-way permutation.

7.5.1 One-time Signatures

In this section, we describe a conceptually simple way to construct a provably secure one-time signature scheme from a one-way function. (A signature scheme is a *one-time* signature scheme if it is secure when only one message is signed. The signature can be verified an arbitrary number of times, of course.) The description of the scheme, which is known as the *Lamport Signature Scheme*, is given in Cryptosystem 7.6.

Cryptosystem 7.6: *Lamport Signature Scheme*

Let k be a positive integer and let $\mathcal{P} = \{0, 1\}^k$. Suppose $f : Y \to Z$ is a one-way function, and let $\mathcal{A} = Y^k$. Let $y_{i,j} \in Y$ be chosen at random, $1 \leq i \leq k$, $j = 0, 1$, and let $z_{i,j} = f(y_{i,j})$, $1 \leq i \leq k$, $j = 0, 1$. The key K consists of the $2k$ y's and the $2k$ z's. The y's are the private key while the z's are the public key.

For $K = (y_{i,j}, z_{i,j} : 1 \leq i \leq k, j = 0, 1)$, define

$$\text{sig}_K(x_1, \ldots, x_k) = (y_{1,x_1}, \ldots, y_{k,x_k}).$$

A signature $(a_1, \ldots, a_k)$ on the message $(x_1, \ldots, x_k)$ is verified as follows:

$$\text{ver}_K((x_1, \ldots, x_k), (a_1, \ldots, a_k)) = \text{true} \Leftrightarrow f(a_i) = z_{i,x_i}, 1 \leq i \leq k.$$

Informally, this is how the system works. A message to be signed is a binary k-tuple. Each bit of the message is signed individually. If the ith bit of the message equals j (where $j \in \{0, 1\}$), then the ith element of the signature is the value $y_{i,j}$, which is a preimage of the public key value $z_{i,j}$. The verification consists simply of checking that each element in the signature is a preimage of the public key element $z_{i,j}$ that corresponds to the ith bit of the message. This can be done using the public function f.

We illustrate the scheme by considering one possible implementation using the exponentiation function $f(x) = \alpha^x \bmod p$, where α is a primitive element modulo p. Here $f : \{0, \ldots, p - 2\} \to \mathbb{Z}_p^*$. We present a toy example to demonstrate the computations that take place in the scheme.

Example 7.6 7879 is prime and 3 is a primitive element in $\mathbb{Z}_{7879}^*$. Define

$$f(x) = 3^x \bmod 7879.$$

Suppose $k = 3$, and Alice chooses the six (secret) random numbers

$$y_{1,0} = 5831$$

$$y_{1,1} = 735$$

$$y_{2,0} = 803$$

$$y_{2,1} = 2467$$

$$y_{3,0} = 4285$$

$$y_{3,1} = 6449.$$

Then Alice computes the images of these six y's under the function f:

$$z_{1,0} = 2009$$
$$z_{1,1} = 3810$$
$$z_{2,0} = 4672$$
$$z_{2,1} = 4721$$
$$z_{3,0} = 268$$
$$z_{3,1} = 5731.$$

These z's are published. Now, suppose Alice wants to sign the message

$$x = (1, 1, 0).$$

The signature for x is

$$(y_{1,1}, y_{2,1}, y_{3,0}) = (735, 2467, 4285).$$

To verify this signature, it suffices to compute the following:

$$3^{735} \bmod 7879 = 3810$$
$$3^{2467} \bmod 7879 = 4721$$
$$3^{4285} \bmod 7879 = 268.$$

Hence, the signature is verified.   ▯

Oscar cannot forge a signature because he is unable to invert the one-way function f to obtain the secret y's. However, the signature scheme can be used to sign only one message securely. Given signatures on two different messages, it is an easy matter for Oscar to construct signatures for another message different from the first two (unless the first two messages differ in exactly one bit).

For example, suppose the messages $(0, 1, 1)$ and $(1, 0, 1)$ are both signed using the same key. The message $(0, 1, 1)$ has as its signature the triple $(y_{1,0}, y_{2,1}, y_{3,1})$, and the message $(1, 0, 1)$ is signed with $(y_{1,1}, y_{2,0}, y_{3,1})$. Given these two signatures, Oscar can manufacture signatures for the messages $(1, 1, 1)$ (namely, $(y_{1,1}, y_{2,1}, y_{3,1})$) and $(0, 0, 1)$ (namely, $(y_{1,0}, y_{2,0}, y_{3,1})$).

The security of the *Lamport Signature Scheme* can be proven if we assume that $f : Y \to Z$ is a bijective one-way function and a public key consists of $2k$ distinct elements of Z. We consider a key-only attack; so an adversary is given only the public key. We will assume that the adversary can carry out an existential forgery. In other words, given the public key, the adversary outputs a message, x, and a valid signature, y. (We are assuming that f, Y, Z and k are fixed.)

The adversary is modeled by an algorithm, say LAMPORT-FORGE. For simplicity, we assume that LAMPORT-FORGE is deterministic: given any particular public key, it always outputs the same forgery. We will describe an algorithm, LAMPORT-PREIMAGE, to find preimages of the function f on randomly chosen elements $z \in Z$. This algorithm is a reduction which uses the algorithm LAMPORT-FORGE as an oracle. The existence of this reduction contradicts the assumed one-wayness of f. Hence, if we believe that f is one-way, then we conclude that a key-only existential forgery is computationally infeasible. LAMPORT-PREIMAGE is presented as Algorithm 7.1.

Algorithm 7.1: LAMPORT-PREIMAGE(z)

external f, LAMPORT-FORGE
comment: we assume $f : Y \rightarrow Z$ is a bijection
choose a random $i_0 \in \{1, \ldots, k\}$ and a random $j_0 \in \{0, 1\}$
construct a random public key $\mathcal{Z} = (z_{i,j} : 1 \leq i \leq k, j = 0, 1)$
   such that $z_{i_0, j_0} = z$
$((x_1, \ldots, x_k), (a_1, \ldots, a_k)) \leftarrow$ LAMPORT-FORGE($\mathcal{Z}$)
if $x_{i_0} = j_0$
  then return (a_{i_0})
  else return (fail)

Let's consider the (average) success probability of Algorithm 7.1, where the average is computed over all $z \in Z$. We are assuming that LAMPORT-FORGE always succeeds in finding a forgery. If $x_{i_0} = j_0$ in the forgery, then it is the case that

$$f(a_{i_0}) = z_{i_0, x_{i_0}} = z_{i_0, j_0} = z,$$

and we have found $f^{-1}(z)$, as desired. Recall that each x_i has the value 0 or 1. We will prove that the probability that $x_{i_0} = j_0$, averaged over all possible runs of the algorithm, is $1/2$. Therefore, the average success probability of Algorithm 7.1 is equal to $1/2$. We give a proof of this in the following theorem.

THEOREM 7.1 *Suppose that $f : Y \rightarrow Z$ is a one-way bijection, and suppose there exists a deterministic algorithm, LAMPORT-FORGE, that will create an existential forgery for the Lamport Signature Scheme using a key-only attack, for any public key $\mathcal{Z}$ consisting of $2k$ distinct elements of Z. Then there exists an algorithm, LAMPORT-PREIMAGE, that will find preimages of random elements $z \in Z$ with average probability at least $1/2$.*

PROOF Let $\mathcal{S}$ denote the set of all possible public keys, and for any $z \in Z$, let $\mathcal{S}_z$ denote the set of all possible public keys that contain z. Denote $s = |\mathcal{S}|$ and, for all $z \in Z$, denote $s_z = |\mathcal{S}_z|$. Let $\mathcal{T}_z$ consist of all public keys in $\mathcal{Z} \in \mathcal{S}_z$ such that LAMPORT-PREIMAGE(z) succeeds when $\mathcal{Z}$ is the public key that is chosen by LAMPORT-PREIMAGE(z), and denote $t_z = |\mathcal{T}_z|$.

We will make use of two equations, which follow from elementary counting techniques. First, we have the following:

$$\sum_{z \in Z} t_z = ks. \tag{7.6}$$

This is seen as follows: There are s possible public keys. For each public key $\mathcal{Z} \in \mathcal{S}$, LAMPORT-FORGE finds inverses of k elements in Z. On the other hand, the total number of inverses computed by LAMPORT-FORGE, over all possible public keys, can be computed to be $\sum t_z$.

Second, for any $z \in Z$, the following equation holds:

$$2ks = s_z |Z|. \tag{7.7}$$

Again, this is easy to prove. Each of the s possible public keys contains $2k$ elements of Z. However, it is obvious that every element $z \in Z$ occurs in the same number of public keys. Therefore s_z is constant (i.e., independent of z) and $2ks = s_z |Z|$.

Now, let p_z denote the probability that LAMPORT-PREIMAGE(z) succeeds. It is clear that $p_z = t_z / s_z$. We compute the average value of p_z, denoted $\overline{p}$, as follows:

$$\begin{aligned}
\overline{p} &= \frac{1}{|Z|} \sum_{z \in Z} p_z \\
&= \frac{1}{|Z|} \sum_{z \in Z} \frac{t_z}{s_z} \\
&= \frac{1}{s_z |Z|} \sum_{z \in Z} t_z \\
&= \frac{1}{2ks} \sum_{z \in Z} t_z \qquad \text{from (7.6)} \\
&= \frac{ks}{2ks} \qquad\qquad \text{from (7.7)} \\
&= \frac{1}{2}.
\end{aligned}$$

∎

The *Lamport Signature Scheme* is quite elegant, but it is not of practical use. One problem is the size of the signatures it produces. For example, if we use the modular exponentiation function to construct f, as in Example 7.6, then a secure implementation would require that p be at least 1024 bits in length. This means that each bit of the message is signed using 1024 bits. Consequently, the signature is 1024 times as long as the message! It might be more efficient to use a one-way bijection whose security is based on the infeasibility of the elliptic curve Discrete Logarithm problem, but the scheme still would not be very practical.

7.5.2 Full Domain Hash

In Section 5.9.2, we showed how to construct provably secure public-key cryptosystems from trapdoor one-way permutations (in the random oracle model). Practical implementations of these systems are based on the *RSA Cryptosystem* and they replace the random oracle by a hash function such as *SHA-1*. In this section, we use a trapdoor one-way permutation to construct a secure signature scheme in the random oracle model. The scheme we present is called *Full Domain Hash*. The name of this scheme comes from the requirement that the range of the random oracle be the same as the domain of the trapdoor one-way permutation used in the scheme. The scheme is presented as Cryptosystem 7.7.

Cryptosystem 7.7: *Full Domain Hash*

Let k be a positive integer; let $\mathcal{F}$ be a family of trapdoor one-way permutations such that $f : \{0,1\}^k \to \{0,1\}^k$ for all $f \in \mathcal{F}$; and let $G : \{0,1\}^* \to \{0,1\}^k$ be a "random" function. Let $\mathcal{P} = \{0,1\}^*$ and $\mathcal{A} = \{0,1\}^k$, and define

$$\mathcal{K} = \{(f, f^{-1}, G) : f \in \mathcal{F}\}.$$

Given a key $K = (f, f^{-1}, G)$, f^{-1} is the private key and (f, G) is the public key.

For $K = (f, f^{-1}, G)$ and $x \in \{0,1\}^*$, define

$$\mathrm{sig}_K(x) = f^{-1}(G(x)).$$

A signature $y = (y_1, \ldots, y_k) \in \{0,1\}^k$ on the message x is verified as follows:

$$\mathrm{ver}_K(x, y) = \text{true} \Leftrightarrow f(y) = G(x).$$

Full Domain Hash uses the familiar hash-then-sign method. $G(x)$ is the message digest produced by the random oracle, G. f^{-1} is used to sign the message digest, and f is used to verify it.

Let's briefly consider an RSA-based implementation of this scheme. The function f^{-1} would be the RSA signing (i.e., decryption) function, and f would be the RSA verification (i.e., encryption) function. In order for this to be secure, we would have to take $k = 1024$, say. Now suppose that the random oracle G is replaced by the hash function *SHA-1*. *SHA-1* constructs a 160-bit message digest, so the range of the hash function, namely $\{0,1\}^{160}$, is a very small subset of

$\{0, 1\}^k = \{0, 1\}^{1024}$. In practice, it is necessary to specify some padding scheme in order to expand a 160-bit message to 1024 bits before applying f^{-1}. This is typically done using a fixed (deterministic) padding scheme.

We now proceed to our security proof, in which we assume that $\mathcal{F}$ is a family of trapdoor one-way permutations and G is a "full domain" random oracle. (Note that the security proofs we will present do not apply when the random oracle is replaced by a fully specified hash function such as *SHA-1*.) It can be proven that *Full Domain Hash* is secure against existential forgery using a chosen-message attack; however, we will only prove the easier result that *Full Domain Hash* is secure against existential forgery using a key-only attack.

As usual, the security proof is a type of reduction. We assume that there is an adversary (i.e., a randomized algorithm, which we denote by FDH-FORGE) which is able to forge signatures (with some specified probability) when it is given the public key and access to the random oracle (recall that it can query the random oracle for values $G(x)$, but there is no algorithm specified to evaluate the function G). FDH-FORGE is allowed to make some number of oracle queries, say q_h. Eventually, FDH-FORGE outputs a valid forgery with some probability, denoted by ϵ.

We construct an algorithm, FDH-INVERT, which attempts to invert randomly chosen elements $z_0 \in \{0, 1\}^k$. That is, given $z_0 \in \{0, 1\}^k$, our hope is that FDH-INVERT$(z_0) = f^{-1}(z_0)$. We now present FDH-INVERT as Algorithm 7.2.

Algorithm 7.2: FDH-INVERT(z_0, q_h)

external f
procedure SIMG(x)
if $j > q_h$
    then return ("failure")
    else if $j = j_0$
    then $z \leftarrow z_0$
    else let $z \in \{0, 1\}^k$ be chosen at random
$j \leftarrow j + 1$
return (z)

main
choose $j_0 \in \{1, \dots, q_h\}$ at random
$j \leftarrow 1$
insert the code for FDH-FORGE(f) here
if FDH-FORGE$(f) = (x, y)$
then $\begin{cases} \textbf{if } f(y) = z_0 \\ \quad \textbf{then return } (y) \\ \quad \textbf{else return ("failure")} \end{cases}$

Algorithm 7.2 is fairly simple. It basically consists of running the adversary, FDH-FORGE. Hash queries made by FDH-FORGE are handled by the function SIMG, which is a simulation of a random oracle. We have assumed that FDH-FORGE will make q_h hash queries, say $x_1, \ldots, x_{q_h}$. For simplicity, we assume that the x_i's are distinct. (If they are not, then we need to ensure that $\text{SIMG}(x_i) = \text{SIMG}(x_j)$ whenever $x_i = x_j$. This is not difficult to do; it just requires some bookkeeping, as was done in Algorithm 5.14.) We randomly choose one query, say the j_0th query, and define $\text{SIMG}(x_{j_0}) = z_0$ (z_0 is the value we are trying to invert). For all other queries, the value $\text{SIMG}(x_j)$ is chosen to be a random number. Because z_0 is also random, it is easy to see that SIMG is indistinguishable from a true random oracle. It therefore follows that FDH-FORGE outputs a message and a valid forged signature, which we denote by (x, y), with probability ϵ. We then check to see if $f(y) = z_0$; if so, then $y = f^{-1}(z_0)$ and we have succeeded in inverting z_0.

Our main task is to analyze the success probability of the algorithm FDH-INVERT, as a function of the success probability, ϵ, of FDH-FORGE. We will assume that $\epsilon > 2^{-k}$, because a random choice of y will be a valid signature for a message x with probability 2^{-k}, and we are only interested in adversaries that have a higher success probability than a random guess. As we did above, we denote the hash queries made by FDH-FORGE by $x_1, \ldots, x_{q_h}$, where x_j is the jth hash query, $1 \leq j \leq q_h$.

We begin by conditioning the success probability, ϵ, on whether or not $x \in \{x_1, \ldots, x_{q_h}\}$:

$$\epsilon = \mathbf{Pr}[\text{FDH-FORGE succeeds} \wedge (x \in \{x_1, \ldots, x_{q_h}\})]$$
$$+ \mathbf{Pr}[\text{FDH-FORGE succeeds} \wedge (x \notin \{x_1, \ldots, x_{q_h}\})]. \quad (7.8)$$

It is not hard to see that

$$\mathbf{Pr}[\text{FDH-FORGE succeeds} \wedge (x \notin \{x_1, \ldots, x_{q_h}\})] = 2^{-k}.$$

This is because the (undetermined) value $\text{SIMG}(x)$ is equally likely to take on any given value in $\{0, 1\}^k$, and hence the probability that $\text{SIMG}(x) = f(y)$ is 2^{-k}. (This is where we use the assumption that the hash function is a "full domain" hash.) Substituting into (7.8), we obtain the following:

$$\mathbf{Pr}[\text{FDH-FORGE succeeds} \wedge (x \in \{x_1, \ldots, x_{q_h}\})] \geq \epsilon - 2^{-k}. \quad (7.9)$$

Now we turn to the success probability of FDH-INVERT. The next inequality is obvious:

$$\mathbf{Pr}[\text{FDH-INVERT succeeds}] \geq \mathbf{Pr}[\text{FDH-FORGE succeeds} \wedge (x = x_{j_0})]. \quad (7.10)$$

Our final observation is that

$$\mathbf{Pr}[\text{FDH-FORGE succeeds} \wedge (x = x_{j_0})]$$
$$= \frac{1}{q_h} \times \mathbf{Pr}[\text{FDH-FORGE succeeds} \wedge (x \in \{x_1, \ldots, x_{q_h}\})]. \quad (7.11)$$

Note that equation (7.11) is true because there is a $1/q_h$ chance that $x = x_{j_0}$, given that $x \in \{x_1, \ldots, x_{q_h}\}$. Now, if we combine (7.9), (7.10) and (7.11), then we obtain the following bound:

$$\mathbf{Pr}[\text{FDH-INVERT succeeds}] \geq \frac{\epsilon - 2^{-k}}{q_h}. \tag{7.12}$$

Therefore we have obtained a concrete lower bound on the success probability of FDH-INVERT. We have proven the following result.

THEOREM 7.2 *Suppose there exists an algorithm* FDH-FORGE *that will output an existential forgery for Full Domain Hash with probability* $\epsilon > 2^{-k}$, *using a key-only attack. Then there exists an algorithm* FDH-INVERT *that will find inverses of random elements* $z_0 \in \{0, 1\}^k$ *with probability at least* $(\epsilon - 2^{-k})/q_h$.

Observe that the usefulness of the resulting inversion algorithm depends on the ability of FDH-FORGE to find forgeries using as few hash queries as possible.

7.6 Undeniable Signatures

Undeniable signatures were introduced by Chaum and van Antwerpen in 1989. They have several novel features. Primary among these is that a signature cannot be verified without the cooperation of the signer, Alice. This protects Alice against the possibility that documents signed by her are duplicated and distributed electronically without her approval. The verification will be accomplished by means of a *challenge-and-response protocol*.

If Alice's cooperation is required to verify a signature, what is to prevent Alice from disavowing a signature she made at an earlier time? Alice might claim that a valid signature is a forgery, and either refuse to verify it, or carry out the protocol in such a way that the signature will not be verified. To prevent this from happening, an undeniable signature scheme incorporates a *disavowal protocol* by which Alice can prove that a signature is a forgery. Thus, Alice will be able to prove "in court" that a forged signature is, in fact, a forgery. (If she refuses to take part in the disavowal protocol, this would be regarded as evidence that the signature is, in fact, genuine.)

Thus, an undeniable signature scheme consists of three components: a signing algorithm, a verification protocol, and a disavowal protocol. First, we present the signing algorithm and verification protocol of the *Chaum-van Antwerpen Signature Scheme* as Cryptosystem 7.8.

Cryptosystem 7.8: *Chaum-van Antwerpen Signature Scheme*

Let $p = 2q + 1$ be a prime such that q is prime and the discrete log problem in $\mathbb{Z}_p$ is intractable. Let $\alpha \in \mathbb{Z}_p^*$ be an element of order q. Let $1 \le a \le q - 1$ and define $\beta = \alpha^a \bmod p$. Let G denote the multiplicative subgroup of $\mathbb{Z}_p^*$ of order q (G consists of the quadratic residues modulo p). Let $\mathcal{P} = \mathcal{A} = G$, and define

$$\mathcal{K} = \{(p, \alpha, a, \beta) : \beta \equiv \alpha^a \pmod{p}\}.$$

The values p, α and β are the public key, and a is the private key.

For $K = (p, \alpha, a, \beta)$ and $x \in G$, define

$$y = \text{sig}_K(x) = x^a \bmod p.$$

For $x, y \in G$, verification is done by executing the following protocol:

1. Bob chooses e_1, e_2 at random, $e_1, e_2 \in \mathbb{Z}_q^*$.
2. Bob computes $c = y^{e_1} \beta^{e_2} \bmod p$ and sends it to Alice.
3. Alice computes $d = c^{a^{-1} \bmod q} \bmod p$ and sends it to Bob.
4. Bob accepts y as a valid signature if and only if

$$d \equiv x^{e_1} \alpha^{e_2} \pmod{p}.$$

We should explain the roles of p and q in this scheme. The scheme lives in $\mathbb{Z}_p$; however, we need to be able to do computations in a multiplicative subgroup G of $\mathbb{Z}_p^*$ of prime order. In particular, we need to be able to compute inverses modulo $|G|$, which is why $|G|$ should be prime. It is convenient to take $p = 2q + 1$ where q is prime. In this way, the subgroup G is as large as possible, which is desirable since messages and signatures are both elements of G.

We first prove that Bob will accept a valid signature. In the following computations, all exponents are to be reduced modulo q. First, we show that

$$d \equiv c^{a^{-1}} \pmod{p}$$

$$\equiv y^{e_1 a^{-1}} \beta^{e_2 a^{-1}} \pmod{p}.$$

Since

$$\beta \equiv \alpha^a \pmod{p},$$

we have that

$$\beta^{a^{-1}} \equiv \alpha \pmod{p}.$$

Similarly,

$$y = x^a \pmod{p}$$

implies that

$$y^{a^{-1}} \equiv x \pmod{p}.$$

Hence,

$$d \equiv x^{e_1} \alpha^{e_2} \pmod{p},$$

as desired.

Here is a small example to illustrate.

Example 7.7 Suppose we take $p = 467$. Since 2 is a primitive element, $2^2 = 4$ is a generator of G, the quadratic residues modulo 467. So we can take $\alpha = 4$. Suppose $a = 101$; then

$$\beta = \alpha^a \bmod 467 = 449.$$

Alice will sign the message $x = 119$ with the signature

$$y = 119^{101} \bmod 467 = 129.$$

Now, suppose Bob wants to verify the signature y. Suppose he chooses the random values $e_1 = 38$, $e_2 = 397$. He will compute $c = 13$, whereupon Alice will respond with $d = 9$. Bob checks the response by verifying that

$$119^{38} 4^{397} \equiv 9 \pmod{467}.$$

Hence, Bob accepts the signature as valid. $\square$

We next prove that Alice cannot fool Bob into accepting a fraudulent signature as valid, except with a very small probability. This result does not depend on any computational assumptions, i.e., the security is unconditional.

THEOREM 7.3 *If $y \not\equiv x^a \pmod{p}$, then Bob will accept y as a valid signature for x with probability $1/q$.*

PROOF   First, we observe that each possible challenge c corresponds to exactly q ordered pairs (e_1, e_2) (this is because y and β are both elements of the multiplicative group G of prime order q). Now, when Alice receives the challenge c, she has no way of knowing which of the q possible ordered pairs (e_1, e_2) Bob used to construct c. We claim that, if $y \not\equiv x^a \pmod{p}$, then any possible response $d \in G$ that Alice might make is consistent with exactly one of the q possible ordered pairs (e_1, e_2).

Since α generates G, we can write any element of G as a power of α, where the exponent is defined uniquely modulo q. So write $c = \alpha^i$, $d = \alpha^j$, $x = \alpha^k$, and $y = \alpha^\ell$, where $i, j, k, \ell \in \mathbb{Z}_q$ and all arithmetic is modulo p. Consider the following two congruences:

$$c \equiv y^{e_1} \beta^{e_2} \pmod{p}$$
$$d \equiv x^{e_1} \alpha^{e_2} \pmod{p}.$$

This system is equivalent to the following system:

$$i \equiv \ell e_1 + a e_2 \pmod{q}$$
$$j \equiv k e_1 + e_2 \pmod{q}.$$

Now, we are assuming that

$$y \not\equiv x^a \pmod{p},$$

so it follows that

$$\ell \not\equiv ak \pmod{q}.$$

Hence, the coefficient matrix of this system of congruences modulo q has non-zero determinant, and thus there is a unique solution to the system. That is, every $d \in G$ is the correct response for exactly one of the q possible ordered pairs (e_1, e_2). Consequently, the probability that Alice gives Bob a response d that will be verified is exactly $1/q$, and the theorem is proved. ∎

We now turn to the disavowal protocol. This protocol consists of two runs of the verification protocol and is presented as Algorithm 7.3.

Algorithm 7.3: DISAVOWAL

1. Bob chooses e_1, e_2 at random, $e_1, e_2 \in \mathbb{Z}_q{}^*$
2. Bob computes $c = y^{e_1} \beta^{e_2} \bmod p$ and sends it to Alice
3. Alice computes $d = c^{a^{-1} \bmod q} \bmod p$ and sends it to Bob
4. Bob verifies that $d \not\equiv x^{e_1} \alpha^{e_2} \pmod{p}$
5. Bob chooses f_1, f_2 at random, $f_1, f_2 \in \mathbb{Z}_q{}^*$
6. Bob computes $C = y^{f_1} \beta^{f_2} \bmod p$ and sends it to Alice
7. Alice computes $D = C^{a^{-1} \bmod q} \bmod p$ and sends it to Bob
8. Bob verifies that $D \not\equiv x^{f_1} \alpha^{f_2} \pmod{p}$
9. Bob concludes that y is a forgery if and only if

$$(d\alpha^{-e_2})^{f_1} \equiv (D\alpha^{-f_2})^{e_1} \pmod{p}.$$

Steps 1 to 4 and steps 5 to 8 comprise two unsuccessful runs of the verification protocol. Step 9 is a "consistency check" that enables Bob to determine if Alice is forming her responses in the manner specified by the protocol.

The following example illustrates the disavowal protocol.

Example 7.8   As before, suppose $p = 467$, $\alpha = 4$, $a = 101$ and $\beta = 449$. Suppose the message $x = 286$ is signed with the (bogus) signature $y = 83$, and Alice wants to convince Bob that the signature is invalid.

Suppose Bob begins by choosing the random values $e_1 = 45$, $e_2 = 237$. Bob computes $c = 305$ and Alice responds with $d = 109$. Then Bob computes

$$286^{45}4^{237} \bmod 467 = 149.$$

Since $149 \neq 109$, Bob proceeds to step 5 of the protocol.

Now suppose Bob chooses the random values $f_1 = 125$, $f_2 = 9$. Bob computes $C = 270$ and Alice responds with $D = 68$. Bob computes

$$286^{125}4^9 \bmod 467 = 25.$$

Since $25 \neq 68$, Bob proceeds to step 9 of the protocol and performs the consistency check. This check succeeds, since

$$(109 \times 4^{-237})^{125} \equiv 188 \pmod{467}$$

and

$$(68 \times 4^{-9})^{45} \equiv 188 \pmod{467}.$$

Hence, Bob is convinced that the signature is invalid.   ▯

We have to prove two things at this point:

1. Alice can convince Bob that an invalid signature is a forgery.

2. Alice cannot make Bob believe that a valid signature is a forgery except with a very small probability.

THEOREM 7.4   *If $y \not\equiv x^a \pmod{p}$ and Bob and Alice follow the disavowal protocol, then*
$$(d\alpha^{-e_2})^{f_1} \equiv (D\alpha^{-f_2})^{e_1} \pmod{p}.$$

PROOF   Using the facts that

$$d \equiv c^{a^{-1}} \pmod{p},$$
$$c \equiv y^{e_1}\beta^{e_2} \pmod{p}, \quad \text{and}$$
$$\beta \equiv \alpha^a \pmod{p},$$

we have the following:

$$(d\alpha^{-e_2})^{f_1} \equiv \left((y^{e_1}\beta^{e_2})^{a^{-1}}\alpha^{-e_2}\right)^{f_1} \pmod{p}$$
$$\equiv y^{e_1 a^{-1}f_1}\beta^{e_2 a^{-1}f_1}\alpha^{-e_2 f_1} \pmod{p}$$
$$\equiv y^{e_1 a^{-1}f_1}\alpha^{e_2 f_1}\alpha^{-e_2 f_1} \pmod{p}$$
$$\equiv y^{e_1 a^{-1}f_1} \pmod{p}.$$

A similar computation, using the facts that $D \equiv C^{a^{-1}} \pmod{p}$, $C \equiv y^{f_1}\beta^{f_2}$ $\pmod{p}$ and $\beta \equiv \alpha^a \pmod{p}$, establishes that

$$(D\alpha^{-f_2})^{e_1} \equiv y^{e_1 a^{-1} f_1} \pmod{p},$$

so the consistency check in step 9 succeeds. ∎

Now we look at the possibility that Alice might attempt to disavow a valid signature. In this situation, we do not assume that Alice follows the protocol. That is, Alice might not construct d and D as specified by the protocol. Hence, in the following theorem, we assume only that Alice is able to produce values d and D which satisfy the conditions in steps 4, 8, and 9 of Algorithm 7.3.

THEOREM 7.5 *Suppose* $y \equiv x^a \pmod{p}$ *and Bob follows the disavowal proto-col. If* $d \not\equiv x^{e_1}\alpha^{e_2} \pmod{p}$ *and* $D \not\equiv x^{f_1}\alpha^{f_2} \pmod{p}$, *then the probability that* $(d\alpha^{-e_2})^{f_1} \not\equiv (D\alpha^{-f_2})^{e_1} \pmod{p}$ *is* $1 - 1/q$.

PROOF Suppose that the following congruences are satisfied:

$$y \equiv x^a \pmod{p}$$

$$d \not\equiv x^{e_1}\alpha^{e_2} \pmod{p}$$

$$D \not\equiv x^{f_1}\alpha^{f_2} \pmod{p}$$

$$(d\alpha^{-e_2})^{f_1} \equiv (D\alpha^{-f_2})^{e_1} \pmod{p}.$$

We will derive a contradiction.

The consistency check (step 9) can be rewritten in the following form: $D \equiv d_0^{f_1}\alpha^{f_2} \pmod{p}$, where $d_0 = d^{1/e_1}\alpha^{-e_2/e_1} \bmod p$ is a value that depends only on steps 1 to 4 of the protocol.

Applying Theorem 7.3, we conclude that y is a valid signature for d_0 with probability $1 - 1/q$. But we are assuming that y is a valid signature for x. That is, with high probability we have $x^a \equiv d_0^a \pmod{p}$, which implies that $x = d_0$. However, the fact that $d \not\equiv x^{e_1}\alpha^{e_2} \pmod{p}$ means that $x \not\equiv d^{1/e_1}\alpha^{-e_2/e_1}$ $\pmod{p}$. Since $d_0 \equiv d^{1/e_1}\alpha^{-e_2/e_1} \pmod{p}$, we conclude that $x \neq d_0$ and we have a contradiction. Hence, Alice can fool Bob in this way with probability $1/q$. ∎

7.7 Fail-stop Signatures

A fail-stop signature scheme provides enhanced security against the possibility that a very powerful adversary might be able to forge a signature. In the event that Oscar is able to forge Alice's signature on a message, Alice will (with high probability) subsequently be able to prove that Oscar's signature is a forgery.

In this section, we describe a fail-stop signature scheme constructed by van Heyst and Pedersen in 1992. Like the *Lamport Signature Scheme*, this is a one-time signature scheme. The system consists of signing and verification algorithms, as well as a "proof of forgery" algorithm. The description of the signing and verification algorithms of the *van Heyst and Pedersen Signature Scheme* are presented as Cryptosystem 7.9.

Cryptosystem 7.9: *van Heyst and Pedersen Signature Scheme*

Let $p = 2q + 1$ be a prime such that q is prime and the discrete log problem in $\mathbb{Z}_p$ is intractable. Let $\alpha \in \mathbb{Z}_p^*$ be an element of order q. Let $1 \leq a_0 \leq q - 1$ and define $\beta = \alpha^{a_0} \bmod p$. The values p, q, α, β, and a_0 are chosen by a central (trusted) authority. p, q, α, and β are public and will be regarded as fixed. The value of a_0 is kept secret from everyone (even Alice).

Let $\mathcal{P} = \mathbb{Z}_q$ and $\mathcal{A} = \mathbb{Z}_q \times \mathbb{Z}_q$. A key has the form

$$K = (\gamma_1, \gamma_2, a_1, a_2, b_1, b_2),$$

where $a_1, a_2, b_1, b_2 \in \mathbb{Z}_q$,

$$\gamma_1 = \alpha^{a_1} \beta^{a_2} \bmod p \quad \text{and} \quad \gamma_2 = \alpha^{b_1} \beta^{b_2} \bmod p.$$

(γ_1, γ_2) is the public key and (a_1, a_2, b_1, b_2) is the private key.

For $K = (\gamma_1, \gamma_2, a_1, a_2, b_1, b_2)$ and $x \in \mathbb{Z}_q$, define

$$\text{sig}_K(x) = (y_1, y_2),$$

where

$$y_1 = a_1 + x b_1 \bmod q \quad \text{and} \quad y_2 = a_2 + x b_2 \bmod q.$$

For $y = (y_1, y_2) \in \mathbb{Z}_q \times \mathbb{Z}_q$, we have

$$\text{ver}_K(x, y) = \text{true} \Leftrightarrow \gamma_1 \gamma_2^x \equiv \alpha^{y_1} \beta^{y_2} \pmod{p}.$$

It is straightforward to see that a signature produced by Alice will satisfy the verification condition, so let's turn to the security aspects of this scheme and how the fail-stop property works. First we establish some important facts concerning the keys of the scheme. We begin with a definition. Two keys $(\gamma_1, \gamma_2, a_1, a_2, b_1, b_2)$ and $(\gamma_1', \gamma_2', a_1', a_2', b_1', b_2')$ are said to be *equivalent* if $\gamma_1 = \gamma_1'$ and $\gamma_2 = \gamma_2'$. It is easy to see that there are exactly q^2 keys in any equivalence class.

We establish several lemmas.

LEMMA 7.6 *Suppose K and K' are equivalent keys and suppose that $\text{ver}_K(x, y) =$ true. Then $\text{ver}_{K'}(x, y) =$ true.*

PROOF Suppose $K = (\gamma_1, \gamma_2, a_1, a_2, b_1, b_2)$ and $K' = (\gamma_1, \gamma_2, a_1', a_2', b_1', b_2')$, where

$$\gamma_1 = \alpha^{a_1} \beta^{a_2} \bmod p = \alpha^{a_1'} \beta^{a_2'} \bmod p$$

and

$$\gamma_2 = \alpha^{b_1} \beta^{b_2} \bmod p = \alpha^{b_1'} \beta^{b_2'} \bmod p.$$

Suppose x is signed using K, producing the signature $y = (y_1, y_2)$, where

$$y_1 = a_1 + x b_1 \bmod q,$$

$$y_2 = a_2 + x b_2 \bmod q.$$

Now suppose that we compute $\text{ver}_{K'}(x, y)$:

$$\alpha^{y_1} \beta^{y_2} \equiv \alpha^{a_1' + x b_1'} \beta^{a_2' + x b_2'} \pmod{p}$$

$$\equiv \alpha^{a_1'} \beta^{a_2'} (\alpha^{b_1'} \beta^{b_2'})^x \pmod{p}$$

$$\equiv \gamma_1 \gamma_2^x \pmod{p}.$$

Thus, y will also be verified using K'. ∎

LEMMA 7.7 *Suppose K is a key and $y = \text{sig}_K(x)$. Then there are exactly q keys K' equivalent to K such that $y = \text{sig}_{K'}(x)$.*

PROOF Suppose (γ_1, γ_2) is the public key. We want to determine the number of 4-tuples (a_1, a_2, b_1, b_2) such that the following congruences are satisfied:

$$\gamma_1 \equiv \alpha^{a_1} \beta^{a_2} \pmod{p}$$

$$\gamma_2 \equiv \alpha^{b_1} \beta^{b_2} \pmod{p}$$

$$y_1 \equiv a_1 + x b_1 \pmod{q}$$

$$y_2 \equiv a_2 + x b_2 \pmod{q}.$$

Since α generates G, there exist unique exponents $c_1, c_2, a_0 \in \mathbb{Z}_q$ such that

$$\gamma_1 \equiv \alpha^{c_1} \pmod{p},$$

$$\gamma_2 \equiv \alpha^{c_2} \pmod{p}, \quad \text{and}$$

$$\beta \equiv \alpha^{a_0} \pmod{p}.$$

Hence, it is necessary and sufficient that the following system of congruences be satisfied:

$$c_1 \equiv a_1 + a_0 a_2 \pmod{q}$$
$$c_2 \equiv b_1 + a_0 b_2 \pmod{q}$$
$$y_1 \equiv a_1 + x b_1 \pmod{q}$$
$$y_2 \equiv a_2 + x b_2 \pmod{q}.$$

This system can be written as a matrix equation in $\mathbb{Z}_q$, as follows:

$$\begin{pmatrix} 1 & a_0 & 0 & 0 \\ 0 & 0 & 1 & a_0 \\ 1 & 0 & x & 0 \\ 0 & 1 & 0 & x \end{pmatrix} \begin{pmatrix} a_1 \\ a_2 \\ b_1 \\ b_2 \end{pmatrix} = \begin{pmatrix} c_1 \\ c_2 \\ y_1 \\ y_2 \end{pmatrix}.$$

Now, the coefficient matrix of this system can be seen to have rank[1] three: Clearly, the rank is at least three because rows 1, 2 and 4 are linearly independent over $\mathbb{Z}_q$. The rank is at most three because

$$r_1 + x r_2 - r_3 - a_0 r_4 = (0, 0, 0, 0),$$

where r_i denotes the ith row of the matrix.

This system of equations has at least one solution, obtained by using the key K. Since the rank of the coefficient matrix is three, it follows that the dimension of the solution space is $4 - 3 = 1$, and hence there are exactly q solutions. The result follows.   ∎

By similar reasoning, the following result can be proved. We omit the proof.

LEMMA 7.8   *Suppose K is a key, $y = \operatorname{sig}_K(x)$, and $\operatorname{ver}_K(x', y') = \text{true}$, where $x' \neq x$. Then there is at most one key K' equivalent to K such that $y = \operatorname{sig}_{K'}(x)$ and $y' = \operatorname{sig}_{K'}(x')$.*

Let's interpret what the preceding two lemmas say about the security of the scheme. Given that y is a valid signature for message x, there are q possible keys that would have signed x with y. But for any message $x' \neq x$, these q keys will produce q different signatures on x'. Thus, the following theorem results.

THEOREM 7.9   *Given that $\operatorname{sig}_K(x) = y$ and $x' \neq x$, Oscar can compute $\operatorname{sig}_K(x')$ with probablity $1/q$.*

---

[1]The *rank* of a matrix is the maximum number of linearly independent rows it contains.

Note that this theorem does not depend on the computational power of Oscar: the stated level of security is obtained because Oscar cannot tell which of q possible keys is being used by Alice. So the security is unconditional.

We now go on to look at the fail-stop concept. What we have said so far is that, given a signature y on message x, Oscar cannot compute Alice's signature y' on a different message x'. It is still conceivable that Oscar can compute a forged signature $y'' \neq \text{sig}_K(x')$ which will still be verified. However, if Alice is given a valid forged signature, then with probability $1 - 1/q$ she can produce a proof of forgery. The proof of forgery is the value $a_0 = \log_\alpha \beta$, which is known only to the central authority.

We assume that Alice possesses a pair (x', y'') such that $\text{ver}_K(x', y'') = \text{true}$ and $y'' \neq \text{sig}_K(x')$. That is,

$$\gamma_1 \gamma_2^{x'} \equiv \alpha^{y_1''} \beta^{y_2''} \pmod{p},$$

where $y'' = (y_1'', y_2'')$. Now, Alice can compute her own signature on x', namely $y' = (y_1', y_2')$, and it will be the case that

$$\gamma_1 \gamma_2^{x'} \equiv \alpha^{y_1'} \beta^{y_2'} \pmod{p}.$$

Hence,

$$\alpha^{y_1''} \beta^{y_2''} \equiv \alpha^{y_1'} \beta^{y_2'} \pmod{p}.$$

Writing $\beta = \alpha^{a_0} \bmod p$, we have that

$$\alpha^{y_1'' + a_0 y_2''} \equiv \alpha^{y_1' + a_0 y_2'} \pmod{p},$$

or

$$y_1'' + a_0 y_2'' \equiv y_1' + a_0 y_2' \pmod{q}.$$

This simplifies to give

$$y_1'' - y_1' \equiv a_0(y_2' - y_2'') \pmod{q}.$$

Now, $y_2' \not\equiv y_2'' \pmod{q}$ since y' is a forgery. Hence, $(y_2' - y_2'')^{-1} \bmod q$ exists, and

$$a_0 = \log_\alpha \beta = (y_1'' - y_1')(y_2' - y_2'')^{-1} \bmod q.$$

Of course, by accepting such a proof of forgery, we assume that Alice cannot compute the discrete logarithm $\log_\alpha \beta$ by herself. This is a computational assumption.

Finally, we remark that the scheme is a one-time scheme since Alice's key K can easily be computed if two messages are signed using K.

We close with an example illustrating how Alice can produce a proof of forgery.

Example 7.9 Suppose $p = 3467 = 2 \times 1733 + 1$. The element $\alpha = 4$ has order 1733 in $\mathbb{Z}_{3467}^*$. Suppose that $a_0 = 1567$, so

$$\beta = 4^{1567} \bmod 3467 = 514.$$

(Recall that Alice knows the values of α and β, but not a_0.) Suppose Alice forms her key using $a_1 = 888$, $a_2 = 1024$, $b_1 = 786$ and $b_2 = 999$, so

$$\gamma_1 = 4^{888} 514^{1024} \bmod 3467 = 3405$$

and

$$\gamma_2 = 4^{786} 514^{999} \bmod 3467 = 2281.$$

Now, suppose Alice is presented with the forged signature $(822, 55)$ on the message 3383. This is a valid signature because the verification condition is satisfied:

$$3405 \times 2281^{3383} \equiv 2282 \ (\text{mod } 3467)$$

and

$$4^{822} 514^{55} \equiv 2282 \ (\text{mod } 3467).$$

On the other hand, this is not the signature Alice would have constructed. Alice can compute her own signature to be

$$(888 + 3383 \times 786 \bmod 1733, 1024 + 3383 \times 999 \bmod 1733) = (1504, 1291).$$

Then, she proceeds to calculate the secret discrete logarithm

$$a_0 = (822 - 1504)(1291 - 55)^{-1} \bmod 1733 = 1567.$$

This is the proof of forgery. □

7.8 Notes and References

For a nice survey of signature schemes, we recommend Mitchell, Piper and Wild [149]. This paper also contains the two methods of forging ElGamal signatures that we presented in Section 7.3. Pedersen [162] is a more recent survey that is also worth reading.

The *ElGamal Signature Scheme* was presented by ElGamal [70], and the *Schnorr Signature Scheme* is due to Schnorr [187]. Another popular scheme, which we have not discussed in this book, is the *Fiat-Shamir Signature Scheme* [86].

The *Digital Signature Algorithm* was first published by NIST in August 1991, and it was adopted as FIPS 186 in December 1994 [79]. There is a lengthy discussion of *DSA* and some controversies surrounding it in the July 1992 issue of the *Communications of the ACM*; for a response by NIST to some of the questions raised, see [200]. FIPS 186-2 [80] is a revised version of the standard, which now includes the *RSA Signature Scheme* as well as the *Elliptic Curve Digital Signature Algorithm*. A complete description of the *ECDSA* is found in Johnson, Menezes and Vanstone [107].

The *Lamport Signature Scheme* is described in the 1976 paper by Diffie and Hellman [67]. A more efficient modification, due to Lamport and (independently) Bos and Chaum, is described in [34]. A more general treatment of the construction of signature schemes from arbitrary one-way functions is given by Bleichenbacher and Maurer [28].

Full Domain Hash is due to Bellare and Rogaway [14, 16]. The paper [16] also includes a more efficient variant, known as the *Probabilistic Signature Scheme* (*PSS*). Provably secure ElGamal-type schemes have also been studied; see, for example, Pointcheval and Stern [169].

The undeniable signature scheme presented in Section 7.6 is due to Chaum and van Antwerpen [50]. The fail-stop signature scheme from Section 7.7 is due to van Heyst and Pedersen [210]; see Pfitzmann [165] for an expanded treatment of this topic.

Some of the Exercises point out some security problems with ElGamal type schemes if the "k" values are reused or generated in a predictable fashion. There are now several works that pursue this theme; see, for example, Bellare, Goldwasser and Micciancio [17] and Nguyen and Shparlinski [154].

Exercises

7.1 Suppose Alice is using the *ElGamal Signature Scheme* with $p = 31847$, $\alpha = 5$ and $\beta = 25703$. Compute the values of k and a (without solving an instance of the Discrete Logarithm problem), given the signature $(23972, 31396)$ for the message $x = 8990$ and the signature $(23972, 20481)$ for the message $x = 31415$.

7.2 Suppose I implement the *ElGamal Signature Scheme* with $p = 31847$, $\alpha = 5$ and $\beta = 26379$. Write a computer program which does the following:

(a) Verify the signature $(20679, 11082)$ on the message $x = 20543$.

(b) Determine my private key, a, by solving an instance of the Discrete Logarithm problem.

(c) Then determine the random value k used in signing the message x, without solving an instance of the Discrete Logarithm problem.

7.3 Suppose that Alice is using the *ElGamal Signature Scheme*. In order to save time in generating the random numbers k that are used to sign messages, Alice chooses an initial random value k_0, and then signs the ith message using the value $k_i = k_0 + 2i \bmod p$ (therefore $k_i = k_{i-1} + 2 \bmod p$ for all $i \geq 1$).

(a) Suppose that Bob observes two consecutive signed messages, say $(x_i, \text{sig}(x_i))$ and $(x_{i+1}, \text{sig}(x_{i+1}))$. Describe how Bob can easily compute Alice's secret key, a, given this information, without solving an instance of the Discrete

Logarithm problem. (Note that the value of i does not have to be known for the attack to succeed.)

(b) Suppose that the parameters of the scheme are $p = 28703$, $\alpha = 5$ and $\beta = 11339$, and the two messages observed by Bob are

$$x_i = 12000 \qquad \text{sig}(x_i) = (26530, 19862)$$
$$x_{i+1} = 24567 \qquad \text{sig}(x_{i+1}) = (3081, 7604).$$

Find the value of a using the attack you described in part (a).

7.4 (a) Prove that the second method of forgery on the *ElGamal Signature Scheme*, described in Section 7.3, also yields a signature that satisfies the verification condition.

(b) Suppose Alice is using the *ElGamal Signature Scheme* as implemented in Example 7.1: $p = 467$, $\alpha = 2$ and $\beta = 132$. Suppose Alice has signed the message $x = 100$ with the signature $(29, 51)$. Compute the forged signature that Oscar can then form by using $h = 102$, $i = 45$ and $j = 293$. Check that the resulting signature satisfies the verification condition.

7.5 (a) A signature in the *ElGamal Signature Scheme* or the *DSA* is not allowed to have $\delta = 0$. Show that if a message were signed with a "signature" in which $\delta = 0$, then it would be easy for an adversary to compute the secret key, a.

(b) A signature in the *DSA* is not allowed to have $\gamma = 0$. Show that if a "signature" in which $\gamma = 0$ is known, then the value of k used in that "signature" can be determined. Given that value of k, show that it is now possible to forge a "signature" (with $\gamma = 0$) for any desired message (i.e., a selective forgery can be carried out).

(c) Evaluate the consequences of allowing a signature in the *ECDSA* to have $r = 0$ or $s = 0$.

7.6 Here is a variation of the *ElGamal Signature Scheme*. The key is constructed in a similar manner as before: Alice chooses $\alpha \in \mathbb{Z}_p^*$ to be a primitive element, $0 \leq a \leq p - 2$ where $\gcd(a, p - 1) = 1$, and $\beta = \alpha^a \bmod p$. The key $K = (\alpha, a, \beta)$, where α and β are the public key and a is the private key. Let $x \in \mathbb{Z}_p$ be a message to be signed. Alice computes the signature $\text{sig}(x) = (\gamma, \delta)$, where

$$\gamma = \alpha^k \bmod p$$

and

$$\delta = (x - k\gamma)a^{-1} \bmod (p - 1).$$

The only difference from the original *ElGamal Signature Scheme* is in the computation of δ. Answer the following questions concerning this modified scheme.

(a) Describe how a signature (γ, δ) on a message x would be verified using Alice's public key.

(b) Describe a computational advantage of the modified scheme over the original scheme.

(c) Briefly compare the security of the original and modified scheme.

7.7 Suppose Alice uses the *DSA* with $q = 101$, $p = 7879$, $\alpha = 170$, $a = 75$ and $\beta = 4567$, as in Example 7.4. Determine Alice's signature on the message $x = 52$ using the random value $k = 49$, and show how the resulting signature is verified.

7.8 We showed that using the same value k to sign two messages in the *ElGamal Signature Scheme* allows the scheme to be broken (i.e., an adversary can determine the secret key without solving an instance of the Discrete Logarithm problem). Show

how similar attacks can be carried out for the *Schnorr Signature Scheme*, the *DSA* and the *ECDSA*.

7.9 Suppose that $x_0 \in \{0, 1\}^*$ is a bitstring such that SHA-1$(x_0) = 0\,0 \cdots 0$. Therefore, when used in *DSA* or *ECDSA*, we have that SHA-1$(x_0) \equiv 0 \pmod q$.

    (a) Show how it is possible to forge a *DSA* signature for the message x_0.

        HINT   Let $\delta = \gamma$, where γ is chosen appropriately.

    (b) Show how it is possible to forge an *ECDSA* signature for the message x_0.

7.10   (a) We describe a potential attack against the *DSA*. Suppose that x is given, let $z = (\text{SHA-1}(x))^{-1} \bmod q$, and let $\epsilon = \beta^z \bmod p$. Now suppose it is possible to find $\gamma, \lambda \in \mathbb{Z}_q{}^*$ such that

$$\left((\alpha \epsilon^\gamma)^{\lambda^{-1} \bmod q} \right) \bmod p \bmod q = \gamma.$$

        Define $\delta = \lambda \, \text{SHA-1}(x) \bmod q$. Prove that (γ, δ) is a valid signature for x.

    (b) Describe a similar (possible) attack against the *ECDSA*.

7.11 In a verification of a signature constructed using the *ElGamal Signature Scheme* (or many of its variants), it is necessary to compute a value of the form $\alpha^c \beta^d$. If c and d are random ℓ-bit exponents, then a straightforward use of the SQUARE-AND-MULTIPLY algorithm would require (on average) $\ell/2$ multiplications and ℓ squarings to compute each of α^c and β^d. The purpose of this exercise is to show that the product $\alpha^c \beta^d$ can be computed much more efficiently.

    (a) Suppose that c and d are represented in binary, as in Algorithm 5.5. Suppose also that the product $\alpha\beta$ is precomputed. Describe a modification of Algorithm 5.5, in which at most one multiplication is performed in each iteration of the algorithm.

    (b) Suppose that $c = 26$ and $d = 17$. Show how your algorithm would compute $\alpha^c \beta^d$, i.e., what are the values of the exponents i and j at the end of each iteration of your algorithm (where $z = \alpha^i \beta^j$).

    (c) Exlpain why, on average, this algorithm requires ℓ squarings and $3\ell/4$ multiplications to compute $\alpha^c \beta^d$, if c and d are randomly chosen ℓ-bit integers.

    (d) Estimate the average speedup achieved, as compared to using the original SQUARE-AND-MULTIPLY algorithm to compute α^c and β^d separately, assuming that a squaring operation takes roughly the same time as a multiplication operation.

7.12 Prove that a correctly constructed signature in the *ECDSA* will satisfy the verification condition.

7.13 Let E denote the elliptic curve $y^2 \equiv x^3 + x + 26 \bmod 127$. It can be shown that $\#E = 131$, which is a prime number. Therefore any non-identity element in E is a generator for $(E, +)$. Suppose the *ECDSA* is implemented in E, with $A = (2, 6)$ and $m = 54$.

    (a) Compute the public key $B = mA$.

    (b) Compute the signature on a message x if SHA-1$(x) = 10$, when $k = 75$.

    (c) Show the computations used to verify the signature constructed in part (b).

7.14 In the *Lamport Signature Scheme*, suppose that two k-tuples, x and x', were signed by Alice using the same key. Let ℓ denote the number of coordinates in which x and x' differ, i.e.,

$$\ell = |\{i : x_i \neq x'_i\}|.$$

Show that Oscar can now sign $2^\ell - 2$ new messages.

7.15 Suppose Alice is using the *Chaum-van Antwerpen Signature Scheme* as in Example 7.7. That is, $p = 467$, $\alpha = 4$, $a = 101$ and $\beta = 449$. Suppose Alice is presented with a signature $y = 25$ on the message $x = 157$ and she wishes to prove it is a forgery. Suppose Bob's random numbers are $e_1 = 46$, $e_2 = 123$, $f_1 = 198$ and $f_2 = 11$ in the disavowal protocol. Compute Bob's challenges, c and d, and Alice's responses, C and D, and show that Bob's consistency check will succeed.

7.16 Prove that each equivalence class of keys in the *Pedersen-van Heyst Signature Scheme* contains q^2 keys.

7.17 Suppose Alice is using the *Pedersen-van Heyst Signature Scheme*, where $p = 3467$, $\alpha = 4$, $a_0 = 1567$ and $\beta = 514$ (of course, the value of a_0 is not known to Alice).

    (a) Using the fact that $a_0 = 1567$, determine all possible keys

$$K = (\gamma_1, \gamma_2, a_1, a_2, b_1, b_2)$$

such that $\text{sig}_K(42) = (1118, 1449)$.

    (b) Suppose that $\text{sig}_K(42) = (1118, 1449)$ and $\text{sig}_K(969) = (899, 471)$. Without using the fact that $a_0 = 1567$, determine the value of K (this shows that the scheme is a one-time scheme).

7.18 Suppose Alice is using the *Pedersen-van Heyst Signature Scheme* with $p = 5087$, $\alpha = 25$ and $\beta = 1866$. Suppose the key is

$$K = (5065, 5076, 144, 874, 1873, 2345).$$

Now, suppose Alice finds the signature $(2219, 458)$ has been forged on the message 4785.

    (a) Prove that this forgery satisfies the verification condition, so it is a valid signature.

    (b) Show how Alice will compute the proof of forgery, a_0, given this forged signature.

Further Reading

There are now many books and monographs on various aspects of cryptography. Here are a few (mostly recent) cryptographic works of fairly general interest:

- Buchmann [41]
- Garrett [89]
- Koblitz [120]
- Menezes, Van Oorschot and Vanstone [145]
- Mollin [150]
- Schneier [185]
- Simmons [196]
- Singh [197]
- Stallings [203]

The International Association for Cryptologic Research (or IACR) sponsors several annual cryptology conferences: CRYPTO, EUROCRYPT, ASIACRYPT and (starting in 2002) Fast Software Encryption.

CRYPTO has been held since 1981 in Santa Barbara. The proceedings of CRYPTO have been published annually since 1982:

CRYPTO '82 [49]	CRYPTO '83 [47]
CRYPTO '84 [27]	CRYPTO '85 [218]
CRYPTO '86 [158]	CRYPTO '87 [171]
CRYPTO '88 [94]	CRYPTO '89 [36]
CRYPTO '90 [144]	CRYPTO '91 [82]
CRYPTO '92 [40]	CRYPTO '93 [205]
CRYPTO '94 [65]	CRYPTO '95 [52]
CRYPTO '96 [121]	CRYPTO '97 [110]
CRYPTO '98 [124]	CRYPTO '99 [216]
CRYPTO 2000 [10]	CRYPTO 2001 [114]

EUROCRYPT has been held annually since 1982, and except for 1983 and 1986, its proceedings have been published, as follows:

EUROCRYPT '82 [18]   EUROCRYPT '84 [19]
EUROCRYPT '85 [166]   EUROCRYPT '87 [48]
EUROCRYPT '88 [100]   EUROCRYPT '89 [176]
EUROCRYPT '90 [57]   EUROCRYPT '91 [59]
EUROCRYPT '92 [183]   EUROCRYPT '93 [103]
EUROCRYPT '94 [64]   EUROCRYPT '95 [99]
EUROCRYPT '96 [139]   EUROCRYPT '97 [88]
EUROCRYPT '98 [157]   EUROCRYPT '99 [204]
EUROCRYPT 2000 [174] EUROCRYPT 2001 [164]

ASIACRYPT (formerly AUSCRYPT) has been held since 1991. Its conference proceedings have also been published:

AUSCRYPT '90 [188]   ASIACRYPT '91 [106]
AUSCRYPT '92 [189]   ASIACRYPT '94 [167]
ASIACRYPT '96 [115]   ASIACRYPT '98 [160]
ASIACRYPT '99 [127]   ASIACRYPT 2000 [161]
ASIACRYPT 2001 [35]

Fast Software Encryption has been held since 1993. Its published conference proceedings are as follows:

FSE '93 [2]   FSE '94 [172]
FSE '96 [97]   FSE '97 [21]
FSE '98 [211]   FSE '99 [118]
FSE 2000 [186]

Bibliography

[1] W. ALEXI, B. CHOR, O. GOLDREICH AND C. P. SCHNORR. RSA and Rabin functions: certain parts are as hard as the whole. *SIAM Journal on Computing*, **17** (1988), 194–209.

[2] R. ANDERSON (ED.) *Fast Software Encryption. Lecture Notes in Computer Science*, vol. 809, Springer-Verlag, 1994.

[3] H. ANTON. *Elementary Linear Algebra, Eighth Edition*. John Wiley and Sons, 2000.

[4] E. BACH AND J. SHALLIT. *Algorithmic Number Theory, Volume 1: Efficient Algorithms*. The MIT Press, 1996.

[5] F. L. BAUER. *Decrypted Secrets, Methods and Maxims of Cryptology, Second Edition*. Springer-Verlag, 2000.

[6] P. BEAUCHEMIN AND G. BRASSARD. A generalization of Hellman's extension to Shannon's approach to cryptography. *Journal of Cryptology*, **1** (1988), 129–131.

[7] P. BEAUCHEMIN, G. BRASSARD, C. CRÉPEAU, C. GOUTIER AND C. POMERANCE. The generation of random numbers that are probably prime. *Journal of Cryptology*, **1** (1988), 53–64.

[8] H. BEKER AND F. PIPER. *Cipher Systems, The Protection of Communications*. John Wiley and Sons, 1982.

[9] M. BELLARE. Practice-oriented provable-security. In *Lectures on Data Security*, pages 1–15. Springer-Verlag, 1999.

[10] M. BELLARE (ED.) *Advances in Cryptology – CRYPTO 2000 Proceedings. Lecture Notes in Computer Science*, vol. 1880, Springer-Verlag, 2000.

[11] M. BELLARE, R. CANETTI AND H. KRAWCZYK. Keying hash functions for message authentication. *Lecture Notes in Computer Science*, **1109** (1996), 1–15. (Advances in Cryptology – CRYPTO '96.)

[12] M. BELLARE, R. GUERIN AND P. ROGAWAY. XOR MACs: new methods for message authentication using finite pseudorandom functions. *Lecture Notes in Computer Science*, **963** (1995), 15–28. (Advances in Cryptology

– CRYPTO '95.)

[13] M. BELLARE, J. KILIAN AND P. ROGAWAY. The security of the cipher block chaining message authentication code. *Journal of Computer and System Sciences*, **61** (2000), 362–399.

[14] M. BELLARE AND P. ROGAWAY. Random oracles are practical: a paradigm for designing efficient protocols. In *First ACM Conference on Computer and Communications Security*, pages 62–73. ACM Press, 1993.

[15] M. BELLARE AND P. ROGAWAY. Optimal asymmetric encryption. *Lecture Notes in Computer Science*, **950** (1995), 92–111. (Advances in Cryptology – EUROCRYPT '94.)

[16] M. BELLARE AND P. ROGAWAY. The exact security of digital signatures: how to sign with RSA and Rabin. *Lecture Notes in Computer Science*, **1070** (1996), 399–416. (Advances in Cryptology – EUROCRYPT '96.)

[17] M. BELLARE, S. GOLDWASSER AND D. MICCIANCIO. "Pseudo-random" number generation within cryptographic algorithms: the DSS case. *Lecture Notes in Computer Science*, **1294** (1997), 277–292. (Advances in Cryptology – CRYPTO '97.)

[18] T. BETH (ED.) *Cryptography Proceedings, 1982. Lecture Notes in Computer Science*, vol. 149, Springer-Verlag, 1983.

[19] T. BETH, N. COT AND I. INGEMARSSON (EDS.) *Advances in Cryptology: Proceedings of EUROCRYPT '84. Lecture Notes in Computer Science*, vol. 209, Springer-Verlag, 1985.

[20] A. BEUTELSPACHER. *Cryptology*. Mathematical Association of America, 1994.

[21] E. BIHAM (ED.) *Fast Software Encryption. Lecture Notes in Computer Science*, vol. 1267, Springer-Verlag, 1997. (FSE '97.)

[22] E. BIHAM AND A. SHAMIR. Differential cryptanalysis of DES-like cryptosystems. *Journal of Cryptology*, **4** (1991), 3–72.

[23] E. BIHAM AND A. SHAMIR. *Differential Cryptanalysis of the Data Encryption Standard*. Springer-Verlag, 1993.

[24] E. BIHAM AND A. SHAMIR. Differential cryptanalysis of the full 16-round DES. *Lecture Notes in Computer Science*, **740** (1993), 494–502. (Advances in Cryptology – CRYPTO '92.)

[25] J. BLACK, S. HALEVI, H. KRAWCZYK, T. KROVETZ AND P. ROGAWAY. UMAC: fast message authentication via optimized universal hash functions. *Lecture Notes in Computer Science*, **1666** (1999), 234–251. (Advances in Cryptology – CRYPTO '99.)

[26] I. BLAKE, G. SEROUSSI AND N. SMART. *Elliptic Curves in Cryptography*. Cambridge University Press, 1999.

[27] G. R. BLAKLEY AND D. CHAUM (EDS.) *Advances in Cryptology: Pro-*

ceedings of CRYPTO '84. Lecture Notes in Computer Science, vol. 196, Springer-Verlag, 1985.

[28] D. BLEICHENBACHER AND U. M. MAURER. Directed acyclic graphs, one-way functions and digital signatures. *Lecture Notes in Computer Science*, **839** (1994), 75–82. (Advances in Cryptology – CRYPTO '94.)

[29] M. BLUM AND S. GOLDWASSER. An efficient probabilistic public-key cryptosystem that hides all partial information. *Lecture Notes in Computer Science*, **196** (1985), 289–302. (Advances in Cryptology – CRYPTO '84.)

[30] D. BONEH. The decision Diffie-Hellman problem. *Lecture Notes in Computer Science*, **1423** (1998), 48–63. (Proceedings of the Third Algorithmic Number Theory Symposium.)

[31] D. BONEH. Twenty years of attacks on the RSA cryptosystem. *Notices of the American Mathematical Society*, **46** (1999), 203–213.

[32] D. BONEH. Simplified OAEP for the RSA and Rabin functions. *Lecture Notes in Computer Science*, **2139** (2001), 275–291. (Advances in Cryptology – CRYPTO 2001.)

[33] D. BONEH AND G. DURFEE. Cryptanalysis of RSA with private key d less than $N^{0.292}$. *IEEE Transactions on Information Theory*, **46** (2000), 1339–1349.

[34] J. N. E. BOS AND D. CHAUM. Provably unforgeable signatures. *Lecture Notes in Computer Science*, **740** (1993), 1–14. (Advances in Cryptology – CRYPTO '92.)

[35] C. BOYD, (ED.) *Advances in Cryptology – ASIACRYPT 2001 Proceedings. Lecture Notes in Computer Science*, vol. 2248, Springer-Verlag, 2001.

[36] G. BRASSARD (ED.) *Advances in Cryptology – CRYPTO '89 Proceedings. Lecture Notes in Computer Science*, vol. 435, Springer-Verlag, 1990.

[37] G. BRASSARD AND P. BRATLEY. *Fundamentals of Algorithmics*. Prentice Hall, 1995.

[38] R. P. BRENT. An improved Monte Carlo factorization method. *BIT*, **20** (1980), 176–184.

[39] D. M. BRESSOUD AND S. WAGON. *A Course in Computational Number Theory*. Springer-Verlag, 2000.

[40] E. F. BRICKELL (ED.) *Advances in Cryptology – CRYPTO '92 Proceedings. Lecture Notes in Computer Science*, vol. 740, Springer-Verlag, 1993.

[41] J. A. BUCHMANN. *Introduction to Cryptography*. Springer-Verlag, 2001.

[42] J. L. CARTER AND M. N. WEGMAN. Universal classes of hash functions. *Journal of Computer and System Sciences*, **18** (1979), 143–154.

[43] S. CAVALLAR, B. DODSON, A. LENSTRA, W. LIOEN, P. MONTGOMERY, B. MURPHY, H. TE RIELE, K. AARDAL, J. GILCHRIST, G. GUILLERM, P. LEYLAND, J. MARCHAND, F. MORAIN, A. MUFFETT,

C. PUTNAM, C. PUTNAM AND P. ZIMMERMANN. Factorization of a 512-bit RSA modulus. *Lecture Notes in Computer Science*, **1807** (2000), 1–18 (Advances in Cryptology – EUROCRYPT 2000.)

[44] F. CHABAUD AND A. JOUX. Differential collisions in SHA-0. *Lecture Notes in Computer Science*, **1462** (1998), 56–71. (Advances in Cryptology – CRYPTO '98.)

[45] F. CHABAUD AND S. VAUDENAY. Links between differential and linear cryptanalysis. *Lecture Notes in Computer Science*, **950** (1995), 356–365. (Advances in Cryptology – EUROCRYPT '94.)

[46] M. CHATEAUNEUF, A. C. H. LING AND D. R. STINSON. Slope packings and coverings, and generic algorithms for the discrete logarithm problem. Preprint.

[47] D. CHAUM (ED.) *Advances in Cryptology: Proceedings of CRYPTO '83.* Plenum Press, 1984.

[48] D. CHAUM AND W. L. PRICE (EDS.) *Advances in Cryptology – EURO-CRYPT '87 Proceedings. Lecture Notes in Computer Science*, vol. 304, Springer-Verlag, 1988.

[49] D. CHAUM, R. L. RIVEST AND A. T. SHERMAN (EDS.) *Advances in Cryptology: Proceedings of CRYPTO '82.* Plenum Press, 1983.

[50] D. CHAUM AND H. VAN ANTWERPEN. Undeniable signatures. *Lecture Notes in Computer Science*, **435** (1990), 212–216. (Advances in Cryptology – CRYPTO '89.)

[51] D. COPPERSMITH. The data encryption standard (DES) and its strength against attacks. *IBM Journal of Research and Development*, **38** (1994), 243–250.

[52] D. COPPERSMITH (ED.) *Advances in Cryptology – CRYPTO '95 Proceedings. Lecture Notes in Computer Science*, vol. 963, Springer-Verlag, 1995.

[53] J. DAEMEN, L. KNUDSEN AND V. RIJMEN. The block cipher Square. *Lecture Notes in Computer Science*, **1267** (1997), 149–165. (Fast Software Encryption '97.)

[54] J. DAEMEN AND V. RIJMEN. The block cipher Rijndael. *Lecture Notes in Computer Science*, **1820** (2000), 288-296. (Smart Card Research and Applications.)

[55] J. DAEMEN AND V. RIJMEN. *The design of Rijndael. AES - The Advanced Encryption Standard.* Springer-Verlag, 2002.

[56] I. B. DAMGÅRD. A design principle for hash functions. *Lecture Notes in Computer Science*, **435** (1990), 416–427. (Advances in Cryptology – CRYPTO '89.)

[57] I. B. DAMGÅRD (ED.) *Advances in Cryptology – EUROCRYPT '90 Proceedings. Lecture Notes in Computer Science*, vol. 473, Springer-Verlag,

1991.

[58] I. DAMGÅRD, P. LANDROCK AND C. POMERANCE. Average case error estimates for the strong probable prime test. *Mathematics of Computation*, **61** (1993), 177–194.

[59] D. W. DAVIES (ED.) *Advances in Cryptology – EUROCRYPT '91 Proceedings. Lecture Notes in Computer Science*, vol. 547, Springer-Verlag, 1991.

[60] J. M. DELAURENTIS. A further weakness in the common modulus protocol for the RSA cryptosystem. *Cryptologia*, **8** (1984), 253–259.

[61] B. DEN BOER AND A. BOSSALAERS. An attack on the last two rounds of MD4. *Lecture Notes in Computer Science*, **576** (1994), 194–203. (Advances in Cryptology – CRYPTO '91.)

[62] B. DEN BOER AND A. BOSSALAERS. Collisions for the compression function of MD5. *Lecture Notes in Computer Science*, **765** (1992), 293–304. (Advances in Cryptology – EUROCRYPT '93.)

[63] D. E. R. DENNING. *Cryptography and Data Security*. Addison-Wesley, 1982.

[64] A. DE SANTIS (ED.) *Advances in Cryptology – EUROCRYPT '94 Proceedings. Lecture Notes in Computer Science*, vol. 950, Springer-Verlag, 1995.

[65] Y. G. DESMEDT (ED.) *Advances in Cryptology – CRYPTO '94 Proceedings. Lecture Notes in Computer Science*, vol. 839, Springer-Verlag, 1994.

[66] W. DIFFIE. The first ten years of public-key cryptography. In *Contemporary Cryptology, The Science of Information Integrity*, pages 135–175. IEEE Press, 1992.

[67] W. DIFFIE AND M. E. HELLMAN. Multiuser cryptographic techniques. *Federal Information Processing Standard Conference Proceedings*, **45** (1976), 109–112.

[68] W. DIFFIE AND M. E. HELLMAN. New directions in cryptography. *IEEE Transactions on Information Theory*, **22** (1976), 644–654.

[69] H. DOBBERTIN. Cryptanalysis of MD4. *Journal of Cryptology*, **11** (1998), 253–271.

[70] T. ELGAMAL. A public key cryptosystem and a signature scheme based on discrete logarithms. *IEEE Transactions on Information Theory*, **31** (1985), 469–472.

[71] A. ENGE. *Elliptic Curves and their Applications to Cryptography: an Introduction*. Kluwer Academic Publishers, 1999.

[72] *Data Encryption Standard (DES)*. Federal Information Processing Standard Publication 46, 1977.

[73] *DES Modes of Operation*. Federal Information Processing Standard Publi-

cation 81, 1980.

[74] *Guidelines for Implementing and Using the NBS Data Encryption Standard.* Federal Information Processing Standard Publication 74, 1981.

[75] *Computer Data Authentication.* Federal Information Processing Standard Publication 113, 1985.

[76] *Secure Hash Standard.* Federal Information Processing Standard Publication 180, 1993.

[77] *Secure Hash Standard.* Federal Information Processing Standard Publication 180-1, 1995.

[78] *Secure Hash Standard.* Federal Information Processing Standard Publication 180-2 (Draft), 2001.

[79] *Digital Signature Standard.* Federal Information Processing Standard Publication 186, 1994.

[80] *Digital Signature Standard.* Federal Information Processing Standard Publication 186-2, 2000.

[81] *Advanced Encryption Standard.* Federal Information Processing Standard Publication 197, 2001.

[82] J. FEIGENBAUM (ED.) *Advances in Cryptology – CRYPTO '91 Proceedings. Lecture Notes in Computer Science*, vol. 576, Springer-Verlag, 1992.

[83] H. FEISTEL. Cryptography and computer privacy. *Scientific American*, **228**(5) (1973), 15–23.

[84] N. FERGUSON, J. KELSEY, S. LUCKS, B. SCHNEIER, M. STAY, D. WAGNER AND D. WHITING. Improved cryptanalysis of Rijndael. *Lecture Notes in Computer Science*, **1978** (2001), 1213–230. (Fast Software Encryption 2000.)

[85] N. FERGUSON, R. SCHROEPPEL AND D. WHITING. A simple algebraic representation of Rijndael. *Lecture Notes in Computer Science*, **2259** (2001), 103–111. (Selected Areas in Cryptography 2001.)

[86] A. FIAT AND A. SHAMIR. How to prove yourself: practical solutions to identification and signature problems. *Lecture Notes in Computer Science*, **263** (1987), 186–194. (Advances in Cryptology – CRYPTO '86.)

[87] E. FUJISAKI, T. OKAMOTO, D. POINTCHEVAL AND J. STERN. RSA-OAEP is secure under the RSA assumption. *Lecture Notes in Computer Science*, **2139** (2001), 260–274. (Advances in Cryptology – CRYPTO 2001.)

[88] W. FUMY (ED.) *Advances in Cryptology – EUROCRYPT '97 Proceedings. Lecture Notes in Computer Science*, vol. 1233, Springer-Verlag, 1997.

[89] P. GARRETT. *Making, Breaking Codes: An Introduction to Cryptology.* Prentice-Hall, 2001.

[90] J. VON ZUR GATHEN AND J. GERHARD. *Modern Computer Algebra*. Cambridge University Press, 1999.

[91] J. K. GIBSON. Discrete logarithm hash function that is collision free and one way. *IEE Proceedings-E*, **138** (1991), 407–410.

[92] E. N. GILBERT, F. J. MACWILLIAMS AND N. J. A. SLOANE. Codes which detect deception. *Bell Systems Technical Journal*, **53** (1974), 405–424.

[93] C. M. GOLDIE AND R. G. E. PINCH. *Communication Theory*. Cambridge University Press, 1991.

[94] S. GOLDWASSER (ED.) *Advances in Cryptology – CRYPTO '88 Proceedings. Lecture Notes in Computer Science*, vol. 403, Springer-Verlag, 1990.

[95] S. GOLDWASSER AND S. MICALI. Probabilistic encryption. *Journal of Computer and Systems Science*, **28** (1984), 270–299.

[96] S. GOLDWASSER, S. MICALI AND P. TONG. Why and how to establish a common code on a public network. In *23rd Annual Symposium on the Foundations of Computer Science*, pages 134–144. IEEE Press, 1982.

[97] D. GOLLMANN (ED.) *Fast Software Encryption. Lecture Notes in Computer Science*, vol. 1039, Springer-Verlag, 1996.

[98] D. M. GORDON AND K. S. MCCURLEY. Massively parallel computation of discrete logarithms. *Lecture Notes in Computer Science*, **740** (1993), 312–323. (Advances in Cryptology – CRYPTO '92.)

[99] L. C. GUILLOU AND J.-J. QUISQUATER (EDS.) *Advances in Cryptology – EUROCRYPT '95 Proceedings. Lecture Notes in Computer Science*, vol. 921, Springer-Verlag, 1995.

[100] C. G. GUNTHER (ED.) *Advances in Cryptology – EUROCRYPT '88 Proceedings. Lecture Notes in Computer Science*, vol. 330, Springer-Verlag, 1988.

[101] J. HÅSTAD, A. W. SCHRIFT AND A. SHAMIR. The discrete logarithm modulo a composite hides $O(n)$ bits. *Journal of Computer and Systems Science*, **47** (1993), 376–404.

[102] M. E. HELLMAN. A cryptanalytic time-memory trade-off. *IEEE Transactions on Information Theory*, **26** (1980), 401–406.

[103] T. HELLESETH (ED.) *Advances in Cryptology – EUROCRYPT '93 Proceedings. Lecture Notes in Computer Science*, vol. 765, Springer-Verlag, 1994.

[104] H. M. HEYS. *A Tutorial on Linear and Differential Cryptanalysis*. Technical report CORR 2001-17, Dept. of Combinatorics and Optimization, University of Waterloo, Waterloo, Canada, 2001.

[105] H. M. HEYS AND S. E. TAVARES. Substitution-permutation networks resistant to differential and linear cryptanalysis. *Journal of Cryptology*, **9**

(1996), 1–19.

[106] H. IMAI, R. L. RIVEST AND T. MATSUMOTO (EDS.) *Advances in Cryptology – ASIACRYPT '91 Proceedings. Lecture Notes in Computer Science*, vol. 739, Springer-Verlag, 1993.

[107] D. JOHNSON, A. MENEZES AND S. VANSTONE. The elliptic curve digital signature algorithm (ECDSA). *International Journal on Information Security*, **1** (2001), 36–63.

[108] P. JUNOD. On the complexity of Matsui's attack. *Lecture Notes in Computer Science*, **2259** (2001), 199–211. (Selected Areas in Cryptography 2001.)

[109] D. KAHN. *The Codebreakers*. Scribner, 1996.

[110] B. KALISKI, JR. (ED.) *Advances in Cryptology – CRYPTO '97 Proceedings. Lecture Notes in Computer Science*, vol. 1294, Springer-Verlag, 1997.

[111] C. KAUFMAN, R. PERLMAN AND M. SPECINER. *Network Security. Private Communication in a Public World.* Prentice Hall, 1995.

[112] L. KELIHER, H. MEIJER AND S. TAVARES. New method for upper bounding the maximum average linear hull probability for SPNs. *Lecture Notes in Computer Science*, **2045** (2001), 420–436. (Advances in Cryptology – EUROCRYPT 2001.)

[113] L. KELIHER, H. MEIJER AND S. TAVARES. Improving the upper bound on the maximum average linear hull probability for Rijndael. *Lecture Notes in Computer Science*, **2259** (2001), 112–128. (Selected Areas in Cryptography 2001.)

[114] J. KILIAN (ED.) *Advances in Cryptology – CRYPTO 2001 Proceedings. Lecture Notes in Computer Science*, vol. 2139, Springer-Verlag, 2001.

[115] K. KIM AND T. MATSUMOTO (EDS.) *Advances in Cryptology – ASIACRYPT '96 Proceedings. Lecture Notes in Computer Science*, vol. 1163, Springer-Verlag, 1996.

[116] R. KIPPENHAHN. *Code Breaking, A History and Exploration*. Overlook Press, 1999.

[117] L. R. KNUDSEN. Contemporary block ciphers. *Lecture Notes in Computer Science*, **1561** (1999), 105–126. (Lectures on Data Security.)

[118] L. R. KNUDSEN (ED.) *Fast Software Encryption. Lecture Notes in Computer Science*, vol. 1636, Springer-Verlag, 1999. (FSE '99.)

[119] N. KOBLITZ. Elliptic curve cryptosystems. *Mathematics of Computation*, **48** (1987), 203–209.

[120] N. KOBLITZ. *A Course in Number Theory and Cryptography, Second Edition*. Springer-Verlag, 1994.

[121] N. KOBLITZ (ED.) *Advances in Cryptology – CRYPTO '96 Proceedings. Lecture Notes in Computer Science*, vol. 1109, Springer-Verlag, 1996.

[122] N. KOBLITZ, A. MENEZES AND S. VANSTONE. The state of elliptic curve cryptography. *Designs, Codes and Cryptography,* **19** (2000), 173–193.

[123] A. G. KONHEIM. *Cryptography, A Primer.* John Wiley and Sons, 1981.

[124] H. KRAWCZYK, (ED.) *Advances in Cryptology – CRYPTO '98 Proceedings. Lecture Notes in Computer Science,* vol. 1462, Springer-Verlag, 1998.

[125] K. KUROSAWA, T. ITO AND M. TAKEUCHI. Public key cryptosystem using a reciprocal number with the same intractability as factoring a large number. *Cryptologia,* **12** (1988), 225–233.

[126] X. LAI, J. L. MASSEY AND S. MURPHY. Markov ciphers and differential cryptanalysis. *Lecture Notes in Computer Science,* **547** (1992), 17–38. (Advances in Cryptology – EUROCRYPT '91.)

[127] K. Y. LAM, E. OKAMOTO AND C. XING (EDS.) *Advances in Cryptology – ASIACRYPT '99 Proceedings. Lecture Notes in Computer Science,* vol. 1716, Springer-Verlag, 1999.

[128] S. LANDAU. Standing the test of time: the data encryption standard. *Notices of the AMS,* **47** (2000), 341–349.

[129] S. LANDAU. Communications security for the twenty-first century: the advanced encryption standard. *Notices of the AMS,* **47** (2000), 450–459.

[130] A. K. LENSTRA. Integer factoring. *Designs, Codes and Cryptography,* **19** (2000), 101–128.

[131] A. K. LENSTRA AND H. W. LENSTRA, JR. (EDS.) *The Development of the Number Field Sieve. Lecture Notes in Mathematics,* vol. 1554. Springer-Verlag, 1993.

[132] A. K. LENSTRA AND H. W. LENSTRA, JR. Algorithms in number theory. In *Handbook of Theoretical Computer Science, Volume A: Algorithms and Complexity,* pages 673–715. Elsevier Science Publishers, 1990.

[133] A. K. LENSTRA AND E. R. VERHEUL. Selecting cryptographic key sizes. *Journal of Cryptology,* **14** (2001), 255–293.

[134] R. LIDL AND H. NIEDERREITER. *Finite Fields, Second Edition.* Cambridge University Press, 1997.

[135] D. L. LONG AND A. WIGDERSON. The discrete log hides $O(\log n)$ bits. *SIAM Jounal on Computing,* **17** (1988), 363–372.

[136] J. C. A. VAN DER LUBBE. *Basic Methods of Cryptography.* Cambridge, 1998.

[137] M. MATSUI. Linear cryptanalysis method for DES cipher. *Lecture Notes in Computer Science,* **765** (1994), 386–397. (Advances in Cryptology – EUROCRYPT '93.)

[138] M. MATSUI. The first experimental cryptanalysis of the data encryption standard. *Lecture Notes in Computer Science,* **839** (1994), 1–11. (Advances in Cryptology – CRYPTO '94.)

[139] U. MAURER (ED.) *Advances in Cryptology – EUROCRYPT '96 Proceedings. Lecture Notes in Computer Science*, vol. 1070, Springer-Verlag, 1996.

[140] U. MAURER AND S. WOLF. The Diffie-Hellman protocol. *Designs, Codes and Cryptography*, **19** (2000), 147–171.

[141] R. MCELIECE. *Finite Fields for Computer Scientists and Engineers.* Kluwer Academic Publishers, 1987.

[142] A. J. MENEZES. *Elliptic Curve Public Key Cryptosystems.* Kluwer Academic Publishers, 1993.

[143] A. J. MENEZES, T. OKAMOTO AND S. A. VANSTONE. Reducing elliptic curve logarithms to logarithms in a finite field. *IEEE Transactions on Information Theory*, **39** (1993), 1639–1646.

[144] A. J. MENEZES AND S. A. VANSTONE (EDS.) *Advances in Cryptology – CRYPTO '90 Proceedings. Lecture Notes in Computer Science*, vol. 537, Springer-Verlag, 1991.

[145] A. J. MENEZES, P. C. VAN OORSCHOT AND S. A. VANSTONE. *Handbook of Applied Cryptography.* CRC Press, 1996.

[146] R. C. MERKLE. One way hash functions and DES. *Lecture Notes in Computer Science*, **435** (1990), 428–446. (Advances in Cryptology – CRYPTO '89.)

[147] G. L. MILLER. Riemann's hypothesis and tests for primality. *Journal of Computer and Systems Science*, **13** (1976), 300–317.

[148] V. MILLER. Uses of elliptic curves in cryptography. *Lecture Notes in Computer Science*, **218** (1986), 417–426. (Advances in Cryptology – CRYPTO '85.)

[149] C. J. MITCHELL, F. PIPER AND P. WILD. Digital signatures. In *Contemporary Cryptology, The Science of Information Integrity*, pages 325–378. IEEE Press, 1992.

[150] R. A. MOLLIN. *An Introduction to Cryptography.* Chapman & Hall/CRC, 2001.

[151] J. H. MOORE. Protocol failures in cryptosystems. In *Contemporary Cryptology, The Science of Information Integrity*, pages 541–558. IEEE Press, 1992.

[152] V. I. NECHAEV. On the complexity of a deterministic algorithm for a discrete logaritum. *Math. Zametki*, **55** (1994), 91–101.

[153] J. NECHVATAL, E. BARKER, L. BASSHAM, W. BURR, M. DWORKIN, J. FOTI AND E. ROBACK. *Report on the Development of the Advanced Encryption Standard (AES).* October 2, 2000. Available from http://csrc.nist.gov/encryption/aes/

[154] P. Q. NGUYEN AND I. E. SHPARLINSKI. The insecurity of the digital signature algorithm with partially known nonces. *Journal of Cryptology*,

to appear.

[155] K. NYBERG. Differentially uniform mappings for cryptography. *Lecture Notes in Computer Science*, **765** (1994), 55–64. (Advances in Cryptology – EUROCRYPT '93.)

[156] K. NYBERG. Linear approximation of block ciphers. *Lecture Notes in Computer Science*, **950** (1995), 439–444. (Advances in Cryptology – EUROCRYPT '94.)

[157] K. NYBERG (ED.) *Advances in Cryptology – EUROCRYPT '98 Proceedings. Lecture Notes in Computer Science*, vol. 1403, Springer-Verlag, 1998.

[158] A. M. ODLYZKO (ED.) *Advances in Cryptology – CRYPTO '86 Proceedings. Lecture Notes in Computer Science*, vol. 263, Springer-Verlag, 1987.

[159] A. M. ODLYZKO. Discrete logarithms: the past and the future. *Designs, Codes, and Cryptography*, **19** (2000), 129–145.

[160] K. OHTA AND D. PEI (EDS.) *Advances in Cryptology – ASIACRYPT '98 Proceedings. Lecture Notes in Computer Science*, vol. 1514, Springer-Verlag, 1998.

[161] T. OKAMOTO (ED.) *Advances in Cryptology – ASIACRYPT 2000 Proceedings. Lecture Notes in Computer Science*, vol. 1976, Springer-Verlag, 2000.

[162] T. P. PEDERSEN. Signing contracts and paying electronically. *Lecture Notes in Computer Science*, **1561** (1999), 134–157. (Lectures on Data Security.)

[163] R. PERALTA. Simultaneous security of bits in the discrete log. *Lecture Notes in Computer Science*, **219** (1986), 62–72. (Advances in Cryptology – EUROCRYPT '85.)

[164] B. PFITZMANN (ED.) *Advances in Cryptology – EUROCRYPT 2001 Proceedings. Lecture Notes in Computer Science*, vol. 2045, Springer-Verlag, 2001.

[165] B. PFITZMANN. *Digital Signature Schemes – General Framework and Fail-Stop Signatures. Lecture Notes in Computer Science*, vol. 1100, Springer-Verlag, 1996.

[166] F. PICHLER (ED.) *Advances in Cryptology – EUROCRYPT '85 Proceedings. Lecture Notes in Computer Science*, vol. 219, Springer-Verlag, 1986.

[167] J. PIEPRYZK AND R. SAFAVI-NAINI (EDS.) *Advances in Cryptology – ASIACRYPT '94 Proceedings. Lecture Notes in Computer Science*, vol. 917, Springer-Verlag, 1995.

[168] S. C. POHLIG AND M. E. HELLMAN. An improved algorithm for computing logarithms over GF(p) and its cryptographic significance. *IEEE Transactions on Information Theory*, **24** (1978), 106–110.

[169] D. POINTCHEVAL AND J. STERN. Security arguments for signature

schemes and blind signatures. *Journal of Cryptology*, **13** (2000), 361–396.

[170] J. M. POLLARD. Monte Carlo methods for index computation (mod p). *Mathematics of Computation*, **32** (1978), 918–924.

[171] C. POMERANCE (ED.) *Advances in Cryptology – CRYPTO '87 Proceedings. Lecture Notes in Computer Science*, vol. 293, Springer-Verlag, 1988.

[172] B. PRENEEL, (ED.) *Fast Software Encryption. Lecture Notes in Computer Science*, vol. 1008, Springer-Verlag, 1995.

[173] B. PRENEEL. The state of cryptographic hash functions. *Lecture Notes in Computer Science*, **1561** (1999), 158–182. (Lectures on Data Security.)

[174] B. PRENEEL (ED.) *Advances in Cryptology – EUROCRYPT 2000 Proceedings. Lecture Notes in Computer Science*, vol. 1807, Springer-Verlag, 2000.

[175] B. PRENEEL AND P. C. VAN OORSCHOT. On the security of iterated message authentication codes. *IEEE Transactions on Information Theory*, **45** (1999), 188–199.

[176] J.-J. QUISQUATER AND J. VANDEWALLE (EDS.) *Advances in Cryptology – EUROCRYPT '89 Proceedings. Lecture Notes in Computer Science*, vol. 434, Springer-Verlag, 1990.

[177] M. O. RABIN. Digitized signatures and public-key functions as intractable as factorization. *MIT Laboratory for Computer Science Technical Report*, LCS/TR-212, 1979.

[178] M. O. RABIN. Probabilistic algorithms for testing primality. *Journal of Number Theory*, **12** (1980), 128–138.

[179] R. L. RIVEST. The MD4 message digest algorithm. *Lecture Notes in Computer Science*, **537** (1991), 303–311. (Advances in Cryptology – CRYPTO '90.)

[180] R. L. RIVEST. The MD5 message digest algorithm. Internet Network Working Group RFC 1321, April 1992.

[181] R. L. RIVEST, A. SHAMIR, AND L. ADLEMAN. A method for obtaining digital signatures and public key cryptosystems. *Communications of the ACM*, **21** (1978), 120–126.

[182] K. H. ROSEN. *Elementary Number Theory and its Applications, Fourth Edition*. Addison-Wesley, 1999.

[183] R. A. RUEPPEL (ED.) *Advances in Cryptology – EUROCRYPT '92 Proceedings. Lecture Notes in Computer Science*, vol. 658, Springer-Verlag, 1993.

[184] A. SALOMAA. *Public-Key Cryptography*. Springer-Verlag, 1990.

[185] B. SCHNEIER. *Applied Cryptography, Protocols, Algorithms and Source Code in C, Second Edition*. John Wiley and Sons, 1995.

[186] B. SCHNEIER (ED.) *Fast Software Encryption. Lecture Notes in Computer Science*, vol. 1978, Springer-Verlag, 2001. (FSE 2000.)

[187] C. P. SCHNORR. Efficient signature generation by smart cards. *Journal of Cryptology*, **4** (1991), 161–174.

[188] J. SEBERRY AND J. PIEPRZYK (EDS.) *Advances in Cryptology – AUSCRYPT '90 Proceedings. Lecture Notes in Computer Science*, vol. 453, Springer-Verlag, 1990.

[189] J. SEBERRY AND Y. ZHENG (EDS.) *Advances in Cryptology – AUSCRYPT '92 Proceedings. Lecture Notes in Computer Science*, vol. 718, Springer-Verlag, 1993.

[190] C. E. SHANNON. A mathematical theory of communication. *Bell Systems Technical Journal*, **27** (1948), 379–423, 623–656.

[191] C. E. SHANNON. Communication theory of secrecy systems. *Bell Systems Technical Journal*, **28** (1949), 656–715.

[192] V. SHOUP. Lower bounds for discrete logarithms and related problems. *Lecture Notes in Computer Science*, **1233** (1997), 256–266. (Advances in Cryptology – EUROCRYPT '97.)

[193] V. SHOUP. OAEP reconsidered. *Lecture Notes in Computer Science*, **2139** (2001), 239–259. (Advances in Cryptology – CRYPTO 2001.)

[194] J. H. SILVERMAN AND J. TATE. *Rational Points on Elliptic Curves*. Springer-Verlag, 1992.

[195] G. J. SIMMONS. A survey of information authentication. In *Contemporary Cryptology, The Science of Information Integrity*, pages 379–419. IEEE Press, 1992.

[196] G. J. SIMMONS (ED.) *Contemporary Cryptology, The Science of Information Integrity*. IEEE Press, 1992.

[197] S. SINGH. *The Code Book*. Doubleday, 1999.

[198] N. P. SMART. The discrete logarithm problem on elliptic curves of trace one. *Journal of Cryptology*, **12** (1999), 193–196.

[199] M. E. SMID AND D. K. BRANSTAD. The data encryption standard: past and future. In *Contemporary Cryptology, The Science of Information Integrity*, pages 43–64. IEEE Press, 1992.

[200] M. E. SMID AND D. K. BRANSTAD. Response to comments on the NIST proposed digital signature standard. *Lecture Notes in Computer Science*, **740** (1993), 76–88. (Advances in Cryptology – CRYPTO '92.)

[201] J. SOLINAS. Efficient arithmetic on Koblitz curves. *Designs, Codes and Cryptography*, **19** (2000), 195–249.

[202] R. SOLOVAY AND V. STRASSEN. A fast Monte Carlo test for primality. *SIAM Journal on Computing*, **6** (1977), 84–85.

[203] W. STALLINGS. *Network and Internetwork Security, Principles and Practice, Second Edition* Prentice Hall, 1999.

[204] J. STERN (ED.) *Advances in Cryptology – EUROCRYPT '99 Proceedings. Lecture Notes in Computer Science*, vol. 1592, Springer-Verlag, 1999.

[205] D. R. STINSON (ED.) *Advances in Cryptology – CRYPTO '93 Proceedings. Lecture Notes in Computer Science*, vol. 773, Springer-Verlag, 1994.

[206] D. R. STINSON. On the connections between universal hashing, combinatorial designs and error-correcting codes. *Congressus Numerantium*, **114** (1996), 7–27.

[207] D. R. STINSON. Some observations on the theory of cryptographic hash functions. Technical report CORR 2001-15, Department of Combinatorics & Optimization, University of Waterloo, 2001.

[208] E. TESKE. On random walks for Pollard's rho method. *Mathematics of Computation*, **70** (2001), 809–825.

[209] E. THOMÉ. Computation of discrete logarithms in $\mathbb{F}_{2^{607}}$. *Lecture Notes in Computer Science*, **2248** (2001), 107–124. (Advances in Cryptology – ASIACRYPT 2001.)

[210] E. VAN HEYST AND T. P. PEDERSEN. How to make efficient fail-stop signatures. *Lecture Notes in Computer Science*, **658** (1993), 366–377. (Advances in Cryptology – EUROCRYPT '92.)

[211] S. VAUDENAY, (ED.) *Fast Software Encryption. Lecture Notes in Computer Science*, vol. 1372, Springer-Verlag, 1998. (FSE '98.)

[212] M. N. WEGMAN AND J. L. CARTER. New hash functions and their use in authentication and set equality. *Journal of Computer and System Sciences*, **22** (1981), 265–279.

[213] D. WELSH. *Codes and Cryptography*. Oxford Science Publications, 1988.

[214] M. J. WIENER. Cryptanalysis of short RSA secret exponents. *IEEE Transactions on Information Theory*, **36** (1990), 553–558.

[215] M. J. WIENER. Efficient DES key search. Technical report TR-244, School of Computer Science, Carleton University, Ottawa, Canada, May 1994 (also presented at CRYPTO '93 Rump Session).

[216] M. J. WIENER, (ED.) *Advances in Cryptology – CRYPTO '99 Proceedings. Lecture Notes in Computer Science*, vol. 1666, Springer-Verlag, 1999.

[217] H. C. WILLIAMS. A modification of the RSA public-key encryption procedure. *IEEE Transactions on Information Theory*, **26** (1980), 726–729.

[218] H. C. WILLIAMS (ED.) *Advances in Cryptology – CRYPTO '85 Proceedings. Lecture Notes in Computer Science*, vol. 218, Springer-Verlag, 1986.

[219] S. Y. YAN. *Number Theory for Computing*. Springer-Verlag, 2000.

Cryptosystem Index

Algorithm Index

Problem Index

Subject Index

$p - 1$ algorithm, 182

abelian group, 4
active S-box, 85
adaptive algorithm, 243
additive identity, 3
additive inverse, 4
adjoint matrix, 17
affine function, 8
algorithm
    (ϵ, q)-, 121
    adaptive, 243
    deterministic, 172
    generic, 239
    Las Vegas, 121
    Monte Carlo, 172
    non-adaptive, 242
    randomized, 121
Alice, 1
almost strongly universal, 149
associative property, 3
    of cryptosystems, 69
    of elliptic curve, 250
attack
    chosen ciphertext, 25
    chosen message, 277
    chosen plaintext, 25
    ciphertext only, 25
    key-only, 277
    known message, 277
    known plaintext, 25
attack model, 25
    for cryptosystem, 25
    for signature scheme, 277

authentication matrix, 142
authentication tag, 118
average-case success probability, 121

balanced, 115
Bayes' theorem, 48
bias, 80
big MAC attack, 138
binomial coefficient, 32
birthday paradox, 123
block cipher, 21
block length, 74
Bob, 1

candidate subkey, 87
challenge-and-response protocol, 300
channel, 1
Chinese remainder theorem, 164
chosen ciphertext attack, 25
chosen message attack, 277
chosen plaintext attack, 25
cipher block chaining mode, 109
cipher feedback mode, 109
ciphertext, 1
    for stream cipher, 21
ciphertext only attack, 25
closed, 3
collision resistant, 119
commutative cryptosystems, 68
commutative property, 3
composition, 138
compression function, 118
concave function, 59
    strictly, 59